DNA cloning
Volume II

a practical approach

Edited by
D M Glover

Cancer Research Campaign, Eukaryotic Molecular
Genetics Research Group, Department of Biochemistry,
Imperial College of Science and Technology, London
SW7 2AZ, UK

IRL PRESS
Oxford · Washington DC

IRL Press Limited
P.O. Box 1,
Eynsham,
Oxford OX8 1JJ,
England

First published July 1985
First reprinting February 1986
Second reprinting April 1986

British Library Cataloguing in Publication Data

DNA cloning : a practical approach.—(Practical
  approach series)
  1. Molecular cloning 2. Recombinant DNA
  I. Title II. Series
  574.87'3282   QH442.2

ISBN 0-947946-19-5

Cover illustration. The design for the cover was based on Figure 2 Chapter 2 Volume I, showing a map of λgt11; Figure 10 Chapter 4 Volume II, showing crown gall tumour on *Nicotiana tabacum*; and Figure 4 Chapter 6 Volume II, showing peroxidase stained cells.

Printed in England by Information Printing, Oxford.

Preface

This is the second volume in this series describing practical aspects of DNA cloning techniques. There can be no doubt of the importance of such techniques in bringing about the information explosion that has occurred in Molecular Biology over the past decade. The first volume concentrated on developments that have occurred with cloning systems using *Escherichia coli* as the host organism, and in particular looked at innovations that have been adopted over the past three or four years. The second volume looks at a diversity of systems that are used alongside *E. coli* to allow the cloning and expression of genes in a variety of other prokaryotic and eukaryotic cells. In the previous book, I recommended that newcomers to the field might first want an overview before reading such a book as this which is primarily a laboratory handbook. The texts that I recommended would be an equally appropriate introduction to this volume, namely 'Recombinant DNA: A Short Course' by Watson, Toose and Kurtz (Scientific American Books, New York, 1983); 'Principles of Gene Manipulation' by Old and Primrose (Blackwell, Oxford, 1985); or the one by myself 'Gene Cloning: The Mechanics of DNA Manipulation' (Chapman and Hall, London and New York, 1984). The laboratory manual 'Molecular Cloning' by Maniatis, Fritsch and Sambrook (Cold Spring Harbor Laboratory, New York, 1982) is an additional, invaluable source of protocols for a variety of molecular cloning techniques using *E. coli*. I hope that the cloning techniques described within the first volume will indeed extend and complement the excellent manual of Maniatis and his co-authors as was our intention. The topics covered in Volume 1 include the use of bacteriophage lambda vectors that permit the direct selection of recombinants in library building and the use of lambda vectors for cDNA cloning. Plasmid vectors that direct the synthesis of fusion proteins are also described. These have recently come into their own as a means of providing antigenic material in order to raise antibodies against the products of cloned genes. This can of course be turned around, and libraries of DNA can be constructed in these vectors for screening using available antibodies as probes. Another imaginative group of vectors, the pEMBL plasmids, can be propagated as single-stranded DNA in bacteria carrying F-factors super-infected with the male-specific phage f1. Two approaches to the mutagenesis *in vitro* of DNA carried in 'single-stranded vectors' are then described. Methods for the high efficiency transformation of *E. coli* with naked DNA are presented in considerable detail. In the final chaper of the first volume we begin to consider other host organisms, but stay with Gram negative bacteria in a discussion of vectors that have a broader host range.

In this second volume, alternative bacterial hosts are explored; the first chapter dealing with *Bacilli* and the second chapter with *Streptomycetes*. Chapter 3 is the first encounter with a eukaryotic system, and contains a succinct account of the powerful methods of cloning in yeast. Molecular biologists interested in introducing foreign genes into plants using vectors derived from Ti plasmids have to master a variety of techniques. In Chapter 4 the techniques for working with *E. coli* and *Agrobacteirum tumefaciens* as well as with plant material are described in detail. The discovery that the phenomenon of hybrid dysgenesis in *Drosophila* can be mimicked by the micro-injection of P-elements has lead to the use of these transposons as vectors. The art of achieving the germ line transformation of *Drosophila* is presented in Chapter 5. In the final three chapters of the book, we turn towards mammalian cell systems. Some tricks for efficient gene transfer into cultured cells are given in Chapter 6, whereas Chapters 7 and 8 look specifically at two viral vector systems using Vaccinia Virus and Bovine Papilloma Virus, respectively. This by no means represents a comprehensive cover of the field of animal cell host-vector systems. This field is evolving so rapidly, however, that in a short while a third volume might be appropriate to cover this area more thoroughly.

As before, I hope that the community of Molecular Biologists will find these volumes useful. My thanks go to all the authors for their willing participation in this project, for producing their manuscripts so promptly, and for being so tolerant of the interfering editor.

David M. Glover

Contributors

M.S.Campo
The Beatson Institute for Cancer Research, Garscube Estate, Switchback Road, Bearsden, Glasgow G61 1BD, UK

J.Draper
Department of Botany, Adrian Building, University of Leicester, Leicester LE1 7RH, UK

C.Gorman
Cancer Research Campaign, Department of Biochemistry, Imperial College of Science and Technology, Imperial College Road, London SW7 2AZ, UK

K.G.Hardy
Biogen S.A., 46, route des Acacias, Geneva 1227, Switzerland

I.S.Hunter
Biotechnology Unit, Department of Genetics, University of Glasgow, Glasgow G12 8QQ, UK

R.Karess
Department of Biochemistry, Imperial College of Science and Technology, Imperial College Road, London SW7 2AZ, UK

C.P.Lichtenstein
Department of Biochemistry, Imperial College of Science and Technology, Imperial College Road, London SW7 2AZ, UK

M.Mackett
Paterson Laboratories, Christie Hospital and Holt Radium Institute, Withington, Manchester M20 9BX, UK

B.Moss
Laboratory of Viral Diseases, Building 5, Room 318, National Institute of Allergy and Infectious Diseases, NIH, Bethesda, MD 20205, USA

R.Rothstein
Department of Human Genetics and Development, Columbia University, College of Physicians and Surgeons, 701 West 168th Street, New York, NY 10032, USA

G.L.Smith
Laboratory of Viral Diseases, Building 5, Room 318, National Institute of Allergy and Infectious Diseases, NIH, Bethesda, MD 20205, USA

Contents

6. HIGH EFFICIENCY GENE TRANSFER INTO MAMMALIAN CELLS   **143**

C.Gorman

Abbreviations

APRT	adenine phosphoribosyl transferase
ars	autonomously replicating segment
BPV-1	bovine papillomavirus type 1
BUdR	5-bromodeoxyuridine
C23O	catechol 2,3-oxygenase
CaMV	cauliflower mosaic virus
CAT	chloramphenicol acetyltransferase
CEF	chick embryo fibroblasts
CHO	Chinese hamster ovary
Cm	chloramphenicol
CRM	cross-reacting material
CTAB	cetyl triethylammonium bromide
DHFR	dihydrofolate reductase
DMEM	Dulbecco's modified Eagle's medium
DMSO	dimethylsulphoxide
d.s.	double-stranded
DTT	dithiothreitol
EtBr	ethidium bromide
FCS	foetal calf serum
βgal	β-galactosidase
gdDNA	gapped duplex DNA
GH	growth hormone
HA	haemagglutinin
HBS	Hepes-buffered saline
HbsAg	hepatitis virus surface antigen
HGT	high gelling temperature
HSV-1	herpes simplex virus type 1
IF	interferon
IMP	inosine monophosphate
IPTG	isopropyl-1-thio-β-D-galactoside
Kn	kanamycin
LGT	low gelling temperature
LTR	long terminal repeat
2ME	2-mercaptoethanol
MEM	Eagle's minimal medium
MES	2[N-morphino]ethone sulphonic acid
MMTV	mouse mammary tumour virus
MSV	Moloney murine sarcoma virus
NPTII	neomycin phosphotransferase
NRDC	National Research and Development Corporation
ONPG	o-nitrophenyl-β-D-galactopyranoside
ORF	open reading frame
PBS	phosphate-buffered saline
PEG	polyethylene glycol
RNP	ribonucleoprotein
RSV	Rous sarcoma virus
SAM	S-adenosyl-L-[methyl-^{3}H]methionine

SB	simple transformation buffer
SDS	sodium dodecylsulphate
Sm	streptomycin
s.s.	single-stranded
SSC	standard saline citrate
Su	sulphonamides
T-DNA	transforming DNA of the Ti-plasmid
TBS	Tris-buffered saline
Tc	tetracycline
TES	N-tris(hydroxymethyl)methyl-2-aminoethanesulphonic acid
TFB	transformation buffer
Ti plasmid	tumour inducing plasmid
TK	thymidine kinase
X-gal	5-bromo-4-chloro-3-indolyl-β-D-galactoside

CHAPTER 1

Bacillus Cloning Methods

KIMBER G. HARDY

1. INTRODUCTION

The potential advantages of *Bacillus* as a bacterial host for cloned DNA include its ability to secrete proteins and the wide range of very different species which can be used. A secreted protein may be easier to recover and purify than an intracellular protein, and in addition, may already be correctly folded, in contrast to an intracellular product for which a step for correct folding must form part of the purification process. The various *Bacillus* species differ greatly from each other in terms of their optimal growth temperatures (there are many thermophilic strains, for example) in whether or not they are obligately aerobic or facultatively anaerobic, and in their proteolytic activities.

Much progress has been made towards the realisation of the potential advantages of *Bacillus* since 1978 when the first *B. subtilis* cloning experiments were reported (1 − 3). Several plasmids have been tried as vectors and a number of promoters have been used to express several mammalian or viral genes. The products of some foreign genes have been secreted from *Bacillus* (4,5). Of course, not all gene cloning experiments in *Bacillus* are designed to express animal or viral genes; cloning experiments can tell us much about the bacterium itself, for example about the mechanisms involved in sporulation (6).

This chapter is written for those who are familiar with basic cloning techniques as applied to *Escherichia coli* and who may wish to begin cloning with *Bacillus*. The difficulties encountered when using *Bacillus* as a host, have included plasmid instability (both loss of the entire plasmid and genetic rearrangements) and the high protease levels of certain strains. In addition, plasmid vectors are not as well developed as they are for *E. coli* and, in particular, the strong controlled promoters, such as *trp, tac* and λpL which have proved so useful for expressing foreign genes in *E. coli* do not have such well developed counterparts in *Bacillus*. Problems associated with plasmid instability may be caused by recombination between homologous regions (7) and can sometimes be overcome by avoiding duplicated sequences in constructs. Degradation of the products of cloned genes may be prevented by using a non-proteolytic species such as *Bacillus sphaericus*.

2. STRAINS AND PLASMIDS

2.1 Bacterial strains

B. subtilis strains commonly used as hosts for recombinant plasmids are shown in *Table 1*. Strain BR151 is perhaps the most commonly used host. Almost all

Table 1. *Bacillus* Strains used for Cloning.

Strain	Characteristics	Reference
BR151	trpC2, metB10, lys-3	15
YB886	trpC2, metB10, xin-1, SPβ^-	16
MI112	leuA8, arg-15, thr-5, recE4, r$^-$, m$^-$	12
MI119	leuB6, trpC2, r$^-$, m$^-$	12
MI120	leuB6, recE4, r$^-$, m$^-$	12
CU403	thyA, thyB, metB, divIVB1	17
BD170	trpC2, thr-5	9
BD224	trpC2, thr-5, recE4	9

the standard laboratory strains of *B. subtilis* 168 have two prophages in their chromosomes, PBSX and SPβ (8). When PBSX is induced, functional phage particles are not produced; the phage particles contain only bacterial DNA (9). SPβ produces functional particles, but it is difficult to induce. Strains have been made which lack these prophages (e.g., YB886, a derivative of BR151, *Table 1*).

Many mutations causing sporulation deficiency (and in certain cases a concommitant reduction in protease activities) also result in the inability to become competent for transformation by DNA. However, although the cells of such strains cannot be transformed, it is possible to transform protoplasts made from them. Mutants of *B. subtilis*, a highly proteolytic organism, have been made which lack certain proteases, but none of these have such low levels of protease as are found in other species such as *B. sphaericus, B. freundenreichii* or *B. coagulans*. In order to transform species of *Bacillus* other than *subtilis* it is necessary to work with sphaeroplasts.

The *recE4* allele (10) greatly reduces but does not completely eliminate homologous recombination; plasmids carrying a chromosomal homologue can be unstable even in a *recE4* strain (11). Both RecE4$^+$ and RecE$^-$ strains transform equally well with plasmid DNA (2).

The restriction system of *B. subtilis* 168 and its derivatives has only a low activity against unmodified plasmid DNA. Although the transformation efficiency of an unmodified plasmid is often the same as that of the modified form, instances have been reported where the transformation frequency in an r$^+$ strain is 10% or 0.2% of that in an r$^-$ strain (11 − 13). In practice many cloning experiments have been successfully carried out using r$^+$ strains of *B. subtilis*. The *B. subtilis* restriction system also operates against phage DNA introduced by transfection, but is not active against chromosomal DNA introduced by transformation (14).

The Bacillus Genetic Stock Centre (Director, D.H. Dean, Department of Microbiology, The Ohio State University, 484 W 12th Avenue, Columbus, OH 43210) provides a valuable service in keeping a collection of strains and plasmids, and also publishes a useful catalogue of strains, plasmids and phages. Another good source of information about strains is the Monograph on *Bacillus* by Gordon *et al.* (15). Many species other than *B. subtilis* are listed together with their reference numbers as they appear in the American Type culture Collection.

Table 2. Plasmids for Cloning in *Bacillus*.

Plasmid	Source	Mol. wt. (x 10^{-6})	Markers[a]	Reference
pBC16	*Bacillus cereus*	3.0	*Tc*	34
pAB124	*B. stearothermophilus*	2.9	*Tc*	35
pPL10	*B. pumilus*	4.4	bacteriocin	36
pPL7065	*B. pumilus*	4.6	bacteriocin	37
pIM13	*B. subtilis*	1.5	*Em*	38
pLS28	*B. subtilis (natto)*	4.1	−	40
pBS1	*B. subtilis*	5.5	−	34
pFTB14	*B. amyloliquefacians*	5.3	−	40
pUB110	*Staphylococcus aureus*	3.0	*Km (Nm)*	41
pE194	*S. aureus*	2.4	*Em*	20,41
pC194	*S. aureus*	1.8	*Cm*	12,21,42
pT127	*S. aureus*	2.9	*Tc*	12
pC221	*S. aureus*	3.0	*Cm*	12
pC223	*S. aureus*	3.0	*Cm*	12
pUB112	*S. aureus*	3.0	*Cm*	12
pSA501 (pS194)	*S. aureus*	2.8	*Sm*	41−43
pSA2100 (pSC194)	*S. aureus*	4.7	*Cm,Sm*	41−43

[a]Resistance markers are *Tc*, tetracycline, *Cm*, chloramphenicol; *Sm*, streptomycin; *Km*, kanamycin; *Tp*, trimethoprim; *Ap*, ampicillin; *Em*, erythromycin; *Nm*, neomycin. Auxotrophic markers are *trp*, tryptophan biosynthesis; *leu*, leucine biosynthesis; *ilv*, isoleucine-valine biosynthesis.

Table 3. Recombinant Plasmids used for *Bacillus* Cloning.

Plasmid	Mol. wt. (x 10^{-6})	Parental plasmids	Markers[a]	Reference
pBD6	5.8	pSA501, pUB110	*Sm, Km*	2,44
pBD8	6.0	pSA2100, pUB110	*Sm, Cm, Km*	2,44
pBD9	5.4	pE194, pUB110	*Em, Km*	2,44
pBD64	3.2	pC194, pUB110	*Cm, Km*	44
pHV11	3.3	pC194, pT127	*Cm, Tc*	3
pHV41	4.5	pC194, pUB110, pBR322	*Cm, Km*	45
pLS103	5.0	pUB110, *B. pumilus trp* gene	*Km, trp*	1
pLS105	5.4	pUB110, *B. licheniformis trp* gene	*Km, trp*	46
pTL10	9.4	pLS28, *B. subtilis leu* and *Tp-r* genes	*leu, Tp*	47
pTL12	6.4	pLS28, *B. subtilis leu* and *Tp-r* genes	*leu, Tp*	47
pTB90	4.4	pTB19	*Km, Tc*	48
pPL608	3.5	pUB110, *B. pumilus cat-86* gene, 0.3 kb SP02 phage promoter	*Km, Cm*	49
pPL708	3.3	pPL608 with linker from phage M13mp7	*Km, Cm*	50

[a]The abbreviations used for markers are given in the footnote to *Table 2*.

2.2 Plasmids

Commonly used plasmids and vectors are listed in *Tables 2−4*. Many vectors are derived from plasmids originally found in *Staphylococcus aureus* and which were subsequently found to replicate in *Bacillus* (12). Some are closely related to plasmids found in *Bacillus* (19); for example, pUB110 and pBC16 have many

Table 4. Shuttle (or Bridge) Vectors Replicating in *E. coli* and *B. subtilis*.

Plasmid	Mol. wt. (x 10⁻⁶)	Parental plasmids	Markers[a]		Reference
			B. subtilis	*E. coli*	
pHV14	4.6	pC194, pBR322	Cm	Ap, Cm	3
pHV33	4.6	pC194, pBR322	Cm	Ap, Tc, Cm	3
pHV23	6.1	pC194, pBR322, pT127	Tc, Cm	Ap, Cm	47
pOG2165	5.0	pC194, pUB110, *B. licheniformis*			
		pen-r gene	Ap, Cm	Ap, Cm	51
pJK3	5.0	pBS161-1, pBR322	Tc	Ap, Tc	52
pCPP-3	3.7	pBR322, pUB110, pC194	Km	Km	53

[a]The abbreviations used for markers are given in the footnote to *Table 2*.

homologous regions (19). The pBC16 plasmid represents a common type in *Bacillus*; pBC16 and pAB124 are almost identical (19). Because of the considerable homology between many plasmids from Gram-positive bacteria, care must be taken when constructing cloning vectors to ensure that homologous regions are not repeated. It should be noted that several plasmids listed in *Table 3* have two or more origins of replication derived from their parent plasmids. This may cause some instability. The complete sequences of plasmids pC194 and pE194 have been published (20,21). The maps of these two plasmids together with that of pUB110, which form components of many of the most commonly used vectors, are shown in *Figures 1–3*.

Hybrid, or chimaeric, plasmids which can replicate in both *E. coli* and *B. subtilis* are especially useful vectors. These are listed in *Table 4*. It is often much simpler to carry out the initial cloning work using *E. coli* as a host, and then to transfer the final construction into *B. subtilis*.

There are numerous examples of plasmid instability, especially amongst plasmids derived from pC194 (2,22–24). The particular region of pC194 associated with instability may be a transposon-like element having an inverted repeat sequence (24). Small regions of homology which may provide foci for a high frequency of recombination (7,25) is another possible explanation for the instability of certain plasmids. Not all recombinant plasmids are unstable (see for example, 4). But because of the high frequency of rearrangements often seen in *B subtilis*, it is perhaps all the more important to ensure that the expression of foreign genes is well controlled, as variants which are not subject to growth inhibition by the foreign protein may quickly outgrow the parental strain. The difficulties arising from the recombination of plasmid promoters with their chromosomal homologues can be avoided by using promoters from unrelated species (the DNA of many *Bacillus* species have little homology with each other, 26) or from phages (see *Table 4*).

A number of *Bacillus* plasmids have been specially designed for cloning and selecting *Bacillus* promoters (for example, pCPP-3, *Table 4*).

2.3 Bacillus Phages and Phasmids

Several phage and plasmid vectors are being developed including those derived

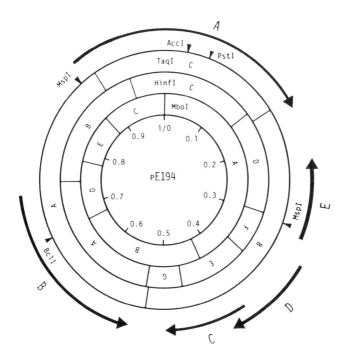

Figure 1. Physical map of pE194 [from Horinouchi and Weisblum (20)]. The arrows indicate open reading frames. Frame B corresponds to the *ermC* gene encoding the methylase responsible for erythromycin resistance. Replication determinants are located in the *Taq*1-B fragment; a single base change in frame E was found in a copy number mutant of pE194. A replication origin (replication proceeding anticlockwise) was mapped by Scheer-Abramowitz and Gryczan (75) to the region 0.35 to 0.45.

from rhoII (27,28), phi105 (28,30), SP02 (31,32) and SPP1 (33). One example of their use has been the cloning of amylase genes (28,30).

2.4 **Storage of Strains**

The instability of certain plasmids in *Bacillus* means that it is very important to store strains immediately after they have been isolated to avoid the possibility of working with rearranged plasmids in subsequent experiments. Lyophilisation is, of course, an excellent method of storing cultures for long periods. A quicker and more convenient method for day-to-day use is to store strains in L-broth containing 10% (v/v) glycerol at $-70°C$. The bacterial growth from a fresh plate is inoculated to a concentration of at least 10^{10} cells/ml into 2 ml of the medium. The tubes are frozen in a dry ice-ethanol bath and then stored in a freezer. A single tube can be used many times to inoculate plates by scraping a small amount of material from the frozen contents of the tube using a loop and streaking this onto a plate. It is important to ensure that the contents of the tube remain frozen when taking samples for inoculation; repeated thawing and freezing of the culture reduces its viability and leads to the selection of variants.

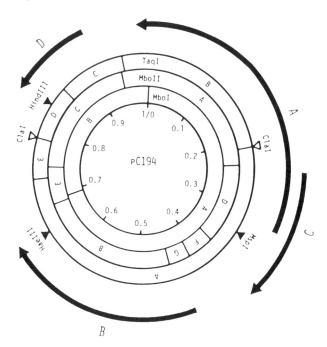

Figure 2. Physical map of pC194 [from Horinouchi and Weisblum (21), as modified by S.D.Ehrlich]. Arrows indicate open reading frames. Frame B encodes the chloramphenicol acetyltransferase, and sequences within frame C are associated with plasmid replication.

2.5 Growth Media

As a complex medium for growing *Bacillus*, L-broth or L-broth agar which is commonly used for growing *E. coli*, can be used. Media which permit somewhat faster growth of some strains are tryptose blood agar base as a solid medium, and VY broth (41) as a liquid medium. Tryptose blood agar base is supplied by Difco and is made up at 3.3% (w/v), sterilised by autoclaving and then poured onto plates. VY Broth is made up from Veal infusion (Difco) 2.5% (w/v) and Yeast extract (Difco) 0.5% (w/v) and is sterilised by autoclaving.

As a defined medium base, Spizizen salts (54) are used; this is usually supplemented with casamino acids and glucose for procedures such as the production of competent cells. Spizizen salts consists of 2 g of $(NH_4)_2SO_4$, 18.3 g of $K_2HPO_4.3H_2O$, 6 g of KH_2PO_4, 1 g of Trisodium citrate.$2H_2O$ and 0.2 g of $MgSO_4.7H_2O$ made up to 1 litre with water and adjusted to pH 7.2.

Many of the strains commonly used for genetic experiments do not grow well in minimal medium unless it is supplemented with casamino acids and yeast extract (see *Table 6*); an exception is strain CU403 (18) which is used to produce mini-cells and which will grow in a minimal medium supplemented only with 0.5% (w/v) glucose, 100 μg/ml methionine and 100 μg/ml thymine.

B. subtilis is an obligate aerobe, in contrast to *E. coli*. It must be grown with a good supply of air at all times. When grown in conical flasks, the volume of a *Bacillus* culture should not exceed 10% of the volume of the flask. Fluted flasks

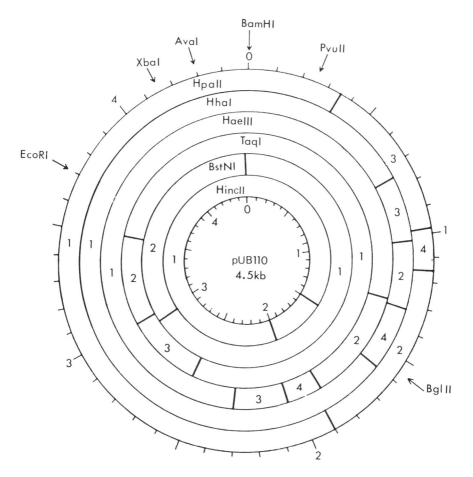

Figure 3. Physical map of pUB110 [redrawn from Jalanko *et al.* (74)]. The *Bgl*II site is within the kanamycin-resistance determinant; the limits of the resistance gene are unknown. An origin of replication (replication proceeding in an anticlockwise direction) was mapped at about 3.2 (75).

which have indented walls are often used to increase the aeration in rotated flasks. If liquid *Bacillus* cultures are left unshaken for more than a few minutes, they often lyse. Cultures are normally grown at 37°C.

3. ISOLATION OF PLASMID DNA

3.1 Large-scale Isolation of Plasmid DNA

Plasmid DNA can be isolated by using modifications of methods which are used for isolating *E. coli* plasmids. The following method for *B. subtilis* plasmids is based on the method of Birnboim and Doly (55) and uses the solutions in *Table 5*.

(i)   Grow 250 ml of cells in VY broth or L-broth at 30°C with vigorous aeration in a 2 litre flask for 18 h. A few colonies from a fresh plate are used as an inoculum.

Table 5. Solutions for Large- and Small-scale Isolation of Plasmid DNA.

Solution I		
Glucose	9.9 g	(50 mM)
EDTA	3.7 g	(10 mM)
Tris	3.0 g	(25 mM)
Make up to 1 litre with water and adjust pH to 8.0 with HCl.		
Solution II		
NaOH	8.0 g	(0.2 M)
SDS	10.0 g	(1% w/v)
Make up to 1 litre with water. Use a freshly prepared solution.		
Solution III		
Sodium acetate, pH 4.8	(3 M)	
246 g/litre. Adjust pH to 4.8 with acetic acid.		
Solution IV		
Sodium acetate	8.2 g	(0.1 M)
Tris	6.05 g	(50 mM)
Make up to 1 litre with water and adjust pH to 8.0 with HCl.		

(ii)   Harvest the cells by centrifugation (10 000 r.p.m. for 10 min in the Sorvall GSA rotor or equivalent).

(iii)   Resuspend the cell pellet in 4.8 ml of Solution I containing 2 mg/ml lysozyme. Incubate at 30°C for 5 min and for 10 min in ice.

(iv)   Add 9.6 ml of Solution II. Vortex and then leave for 5 min in ice.

(v)   Add 7.2 ml of Solution III. Vortex and then leave for 20 min in ice.

(vi)   Centrifuge the mixture at 12 000 r.p.m. for 15 min at 4°C in the Sorvall SS34 rotor or its equivalent.

(vii)   Transfer the supernatant to a fresh tube and add 13 ml of isopropanol. Mix and then leave at −20°C for 15 min to precipitate the nucleic acids.

(viii) Pellet the nucleic acids by centrifugation at 15 000 r.p.m. for 5 min at 4°C in the Sorvall SS34 rotor or its equivalent. Discard the supernatant.

(ix)   Dry the pellet in a vacuum desiccator and then resuspend it in 4.3 ml TE Buffer (10 mM Tris-HCl, 1 mM EDTA, pH 7.5).

(x)   To 3.9 ml of resuspended pellet, add 3.85 g CsCl. Allow the CsCl to dissolve and add 160 μl of a 1% (w/v) solution of ethidium bromide.

(xi)   Add the solution to a Beckmann quickseal tube, adding paraffin oil to fill the tube. Seal the tube following the manufacturer's instructions and centrifuge at 40 000 r.p.m. for at least 18 h at 15°C in the Beckmann Ti70 rotor. (The equivalent fixed-angle or vertical rotors from other manufacturers could also be used).

3.2 Method for the Isolation of Small Quantities of Plasmid DNA

Small amounts of DNA can be prepared using the procedure of Birnboim and Doly (55). However, poor results are obtained, especially with *Bacillus* species other than *B. subtilis*, if overnight cultures are used. Higher yields of plasmid

DNA are obtained if fresh colonies are inoculated into broth and incubated for only $2-3$ h to obtain an O.D. of about 1.0.

(i)   Grow a 1.5 ml culture in VY broth or L broth in a 5 ml or 10 ml flask to O.D.$_{600}$ of about 1.0. The culture should ideally still be in the logarithmic phase of growth.

(ii)   Harvest the cells by centrifuging for 1 min in a microcentrifuge tube.

(iii)   Resuspend the pellet in 100 μl of solution I containing 2 mg/ml lysozyme. (The lysozyme should be added to solution I shortly before use). Leave in ice for 15 min.

(iv)   Add 200 μl of solution II. Vortex and then leave in ice for 5 min.

(v)   Add 150 μl of solution III. Vortex and then leave for 1 h in ice.

(vi)   Centrifuge for 5 min at 4°C at full speed in a microcentrifuge.

(vii)   Remove the supernatant into a fresh tube and add 1 ml of ethanol. Leave at -70°C in a dry ice/ethanol bath for 10 min.

(viii) Centrifuge for 10 min at 4°C at full speed in a microcentrifuge.

(ix)   Carefully remove the supernatant and resuspend the pellet in 100 μl of Solution IV.

(x)   Add 250 μl of ethanol. Vortex and centrifuge for a further 10 min at 4°C in a microcentrifuge.

(xi)   Add 400 μl of ethanol to the pellet. Centrifuge for 2 min at 4°C in a microcentrifuge.

(xii)   Discard the supernatant, dry the pellet in a dessicator, and then resuspend it in 30 μl of TE Buffer. Store at -20°C.

4. PREPARATION AND STORAGE OF COMPETENT CELLS

Plasmids can be introduced into *Bacillus* by four different methods: transformation of competent cells, transformation of protoplasts, transduction using bacteriophage AR9 (see reference 41 for an example), and mobilisation using the conjugative wide host-range plasmid, pAMβ1 (56).

Transformation of competent cells or protoplasts can be used with *B. subtilis*. Methods for preparing competent cells of species other than *B. subtilis* have not yet been devised, so protoplasts are generally used.

An important difference between *E. coli* and *B. subtilis* is that monomeric closed circular plasmid DNA does not effectively transform *B. subtilis*. Only multimeric plasmids can be introduced into *B. subtilis* by the transformation of competent cells (57,58). Complete duplication of the plasmid in the form of a dimer or trimer is not actually required. It is sufficient to duplicate a part of the molecule (25). On the other hand, protoplasts can be effectively transformed by monomeric plasmids. Indeed they are somewhat more effective than multimers. Because multimers are required for transformation of competent cells of *Bacillus*, transformation efficiencies can usually be greatly increased if the plasmid preparation is first cut with an endonuclease which cleaves the plasmid at only one position, and then religated to form multimers.

Table 6. Medium for Preparing Competent Cells.

4 x SP medium (for SPI and SP2)[a]	
$(NH_4)_2SO_4$	8 g
K_2HPO_4	56 g
KH_2PO_4	24 g
Trisodium citrate.$2H_2O$	4 g
$MgSO_4.7H_2O$	0.8 g
Casamino acids (Difco)	0.8 g
Yeast extract (Difco)	4 g

1. Dissolve the above ingredients in 1 litre water, adjusting the pH to 7.2 with NaOH.
2. Autoclave for 15 min at 10 lb/sq. in.
3. For SPI, dilute the medium four-fold with sterile water, add glucose to 0.5% (w/v), thymine at 100 μg/ml and the required amino acids at 50 μg/ml.
4. SPII is the same as SPI except that required amino acids are omitted.

[a]We have obtained best results with salts supplied by B.D.H. and have had poor results with chemicals from certain other suppliers.

Competent cells are easier to prepare and store than protoplasts. In addition, since the rich medium used for regeneration of protoplasts does not allow the direct selection of nutritional markers, replica plating must be used.

The competence of *Bacillus* cultures is a function of the stage of growth. Cells are induced to take up DNA by growing them in a rich medium and then transferring them to a less rich medium lacking the amino acids required by auxotrophic mutants. The time required to reach maximum competence varies according to strain (54). Only a small proportion of cells (~0.1%) in a culture grown to competence are actually able to take up DNA at saturating DNA concentrations. Several modifications to the original procedure of Anagnostopoulos and Spizizen (54) have been published (see, for example, reference 61); the method described below is derived from that described by Dubnau and Davidoff-Abelson (60).

4.1 Method for Preparing Competent Cells

(i)   Inoculate a colony of the recipient strain into a 10 ml culture of SPI medium (*Table 6*). The medium should contain the amino acids (at a concentration of 50 μg/ml) corresponding to the auxotrophic requirements of the strain and thymine (at a concentration of 100 μg/ml if the strain is Thy$^-$). Glucose is added at a concentration of 0.5% (w/v). Culture overnight in a 200 ml flask at 30°C.

(ii) Inoculate 100 ml of the same pre-warmed medium in a 1l flask with the overnight culture to give an O.D.$_{600}$ 0.1 to 0.2. Grow with vigorous aeration at 37°C, taking samples every 30 min to measure the O.D. It is important to stop agitating the culture for the minimum of period of time when taking samples.

(iii) At the end of logarithmic growth, as judged by a decrease in the rate of increase of O.D., add 10 ml of the culture to 90 ml of SPII medium containing 0.5% glucose (and 100 μg/ml thymine if the strain is Thy$^-$, but not the amino acids which were added to the previous cultures grown in SPI medium.

(iv)   Incubate the 100 ml culture in SPII medium for 90 min. The precise time for maximum competence depends to some extent on the individual strain and so samples can be taken at 30 min intervals if maximum competence is essential for the experiment.

(v)   Centrifuge the cells at 5000 r.p.m. for 10 min at 20°C in the Sorvall GSA rotor or equivalent.

(vi)   Resuspend the cells in 10 ml of the supernatant to which glycerol has been added to a final concentration of 10% (v/v). Quickly dispense 0.5 ml volumes into Eppendorf tubes, freeze them in a dry ice-ethanol bath and store them at -70°C. The cells remain competent for at least 6 months.

4.2 Transformation of Competent Cells

(i)   Thaw 0.5 ml of frozen competent cells quickly in a 37°C water bath.

(ii)   Add the cells to a 50 ml conical flask containing $1-5$ μg of DNA in $5-10$ μl.

(iii)   Incubate for 1 h at 37°C with gentle agitation.

(iv)   Add 5 ml of VY broth or L broth and incubate for 1.5 h at 37°C with vigorous agitation.

(v)   Spread aliquots onto plates (Tryptose blood agar base or L agar) containing the appropriate antibiotic to select for plasmid-determined resistance. Tetracycline, kanamycin, erythromycin, and chloramphenicol are used at a concentration of 5 μg/ml in the medium; streptomycin is used at 30 μg/ml.

Higher frequencies of transformation are obtained if bacteria are plated in an agar overlay, incubated for a while, and then overlayed with antibiotic (59,61). This alternative procedure for step (v) is carried out as follows:

Incubate for 30 min (not 1.5 h as in section (iv) above) and add dilutions of the culture to 5 ml of molten Tryptose blood agar base or L agar at 45°C. Pour onto a plate containing 10 ml of the same agar and incubate for 3 h at 37°C. Add a further 5 ml of agar containing the same concentrations of antibiotics listed in section (v) above.

4.3 Preparation of Protoplasts

The transformation of Bacillus protoplasts was first reported by Chang and Cohen (62). An advantage of protoplasts is that they accept monomeric plasmid DNA molecules. In addition, a high proportion of the protoplasts ($>50\%$) can take up plasmid DNA and the method can be used with a wide variety of species for which methods of preparing competent cells have not been developed.

The following method is based on a modification of the procedure as described by Imanaka *et al.* (48). The advantage of this particular method is that it can not only be used for *B. subtilis*, but also for other speices such as *B. sphaericus, B. licheniformis, B. freundenreichii* and *B. circulans*. The only change required for use with these various species is the length of incubation in lysozyme; some

Table 7. Reagents for Protoplast Transformation.

1. *SMMLBV* is made by mixing equal volumes of
   2 x SMM and LBV
   2 x SMM

Sucrose	342 g	(1M)
Sodium maleate	4.72 g	(40 mM)
MgCl$_2$	8.12 g	(40 mM)

   Make up to 1 litre with water and adjust pH to 6.5 with NaOH.
   Autoclave for 15 min at 10 lb/sq. in.
   LBV comprises 2 x L-broth containing polyvinylpyrrolidone at a concentration of 80 g/litre.

Tryptone (Difco)	20 g
Yeast extract (Difco)	10 g
NaCl	10 g
Polyvinylpyrrolidone	80 g

   Make up to 1 litre with water and autoclave for 15 min at 10 lb/sq. in.

2. *Polyethylene glycol*
   To 40 g of polyethyleneglycol (molecular weight ~6000; Sigma Chemical Co.) add 50 ml of 2 x SMM and make up to 100 ml with water. Autoclave for 15 min at 10 lb/sq. in.

3. *DMP regeneration plates.*
   These are made by mixing the various components listed below which are autoclaved separately.

(i)	Sodium succinate	
	Succinic acid (disodium salt)	81 g
	Polyvinylpyrrolidone	80 g

   Make up to 500 ml with water and adjust pH to 7.3 with NaOH.
   Autoclave for 10 min at 15 lb/sq. in.

(ii)	Nutrient agar	
	Agar (Difco)	10 g
	Casamino acids (Difco)	5 g
	Yeast extract (Difco)	5 g

   Make up to 350 ml with water.
   Autoclave for 15 min at 10 lb/sq. in.

(iii)	Potassium phosphate	
	KH$_2$PO$_4$	1.5 g
	K$_2$HPO$_4$	3.5 g

   Make each up to 50 ml with water, mix and autoclave for 15 min at 15 lb/sq. in.

(iv)	Glucose	20 g

   Make up to 100 ml with water. Autoclave for 10 min at 10 lb/sq. in.

(vi)	MgCl$_2$.6H$_2$O	20.3 g

   Make up to 100 ml with water. Autoclave for 15 min at 15 lb/sq. in.

   The autoclaved reagents are mixed together in the following volumes:

(i)	Sodium succinate + PVP	500 ml
(ii)	Nutrient agar	350 ml
(iii)	Potassium phosphate	100 ml
(iv)	Glucose	25 ml
(v)	MgCl$_2$	20 ml

   The components are mixed together soon after autoclaving, while they are still hot, and appropriate antibiotics are added when the medium has cooled to about 50°C.

species, such as *B. sphaericus* are lysozyme-resistant, whereas others such as *B. freundenreichii* are much more sensitive.

(i)   Use plastic vessels or clean (detergent-free) glassware throughout the procedure.

(ii)   Inoculate 50 ml of VY broth or L broth in a 500 ml flask with a few colonies from a fresh (18 h) plate of the strain.

(iii)   Grow with vigorous aeration to O.D.$_{600}$ 0.4.

(iv)   Centrifuge the culture for 10 min at 5000 r.p.m. in the Sorvall GSA rotor or its equivalent. Pour off the culture supernatant and resuspend the cells in 5 ml of SMMLBP (*Table 7*).

(v)   Add 260 μl of 20 mg/ml lysozyme. (The lysozyme solution should be freshly prepared in SMMLB and filter-sterilised).

(vi)   Incubate at 37°C. Take samples at intervals and examine by phase-contrast microscopy to check how many protoplasts have been formed. When 99% or more of the cells have been converted to protoplasts, add $1-5$ μg of DNA dissolved in 10 μl of TE or water and 1.5 ml of 40% (w/v) polyethylene glycol. Mix gently and leave for 2 min at room temperature.

(vii)   Add 5 ml of SMMLBP and mix gently.

(viii) Centrifuge at 3000 r.p.m. for 10 min in the Sorvall SS34 rotor, or its equivalent.

(ix)   Resuspend the protoplasts in 1 ml SMMLBP and incubate at 37°C for 90 min with gentle shaking.

(x)   Spread 0.1 ml aliquots onto protoplast regeneration medium, DMP (*Table 7*), containing the appropriate antibiotics. Kanamycin is much less effective when incorporated into DMP; it must be added at a concentration of 1 mg/ml. Tetracycline, chloramphenicol and erythromycin are added at a concentration of 5 μg/ml.

5. TECHNIQUES FOR STUDYING GENE EXPRESSION IN BACILLUS

Many of the techniques used for studying gene expression in *E. coli* can be used with little or no modification when studying *Bacillus*. For instance, the Broome-Gilbert procedure (63) for detecting antigens produced by bacteria can be modified for *Bacillus* by using lysozyme instead of virulent bacteriophage to lyse the cells. This is achieved by adding 5 μl of a lysozyme solution (2 mg/ml in water), to the bacteria, which have grown as a 5 mm streak on an agar plate, and then incubating for 30 min at 37°C (64).

A method for preparing extracts of *B. subtilis* which can be used for *in vitro* transcription and translation of DNA has been described in detail by Chambliss *et al.* (65). Mini-cells can also be useful for studying gene expression in *Bacillus* (18). Another technique is based on the ability of *Bacillus* to release or secrete products. Proteins released from bacteria during growth on nitrocellulose filters placed on L agar can be readily detected by washing off the cells and then processing the filter by standard 'Western blotting' procedures (66) to reveal the proteins.

Table 8. The Preparation of Minicells.

1.	Grow strain CU403 at 30°C overnight in 100 ml L-broth or VY-broth in a 1 litre fluted flask.
2.	Add an appropriate volume to 400 ml of minimal medium[a] containing 0.5% w/v glucose, 100 μg/ml methionine and 100 μg/ml thymine to give O.D.$_{600}$ 0.1 – 0.2. Grow with vigorous agitation in a two litre flask to O.D.$_{600}$ 1.0. Put the flasks in ice to chill the cells.
3.	Centrifuge at 10 000 r.p.m. for 10 min at 0°C in a Sorvall GSA rotor or its equivalent. Pour off the supernatant and resuspend the pellet vigorously in 5 ml of cold minimal medium lacking glucose, methionine and thymine.
4.	Carefully layer the suspension on two 25 ml 5 – 35% sucrose gradients[a] in Corex tubes. Centrifuge at 0°C in a Sorvall HB-4 swing out rotor or its equivalent at 5000 r.p.m. for 20 min.
5.	Remove the band of minicells, which should be at the centre of the gradient, using a Pasteur pipette bent into a U-shape at its tip.
6.	Pellet the isolated mini-cells by centrifugation at 0°C for 20 min at 5000 r.p.m. in a Sorvall SS34 rotor or its equivalent.
7.	Resuspend the mini-cells in two, 5 ml aliquots of cold minimal medium and repeat steps 4 – 6.
8.	Wash the minicells twice by resuspending them in 5 ml of cold minimal medium and centrifuging at 0°C for 10 min at 12 500 r.p.m. in a Sorvall SS34 rotor or its equivalent.
9.	If the minicells are not to be used immediately for labelling, they can be stored for future use by resuspending them in 3 ml of cold minimal medium containing 10% (v/v) glycerol, freezing the tubes in a dry ice-ethanol bath and storing them at −70°C.

[a]The minimal medium is Spizizen salts (section 2.5) containing 0.5 mM $CaCl_2$, 0.1 mM $MnSO_4$, 0.005 mM $FeCl_3$, sterilised by autoclaving.
[b]Sucrose gradients can be conveniently made using the method of Stone (68). Layer 12.5 ml of 5% (w/w) sucrose in minimal medium onto 12.5 ml of 35% (w/w) sucrose in a Corex centrifuge tube. Leave the tube at an angle of 30° for 18 h at 4°C.

5.1 Preparation and Labelling of Bacillus Mini-cells

Strain CU403 (*Table 1*) is a minicell-producing strain which has been used in several investigations of phage- and plasmid-encoded proteins (18). It grows in minimal medium supplemented only with glucose, thymine and methionine. It does not lyse when centrifuged in sucrose gradients at 4°C, and it does not produce spores. Like most Spo⁻ strains, CU403 is difficult to make competent. Plasmids are normally introduced into the strain by transformation of protoplasts.

Minicells can be prepared and stored as described in *Table 8*. Beginning with 400 ml of culture, one can normally get 2 ml of minicells at O.D.$_{600}$ 0.5, which corresponds to about 5 x 10^9 minicells/ml. Satisfactory results can be obtained in labelling experiments if there are less than 5 x 10^3 viable cells/10^9 minicells. The labelling of the mini-cells can be carried out as follows:

(i) Thaw the minicells by placing them in water bath at 48°C.

(ii) Centrifuge the minicells for 90 sec in microcentrifuge at full speed.

(iii) Resuspend the mini-cells in 0.2 ml of medium comprising 1 volume of 0.4% (w/v) methionine assay medium (Difco), 3 volumes of minimal medium containing 100 μg/ml thymine and 0.5% (w/v) glucose.

(iv) Add 0.1 ml of 500 μg/ml D-cycloserine and incubate at 37°C for 10 min.

(v) Add 100 μCi of [^{35}S]methionine (70 Ci/mmol) and incubate at 37°C for 30 min with shaking in a 5 ml flask.

(vi) Add 0.1 ml of 0.2 mg/ml methionine and incubate at 37°C for 10 min.

(vii)   Centrifuge in a microcentrifuge at full speed for 90 sec. Discard the radioactive supernatant.

(viii)   Resuspend the minicells in 1 ml of water. Centrifuge in a microcentrifuge at full speed for 90 sec and discard the supernatant.

(ix)   Resuspend the minicells in 50 μl of a 0.5 mg/ml solution of lysozyme in water and incubate for 2 min at 37°C.

(x)   Add 100 μl of sample buffer (0.0625 M Tris-HCl pH 6.8, 2% SDS, 5% 2-mercaptoethanol, 10% glycerol, 0.002% bromophenol blue).

(xi)   Heat for 3 min at 100°C and load onto a 12% polyacrylamide gel (or gel of another percentage as appropriate to the proteins under study) prepared as described in reference 69, for example. Load about 50 000 c.p.m. per track.

(xii)   Following electrophoresis (69) stain the gel in a solution comprising 40% (v/v) acetic acid, 20% (v/v) methanol 0.1%, (w/v) coomassie blue in water for 4−8 h. Destain the gel in a solution of 10% (v/v) acetic acid 30% (v/v) methanol in water and then place the gel in Enhance (New England Nuclear) for 1 h. Wash the gel in two changes of water for an hour. Dry the gel in a vacuum drying apparatus and expose it to X-ray film (Kodak X-Omat R) at −70°C (70).

5.2 Use of the Erythromycin Resistance Gene, ermC, for Expression of Cloned Genes

The *ermC* gene of plasmid pE194 (71) encodes an RNA methylase which catalyses a specific N^6,N^6-dimethylation of ribosomal RNA, causing resistance to erythromycin (72). Synthesis of the reductase can be induced by sub-inhibitory concentrations of erythromycin (72). Induction occurs at the level of translation of mRNA, so it is not therefore the ideal transcriptionally controlled promoter which would be particularly useful for the expression of cloned genes. Nevertheless the gene is well-controlled and provides a means of obtaining good yields of foreign proteins (to at least 1% of the total cell protein) in *Bacillus* (64).

It is often more convenient to obtain recombinant plasmids initially in *E. coli*, and then to transfer them to *B. subtilis*. In common with many other *Bacillus* or staphylococcal drug-resistance genes, the staphylococcal *ermC* gene of pE194 also functions in *E. coli* (73). However, because of the intrinsic resistance of *E. coli* to erythromycin, it is necessary to use very high concentrations of erythromycin, or a highly sensitive strain (73) in order to select against *E. coli* not carrying the *ermC* gene. Inactivation of the *ermC* gene in the highly sensitive strain DB11 can be detected by sensitivity to erythromycin at a concentration of 5 μg/ml in L agar. Recombinant plasmids having insertions within the *ermC* gene can thus be readily detected in *E. coli*. Chimeric plasmids carrying *ermC* which can replicate in both *E. coli* and *Bacillus* include pKH80 which comprises pBD9 and pBR322 (63). The *E. coli* part of this plasmid can easily be removed before transfer into *B. subtilis*.

In a typical experiment to examine production of a foreign protein whose synthesis is controlled by the *ermC* gene (64), cells are grown in VY broth to O.D._{600}

0.2. Erythromycin is then added to a concentration of 0.05 μg/ml (a sub-inhibitory concentration) and the culture is grown for a further 2 h with vigorous aeration. At this time the cells are broken by sonication and extracts examined for production of the foreign protein by Western blotting (66) or other assays. The *B. pumilus* chloramphenicol resistance determinant also appears to be controlled at the level of translation; the use of this promoter for expressing cloned genes has been investigated in detail by Lovett and co-workers (49,50).

6. ACKNOWLEDGEMENT

I am grateful to S.D. Ehrlich for an introduction to cloning methods for *Bacillus*.

7. REFERENCES

1. Keggins,K.M., Lovett,P.S. and Duvall,E.J. (1978) *Proc. Natl. Acad. Sci. USA*, **75**, 1423.
2. Gryczan,T.J. and Dubnau,D. (1978) *Proc. Natl. Acad. Sci. USA*, **75**, 1428.
3. Ehrlich,S.D. (1978) *Proc. Natl. Acad. Sci. USA*, **75**, 1433.
4. Palva,I., Lehtovaara,P., Kääriäinen,L., Sibakov,M., Cantell,K., Schein,C.H., Kashiwagi,K. and Weissmann,C. (1983) *Gene*, **22**, 229.
5. Mosbach,K., Birnbaum,S., Hardy,K., Davies,J. and Bulow,L. (1983) *Nature*, **302**, 543.
6. Losick,R. (1982) in *Molecular Biology of the Bacilli*, Dubnau,D. (ed.), Academic Press, New York, p.179.
7. Ehrlich,S.D., Niaudet,B. and Michel,B. (1982) *Curr. Top. Microbiol. Immunol.*, **96**, 19.
8. Hemphill,H.E. and Whiteley,H.R. (1975) *Bacteriol. Rev.*, **39**, 257.
9. Okamoto,K., Mudd,J.A. and Marmur,J. (1968) *J. Mol. Biol.*, **34**, 429.
10. Dubnau,D. and Cirigliano,C. (1974) *J. Bacteriol.*, **117**, 488.
11. Tanaka,T. (1979) *J. Bacteriol.*, **139**, 775.
12. Ehrlich,S.D. (1977) *Proc. Natl. Acad. Sci. USA*, **74**, 1680.
13. Tanaka,T. (1979) *Mol. Gen. Genet.*, **175**, 235.
14. Trautner,T.A., Pawlek,B., Bron,S. and Anagnostopoulos,C. (1974) *Mol. Gen. Genet.*, **131**, 181.
15. Gordon,R.E., Haynes,W.C. and Pang,C.H-N. (1973) *The Genus Bacillus*, U.S. Dept of Agriculture, Washington, DC.
16. Wilson,G.A. and Young,F.E. (1972) *J. Bacteriol.*, **111**, 705.
17. Yasbin,R.E., Fields,P.I. and Anderson,B.J. (1980) *Gene*, **12**, 155.
18. Reeve,J. (1979) in *Methods in Enzymology*, Vol.68, Wu,R. (ed.), Academic Press Inc., London and New York, p.493.
19. Polak,J. and Novick,R.P. (1982) *Plasmid*, **7**, 152.
20. Horinouchi,S. and Weisblum,B. (1982) *J. Bacteriol.*, **150**, 804.
21. Horinouchi,S. and Weisblum,B. (1982) *J. Bacteriol.*, **150**, 815.
22. Goze,A. and Ehrlich,S.D. (1980) *Proc. Natl. Acad. Sci. USA*, **77**, 7333.
23. Kreft,J. and Hughes,C. (1982) *Curr. Top. Microbiol. Immunol.*, **96**, 1.
24. Uhlen,M., Flock,J.I. and Philipson,L. (1981) *Plasmid*, **5**, 161.
25. Michel,B., Niaudet,B. and Ehrlich,S.D. (1982) *EMBO J.*, **1**, 1565.
26. Copeland,J.C. and Marmur,M.J. (1968) *Bacteriol. Rev.*, **32**, 302.
27. Kawamura,F., Saito,H. and Ikeda,Y. (1979) *Gene*, **5**, 89.
28. Yamazaki,H., Ohmura,K., Nakayama,A., Takeichi,Y., Otazaki,K., Yamasaki,M., Tamura,G. and Yamane,K. (1983) *J. Bacteriol.*, **156**, 327.
29. Iijima,T., Kawamura,J., Saito,H. and Ikeda,Y. (1979) *Gene*, **5**, 87.
30. Flock,J.I. (1983) *Mol. Gen. Genet.*, **189**, 304.
31. Marrero,R. and Lovett,P.S. (1980) *J. Bacteriol.*, **143**, 879.
32. Yoneda,Y., Graham,S. and Young,F.E. (1979) *Gene*, **7**, 51.
33. Heilmann,H. and Reeve,J.N. (1982) *Gene*, **17**, 91.
34. Bernhard,K., Schrempf,H. and Goebel,W. (1978) *J. Bacteriol.*, **133**, 879.
35. Bingham,A.H.A., Bruton,C.J. and Atkinson,T. (1979) *J. Gen. Microbiol.*, **114**, 401.
36. Lovett,P.S., Duvall,E.J. and Keggins,K.M. (1979) *J. Bacteriol.*, **127**, 817.
37. Lovett,P.S., Duvall,E.J., Bramucci,M.G. and Taylor,R. (1977) *Antimicrob. Agents Chemother.*, **12**, 435.
38. Mahler,I. and Halvorson,H.O. (1980) *J. Gen. Microbiol.*, **120**, 259.

39. Tanaka,T. and Koshikawa,T. (1977) *J. Bacteriol.*, **131**, 699.
40. Yoshimura,K., Yamamoto,O., Seki,T. and Oshima,Y. (1983) *Appl. Environ. Microbiol.*, **45**, 1733.
41. Gryczan,T.J., Contente,S. and Dubnau,D. (1978) *J. Bacteriol.*, **134**, 318.
42. Iordenescu,S. (1975) *J. Bacteriol.*, **124**, 597.
43. Löfdahl,S., Sjöström,J.E. and Philipson,L. (1978) *Gene*, **3**, 161.
44. Gryczan,T.J., Shivakumar,A.G. and Dubnau,D. (1980) *J. Bacteriol.*, **141**, 246.
45. Michel,B., Palla,E., Niaudet,B. and Ehrlich,S.D. (1980) *Gene*, **12**, 147.
46. Keggins,K.M., Lovett,P.S., Marrero,R. and Hoch,S.O. (1979) *J. Bacteriol.*, **139**, 1001.
47. Tanaka,T. and Kawano,N. (1980) *Gene*, **10**, 131.
48. Imanaka,T., Fujii,M., Aramori,I. and Aiba,S. (1982) *J. Bacteriol.*, **149**, 824.
49. Schoner,R.G., Williams,D.M. and Lovett,P.S. (1983) *Gene*, **22**, 47.
50. Duvall,E.J., Williams,D.M., Lovett,P.S., Rudolph,C., Vasantha,N. and Guyer,M. (1983) *Gene*, **24**, 171.
51. Gray,O. and Chang,S. (1981) *J. Bacteriol.*, **145**, 422.
52. Kreft,J., Bernhard,K. and Goebel,W. (1978) *Mol. Gen. Genet.*, **162**, 59.
53. Band,L., Yansura,D.G. and Henner,D.J. (1983) *Plasmid*, **26**, 313.
54. Anagnostopoulos,C. and Spizizen,J. (1961) *J. Bacteriol.*, **81**, 741.
55. Birnboim,H.C. and Doly,J. (1979) *Nucleic Acids Res.*, **7**, 1513.
56. Clewell,D. (1981) *Microbiol. Rev.*, **45**, 409.
57. Canosi,U., Morelli,G. and Trautner,T.A. (1978) *Mol. Gen. Genet.*, **166**, 259.
58. Mottes,M., Grandi,G., Sgaramella,V., Canosi,U., Morellei,G. and Trautner,T.A. (1979) *Mol. Genet.*, **174**, 281.
59. Niaudet,B. and Ehrlich,S.D. (1979) *Plasmid*, **2**, 48.
60. Dubnau,D. and Davidoff-Abelson,R. (1971) *J. Mol. Biol.*, **56**, 209.
61. Contente,S. and Dubnau,S.D. (1979) *Mol. Gen. Genet.*, **167**, 251.
62. Chang,S. and Cohen,S.N. (1979) *Mol. Gen. Genet.*, **168**, 111.
63. Broome,S. and Gilbert,W. (1978) *Proc. Natl. Acad. Sci. USA*, **75**, 2746.
64. Hardy,K., Stahl,S. and Küpper,H. (1981) *Nature*, **293** 481.
65. Chambliss,G.H., Henkin,T.M. and Levinthal,J.M. (1983) *Methods in Enzymology*, **101**, 598.
66. Burnette,W.N. (1981) *Anal. Biochem.*, **112**, 195.
67. Mertens,G. and Reeve,J.N. (1977) *J. Bacteriol.*, **129**, 1198.
68. Stone,A.B. (1974) *Biochem. J.*, **137**, 117.
69. Laemmli,U.K. (1970) *Nature, New Biol.*, **227**, 680.
70. Bonner,W.H. and Laskey,R.A. (1974) *Eur. J. Biochem.*, **46**, 83.
71. Shivakumar,A.G. and Dubnau,D. (1981) *Nucleic Acids Res.*, **11**, 2549.
72. Weisblum,B. (1975) in *Microbiology-1974*, Schlessinger,D. (ed.), American Society for Microbiology, Washington, DC, p.199.
73. Hardy,K. and Haefeli,C. (1982) *J. Bacteriol.*, **152**, 524.
74. Jalanko,A., Palva,I. and Soderlund,H. (1981) *Gene*, **14**, 325.
75. Scheer-Abramowitz,J., Gryczan,T.J. and Dubnau,D. (1981) *Plasmid*, **6**, 67.

CHAPTER 2

Gene Cloning in Streptomyces

IAIN S.HUNTER

1. INTRODUCTION

Gene cloning techniques for *Streptomyces* are relatively new. Indeed, it is less than 4 years since the first report (in 1980) of the cloning of a streptomycete gene (1). Gene cloning now allows us to perform experiments with streptomycetes which were impossible previously. The foundations of streptomycete molecular biology were laid in Hopwood's laboratory and depended on a vast background of genetical information obtained from that laboratory over 25 years.

Streptomyces are gram-positive soil microorganisms. They have a genome of size around 10^4 kb, consisting of a single circular chromosome. *S. coelicolor* A3(2) is the best characterised example. A good genetic map exists (2) with map positions for auxotrophic, fermentative and drug-resistant mutations, and for mutations involved in differentiation. *S. coelicolor* serves as a model for a differentiating prokaryote. In the future, cloning of the genes involved in differentiation will undoubtedly help to unravel the complexities of this process.

In the last few years, much of the molecular biological research on the genus has been concerned solely with the development of molecular techniques. With that phase almost over, these techniques are now being used in investigations focussed on other fascinating aspects of the biology of *Streptomyces* including the production of antibiotics. *Streptomyces* produce a wide variety of antibiotic structures, many of which are of commercial importance. Each species may produce more than one antibiotic. For example *S. coelicolor* is known to produce four different antibiotics (3), whereas some species (e.g. *S. clavuligerus*) produce over 30 structures.

Over 60% of known naturally-occurring antibiotics come from *Streptomyces* (4), making the genus the major taxonomic class sought by pharmaceutical companies in screening programmes for new naturally-occurring isolates. As the search continues for novel structures, recombinant DNA techniques will be used both in screening strategies and to improve the titres of fresh isolates producing new antibiotics. The recombinant techniques may also be used to generate novel structures, which are not made in naturally-occurring isolates, by cloning antibiotic production genes from one species into another. In the 'biotechnology era' the commercial aspects of antibiotics have helped focus attention on the subject. To date, comparatively little is known about expression of the genes involved in antibiotic production, but reports of cloning of the genes for these biosynthetic pathways are beginning to appear (5,6). Recently (7) the entire biosynthetic pathway for actinorhodin, an antibiotic made by *S. coleicolor*, has been cloned in blocked mutants of *S. coelicolor*. The cloned genes were transformed into *S. parvulus*, a heterologous host which does not make the antibiotic. The transformed *S. parvulus* host produced actinorhodin. This elegant experiment demonstrates the feasibility of producing antibiotics by interspecific DNA cloning.

2. A STRATEGY FOR DNA CLONING IN STREPTOMYCES

2.1 **The Host Strain**

The first decision to take in planning any cloning experiment is to choose the streptomycete host. The point may seem trivial, but is often overlooked, with the result that some cloning experiments are unsuccessful. A suitable host should be selected, based on the critiera indicated below.

2.1.1 *Host Phenotype*

Most strategies for screening gene banks of streptomycete DNA depend on phenotypic expression of the cloned gene, i.e., the recipient has to be deficient in an enzyme activity which is complemented by the product of the gene when it is cloned into the vector. For example, to clone a gene for resistance to a drug, a host is required which is sensitive to that drug; to clone a gene involved in amino acid biosynthesis, a host is required which is auxotrophic for the amino acid specifically because it lacks that gene product. In the former case, a recipient which is normally sensitive to the drug is preferred because it will not revert; a mutant of a parental strain which originally had the resistant phenotype but was now sensitive, might revert at a high frequency to resistance and is less acceptable. Inevitably, when an auxotrophic host is required a mutant has to be used. But its reversion frequency to prototrophy must be shown to be low. Standard techniques for mutagenising *Streptomyces* to obtain auxotrophs are available (9). Potential hosts should be tested for sensitivity to the drug markers used commonly with streptomycete vectors (thiostrepton, neomycin, viomycin, erythromycin).

2.1.2 *Host Transformation Efficiency/Regeneration Efficiency*

Section 2.3 deals with the statistical analysis of gene banks. From that it is clear that, if transformation of streptomycete protoplasts is less than $10^5/\mu g$ for uncut vector (i.e., $\sim 10^3/\mu g$ vector in ligation reactions), unacceptably large quantities of DNA will be required to construct gene banks.

(i)   The poor apparent transformation efficiency may be due not to poor transformation *per se*, but to poor regeneration of protoplasts once they have been transformed. In that case, some of the changes suggested in Section 4.6 should be tried to enhance regeneration.

(ii)   If transformation itself is poor, a change of vector may be required, e.g., *S. rimosus* is tranformed with uncut SCP2*-based vectors at $<10^3/\mu g$, whereas with pIJ101 replicons transformation is $>10^6/\mu g$.

(iii)   It may be necessary to use another host if both (i) and (ii) do not result in transformation efficiency approaching $10^5/\mu g$.

2.1.3 *Host Restriction Activity*

If because of the considerations outlined in Section 2.1.1 a heterologous strep-tomycete host is to be used, its restriction activity has to be evaluated. Many *Streptomyces* have active restriction systems. This is particularly true of the strains used by industry which are selected for robust performance in large fermenters. Their high restriction activities protect them from infection by phage. When they are used as recipients in cloning experiments, the insert DNA from the donor can be recognised as 'foreign' by the recipient and restricted, so that a good gene bank is not obtained.

The best way of testing if a restriction barrier exists between donor and recipient is to transform a plasmid into the donor, isolate plasmid DNA from a transformant, and then use this DNA to transform the recipient. If no restriction barrier exists, the number of transformants per μg of the DNA will be equivalent to that obtained in the control experiment when the plasmid is isolated from the recipient and tranformed back into protoplasts of the recipient. A lower transformation efficiency than in the control experiment indicates a restriction problem. However, if the donor strain is not well-characterised, the technical problems associated with transformation of a plasmid initially into the donor (i.e., making and regenerating protoplasts, etc.) make this approach untenable.

A second, less rigorous, test can be made using a bifunctional (*Streptomyces/E. coli*) vector with plasmid DNA isolated from *E. coli*. A streptomycete host totally devoid of restriction activity will be transformed with DNA from *E. coli* at an efficiency equal to that for plasmid DNA isolated from the streptomycete host. *S. ambofaciens* was shown to lack restriction activity in this way (14).

If restriction activity is suspected in a host, three options are open:

(i)   choose an alternative host which does not have the problem;
(ii)   attempt to mutate the host to reduce its restriction activity;
(iii)   'heat shock' the protoplasts to attenuate the restriction activity.

The second approach (*Figure 1*) has been used successfully with an industrial strain of *S. rimosus* to obtain a mutant which is phenotypically 'restriction-deficient', although the restriction enzymology of the mutant has not been investigated further (10). Before the mutation treatment, pPZ33 [a bifunctional *E. coli*/streptomycete vector based on pIJ101 (see Section 2.2.1)] transformed *S. rimosus* poorly with the plasmid DNA isolated from *E. coli* ($2 \times 10^2/\mu$g), but well ($7 \times 10^5/\mu$g) when the plasmid DNA came from *S. rimosus*. The mutant was transformed with plasmid DNA from *E. coli* at an efficiency of $6 \times 10^5/\mu$g, indicating that the restriction barrier between *E. coli* and *S. rimosus* no longer existed.

In the third approach, protoplasts are incubated at $45-50°C$ for $10-15$ min to denature the restriction enzyme. Protoplasts remain viable and can still be transformed after mild 'heat-shock'. The exact conditions of temperature and time have to be arrived at empirically for each strain. Unfortunately, not all strains have thermolabile restriction activities.

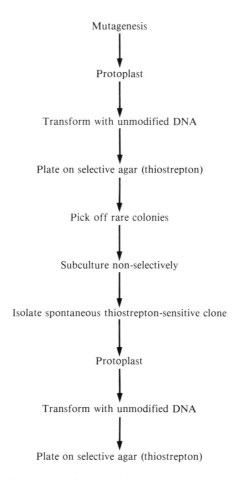

Mutagenesis

↓

Protoplast

↓

Transform with unmodified DNA

↓

Plate on selective agar (thiostrepton)

↓

Pick off rare colonies

↓

Subculture non-selectively

↓

Isolate spontaneous thiostrepton-sensitive clone

↓

Protoplast

↓

Transform with unmodified DNA

↓

Plate on selective agar (thiostrepton)

Figure 1. Strategy for selection of restriction-deficient *Streptomyces*.

2.1.4 *Host DNase Activity*

Some streptomycetes have active extracellular enzymes for degradation of DNA (11), which may either leach out of protoplasts or be present in protoplast suspensions in which some protoplasts have lysed. Thus, if plasmid DNA is added to the protoplasts, it may be partially degraded and rendered inactive for transformation.

DNase activity can be tested by adding a known amount of plasmid to the protoplasts, incubating for a short time, and centrifuging out the protoplasts in a microfuge. The supernate is loaded onto an agarose gel to check for degradation of the DNA. If DNase activity is present, it is worth testing if 'heat shock' (as described in Section 2.1.3) will inactivate it, although this is unlikely. Addition of heterologous DNA (e.g., calf thymus DNA) to protoplasts before addition of plasmid DNA affords partial protection against DNase. At least 100 μg of calf thymus DNA can be added to 100 μl of protoplasts without affecting adversely the transformation of plasmid DNA.

2.1.5 *Host Recombination Activity*

Streptomyces are highly recombinogenic. Ideally, a recombination-deficient streptomycete host would minimize the likelihood of deletion or rearrangement of some of the cloned DNA sequences. Although some u.v.-sensitive mutants of *Streptomyces* are available, there is no well substantiated report of selecting a 'recombination-deficient' strain. The acquisition of such a strain is awaited eagerly by many workers in the field. If available, this attribute would be given high priority in the ranking of the above criteria.

2.2 **The Vector**

A large number of streptomycete plasmids are described in the scientific and patent literature. Therefore, there is no longer a limited choice of vector, although some are better developed than others.

A decision has to be made on the copy number of the vector to be used, i.e. high-copy plasmid, low-copy plasmids or single-copy lysogenic phage. Vectors can be either self-transmissible or non-transmissible. They may be cryptic and recognised by 'lethal zygosis' (where a colony containing the plasmid produces a translucent zone when replicated to a lawn of growing cells which do not contain the plasmid — see Section 4.7.1), or contain a drug-resistance marker.

2.2.1 *High Copy Number Vectors*

Several high-copy number vectors have been reported (see below). Vectors based on pIJ101 are preferable, since they are well characterized and have high transformation efficiencies.

High copy-number vectors are easy to isolate, and have been used in many recent cloning experiments. It is possible that streptomycete vectors may have such a high copy-number that when some genes are cloned onto them the inevitable overexpression is deleterious to the cell. Malpartida and Hopwood (7) have cloned into a low-copy vector two DNA segments coding for an antibiotic pathway which had proved refractory to cloning into a high-copy vector. It has not been possible subsequently to subclone parts of these DNA sequences from the low-copy to the high-copy vector, suggesting (although not proving) that overexpression of some of the pathway genes is lethal. This ongoing aspect of streptomycete molecular biology should be kept under close scrutiny when devising a suitable cloning strategy.

pIJ101. Plasmids based on pIJ101 are used frequently for shotgun cloning. pIJ101 was one of a family of plasmids isolated from a *S. lividans* strain (12). The strain contained three other plasmids of smaller size which, on restriction analysis, were shown to be deletion versions of the parent (pIJ101) plasmid that had arisen *in vivo*. pIJ101 is self-transmissible and exhibits the 'lethal zygosis' response (Section 4.7.1). Analysis of the behaviour of *in vivo* and *in vitro* deletion variants has allowed tentative assignment of loci for these phenotypes on the plasmid map (12). pIJ101 (9 kb) has the widest host range of the streptomycete plasmids documented to date. It was maintained in 13 of the 18 strains tested in the original paper (12). Since then the replicon has been used extensively in ex-

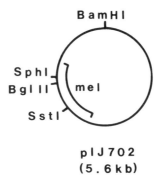

Figure 2. Restriction map of pIJ702. Cloning into the unique *Sph*I, *Bgl*II or *Sst*I sites results in inactivation of the *mel* gene.

periments with many other strains.

The copy number in *S. lividans* is 40 – 300. Deletion variants have even higher copy numbers. Copy number is somewhat strain-dependent, e.g., *S. rimosus* gives higher yields of plasmid than *S. lividans*, presumably reflecting a higher copy number in *S. rimosus*.

Smaller versions of pIJ101 have been constructed *in vitro* and marked with several drug resistant determinants (12). Resistance to thiostrepton (cloned from *S. azureus*) is the most useful, since most streptomycetes are sensitive to that drug. pIJ350 (12) is a 4.1 kb thiostrepton-resistant derivative of pIJ101.

A more useful cloning vector (pIJ702) was made by subcloning a DNA fragment encoding the gene for tyrosinase of *S. antibioticus* into pIJ350 (13). Colonies of *S. lividans* containing active tyrosinase (Mel$^+$) are coloured black on medium supplemented with tryptone and copper ions, which provides a good visual screen. pIJ702 contains several unique restriction sites within the *mel* gene (*Figure 2*). When other DNA sequences are cloned into these sites, colonies of transformants are colourless, since active tyrosinase is no longer made. This provides an 'insertional inactivation' test for DNA cloned into the *mel* gene. The *Bgl*II target within the *mel* gene of pIJ702 is particularly useful in shotgun cloning experiments. The cohesive termini of the vector DNA after *Bgl*II digestion are compatible with the termini generated by many restriction enzymes (*Bam*HI, *Xho*II, *Bcl*I, *Sau*3A or *Mbo*I). *Sau*3A or *Mbo*I recognise tetrameric sequences (GATC) which occur frequently in streptomycete DNA. Thus, partial digestion of donor DNA with *Sau*3A or *Mbo*I generates random DNA fragments to ligate into pIJ702 cut with *Bgl*II.

pFJ103. pFJ103 has been developed by the Lilly group (14). It was isolated originally as a 20 kb plasmid from *S. granuloruber*. After the thiostrepton-resistance determinant was introduced, an *in vitro* variant consisting of the small replicon segment was obtained (pFJ105, 4.7 kb). No definitive copy number data has been reported, although the ease with which the plasmid was isolated suggested it was present in high copy number. *S. ambofaciens, S. fradiae* and *S. aureofaciens* are hosts for the plasmid. The reported transformation efficiency

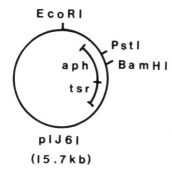

Figure 3. Restriction map of pIJ61. The unique *Pst*I and *Bam*HI sites are within the neomycin resistance gene (aph).

(4.6 x 10^3/μg) is somewhat low for it to be a really good cloning vector (see Section 2.3).

pUC6. pUC6, was isolated by Upjohn as a 9.2 kb cryptic plasmid from *S. espinosus* (15). It has a copy number of 30−40. pUC1061 (7.2 kb) is a copy number mutant (copy number 500−600) generated by *in vitro* deletion mutagenesis of pUC6. It has the highest copy number of streptomycete plasmids reported to date, with over 40% of the total cellular DNA as plasmid. It is difficult to screen for pUC1061 since it does not contain a drug-resistance marker, nor does it display the true 'lethal zygosis' phenotype. Introduction of a drug-resistance determinant would appear to be a trivial task in developing it further as a vector.

2.2.2 *Low Copy Number Vectors*

In the early days low-copy vectors had the disadvantage that sufficient quantities of DNA were difficult to isolate. The methodology for DNA plasmid isolation is now so good that this is less of a problem. If the streptomycete host has no restriction activity (Section 2.1.3) a chimera of the low-copy streptomycete vector and a high-copy *E. coli* plasmid can be made, so that large quantities of DNA can be isolated from *E. coli*. The chimera can be designed so that digestion with a suitable restriction enzyme gives a DNA fragment which contains only the streptomycete vector. It can be purified away from the sequences of *E. coli* plasmid and used in ligations directly.

SLP1.2. SLP1.2 (14.5 kb) was used in the early gene cloning experiments with *Streptomyces* (16). The plasmid was first detected in a 'lethal zygosis' or 'pocking' variant of *S. lividans* after an interspecific mating with *S. coelicolor*. Subsequently, SLP1.2 was shown to be derived from the chromosome of *S. coelicolor*, where it exists as an integrated plasmid (17). In *S. lividans*, the autonomous plasmid has a copy number of 4−5.

Versions of SLP1.2 are available which have single markers or combinations of genes for drug resistances (18). Amongst them, pIJ61 is probably the most versatile (*Figure 3*). pIJ61 (15.7 kb) contains the genes for resistance to thiostrepton

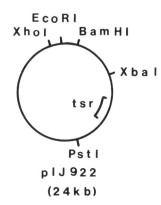

Figure 4. Restriction map of pIJ922 (20). All the restriction sites shown are available for cloning.

(*tsr*) and neomycin (*aph*). Within the *aph* gene, there are unique restriction sites for *Bam*HI and *Pst*I. Cloning into either site results in sensitivity to neomycin, providing a convenient 'insertional inactivation' screen. The *Bam*HI site can be used to clone random DNA fragments generated by *Sau*3A or *Mbo*I, in the same way as described for pIJ702 (Section 2.2.1).

The National Research and Development Corporation (NRDC) of the UK hold a patent on the SLP1.2 replicon. Workers interested in using SLP1.2 for research experiments should not be discouraged from doing so because of the NRDC patent. Complications only arise with patented vectors when a profitable product or process results from their use; all patented vectors are available for research purposes without any penalty.

It has been reported that the host range of SLP1.2-based replicons is limited (18). The apparent limited host range may be due to incompatibility with resident plasmids of the potential host, or due to restriction activity (Section 2.1.3) of the host. It is possible that the host range of the replicon is wider than first thought.

SCP2. Although SCP2 was the first streptomycete plasmid to undergo systematic physical investigation (19), its use as a cloning vector was confined to methylenomycin resistance (1) and (until recently) it has been rather overshadowed by SLP1.2. The SCP2 plasmid (30 kb) is a resident sex factor of *S. coelicolor*. It is present in copy number of 1−3. It is transmissible and displays the 'lethal zygosis' phenotype.

pIJ922 (*Figure 4*; 20) is the most useful available version of SCP2. It is a 24.5 kb plasmid and is marked with thiostrepton resistance. Several unique restriction sites are available for cloning. The success of the cloning of the actinorhodin pathway with this vector (7) will undoubtedly result in reconsideration of its utility in experiments of this type.

Like SLP1.2, SCP2 was formerly considered to have a narrow host range. With a drug-resistance selection (such as exits in pIJ922) rather than 'lethal zygosis', and using hosts which are 'restriction deficient', there is a good chance that, like SLP1.2, the host range of SCP2-based replicons will prove to be much wider than thought previously.

2.2.3 *Phage Vectors*

The broad-host range actinophage ϕC31 has been developed as a vector for strep-tomycetes (21,22). The phage/host interactions are reminiscent of those of phage lambda with *E. coli*. The genome is a linear 41 kb DNA molecule with cohesive ends. It can be lysogenised, when the prophage state is under the control of a repressor (the *C* gene product). Lysogenisation occurs by recombination of the phage attachment site with a chromosomal attachment sequence. Therefore ϕC31 can exist in a host as a single-copy lysogen.

Variants of ϕC31 have been engineered with either thiostrepton- or viomycin-resistance markers. Chimeras with pBR322, which exist in *E. coli* as plasmids, are also available and are good sources of phage DNA. Like phage lambda, there are limits for packaging of DNA into the head of ϕC31. The phage vectors currently available permit a maximum of 7 kb of insert DNA to be cloned. Undoubtedly, the phage vectors will be developed in the future to accept larger DNA inserts.

Chater has developed a 'mutational cloning' system based on ϕC31 (23). With an uncharacterised streptomycete, it is possible to construct gene banks of its DNA using a ϕC31 derivative which is deleted in the normal phage attachment site. The bank is screened for the impaired phenotype (e.g., inability to produce an antibiotic), which is due to gene disruption by the phage lysogenising using the cloned DNA as substitute attachment site. The impaired phenotype (blocked mutant) is only constructed if the cloned DNA contains an incomplete unit of transcription for the gene in question. This technique promises fast genetic analysis of new pathways in uncharacterised strains.

2.3 Statistical Analysis

In order to determine whether there are sufficient clones with inserts of given length to represent the entire genome of a donor streptomycete, or in order to determine how much DNA is necessary to isolate a representative gene, it is necessary to use the binomial theorem for a statistical analysis. The genome size of *Streptomyces* is 10^4 kb. *Table 1* shows the number of clones required to obtain high probabilities of genome representation in streptomycete gene banks of given average insert size. The information can be used in two ways:

(i) *To determine how much vector DNA is needed to make a gene bank.* The transformation efficiency of primary gene banks is invariably 100-fold less

Table 1. Number of Clones of Given Average Size Required for a Gene Bank of *Streptomyces* DNA at Various Confidence Levels.

Average length of cloned fragments (kb)	*Percent probability of complete representation in bank*			
	80	*90*	*95*	*99.99*
5	3218	4604	5998	18 414
7.5	2414	3453	4491	13 810
10	1609	2302	2994	9206
15	1072	1534	1998	6136
35	460	657	854	2627

than of uncut vector. Even with vector DNA which has been phosphatased (4.2.1), the frequency of transformants that contain recombinant molecules is often only 50%. A first approximation of the amount of DNA necessary to make a gene bank can then be calculated.

Example: A streptomycete protoplast preparation is transformed with uncut pIJ702 at a frequency of 5 x $10^5/\mu$g DNA. How many colonies are likely to constitute a gene bank of average insert size 7.5 kb?

For the bank to represent 99.9% of the genome, 13 810 *insert*-containing colonies are necessary (*Table 1*). If the frequency of transformants containing recombinant molecules is 50%, then 27 620 (i.e., 2 x 13 810) *plasmid*-containing colonies are required. Assuming that the primary gene bank will have a transformation efficiency of 5 x $10^3/\mu$g (i.e., 1/100 of 5 x 10^5) then $(2.75 \times 10^4)/(5 \times 10^3)$ = 5.5 μg of vector will be required. By a similar calculation, for 95% representation of the genome, 1.6 μg of vector will be required.

It is usually safer to carry out the experiment at the higher level of DNA, and obtain too many colonies, than to use the minimum amount of DNA and end up with an incomplete bank because of poor transformation on the day.

(ii) *To determine whether the colonies obtained after transformation constitute a representative 'gene bank'.* Pick 30 − 40 colonies from the putative bank, and grow them up as fresh colonies patched onto plates containing the drug required to maintain selection of the plasmids. For each, prepare a colony lysate by rapid technique (*Table 2*) and digest the lysate with suitable restriction enzymes to determine the length of the insert DNA. From these data, calculate the frequency of colonies containing recombinant molecules and the average insert size in order to test if enough insert sequences are present. Although a sample of 40 is probably too small on which to base a statistically-significant conclusion, a good idea of the

---

Table 2. Isolation of plasmid DNA from colonies of *Streptomyces*. Developed from a method originally devised by Kieser (29).

---

1. Remove the colony from the plate into a microcentrifuge tube, taking care not to remove any agar with the colony.
2. Add 30 μl TE[a] containing 2 mg/ml lysozyme. Crush the colony with a micropipette tip. Incubate at 30°C for 15 min.
3. Add 15 μl of a solution containing 0.3 M NaOH and 2% w/v SDS. Agitate immediately on a vortex mixer.
4. Incubate for 15 min at 70°C and cool to room temperature.
5. Add 30 μl phenol/chloroform[b]. Agitate on a vortex mixer for 10 sec. Spin in a microfuge for 2 min to separate the phases.
6. The aqueous phase can be loaded on a gel at this stage. Alternatively, for restriction analysis of the DNA, take the aqueous phase into a clean microcentrifuge tube. Add 4 μl of 3 M sodium acetate and 45 μl isopropanol. Mix by inversion. Stand the tube at room temperature for 5 min. Centrifuge in a microfuge for 2 min, decant the supernatant, and wash the pellet several times in 40% (v/v) isopropanol. Dissolve the pellet in 10 μl of distilled water ready for restriction analysis.

---

[a]TE is 10 mM Tris, 1 mM EDTA, pH 8.0.
[b]5 g phenol, 5 g chloroform, 5 mg hydroxyquinoline.

status of a putative gene bank can be obtained in this way.

Example: A gene bank contained 8550 drug resistant colonies. 40 random isolates were analysed. 28 (i.e., 70%) had inserts of average length 6.2 kb.

Of the 8500 colonies, 5985 (i.e., 70%) will probably have inserts. By calculation from *Table 1*, the 95% confidence level is 4837, and 99.99% level is 14 850. Therefore, the number of insert-containing colonies in the bank is significant at the 95% level, but not significant at the 99.99% level.

3. GROWTH AND MAINTENANCE OF STRAINS

Streak out the streptomycete strain on solid medium [TSB or YEME (see Section 5)] solidified with 1.5% w/v agar. Incubate the plate at 30°C for several days until sporulation occurs. Identify a single colony which is well separated from the others. Using a sterile microbiological loop which has been dipped into sterile distilled water (this helps the spores to stick to the loop) remove the spores from the colony onto the loop. Suspend the spores in 200 µl sterile water.

Store the spores either by inoculating an agar slope which is stored at −20°C after it has grown up, or in aqueous glycerol (20% v/v) suspensions at −20°C. For most strains, spores are maintained stably under these conditions for several years.

Inoculate liquid medium (YEME or TSB) in conical flasks with spores from agar slopes. The spores are scraped from the slopes into a suspension with a few millilitres of sterile distilled water using a sterile pipette or long needle on a syringe. Remove the mycelial debris from spore suspensions by filtering through plugs of non-adsorbent cotton wool (*Figure 5*). To inoculate the liquid medium with spores which have been stored in glycerol, pellet the spores in a bench top centrifuge (3000 *g*, 10 min) and resuspend them in sterile water before inocula-

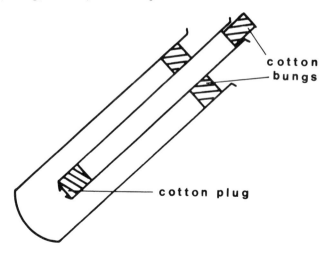

Figure 5. Filter tubes for removing mycelial debris from the spore and protoplast suspensions. A hole is made in the end of a ½ in diameter test tube with a red-hot needle. The bottom of the tube is plugged with non-absorbent cotton wool and held in a 1 in diameter test tube with a cotton bung. The spore (or protoplast) suspension is placed in the inner tube and allowed to filter through the plug to the outer tube.

tion. Glycerol can inhibit the germination of spores, so it is better to remove it before germination. For some strains, spore germination is accelerated by 'heat-shock'. This entails incubating the flask of inoculated media at $45-50°C$ for 10 min before placing it on the shaker at 30°C.

Conical flasks are shaken at 300 rev min^{-1} to give vigorous aeration. Most strains grow reasonably well at 30°C although some are fastidious, e.g., *S. clavuligerus* grows very poorly at 30°C but well at 26°C. It usually takes 2 days for 200 ml of culture to grow well from a spore inoculum.

In some laboratories, fungal contamination is a problem. Adding cyclohex-imide (10 μg/ml) to both liquid and solid medium prevents growth of fungal con-taminants without affecting growth of *Streptomyces*. TSB is preferable to YEME as a growth medium because it is easier to prepare. Centrifugation of TSB is easier because it is not as viscous as YEME which contains a high level of sucrose. However, some strains do not grow well in TSB, and for them YEME has to be used.

4. PRACTICAL APPROACHES TO DNA CLONING

4.1 An overview of DNA Cloning in Streptomyces

The steps involved in the preparation of a 'gene bank' of streptomycete genomic DNA are summarized in *Figure 6.*

(i)   Prepare chromosomal DNA from the donor (see Section 4.2.1) and partial-ly digest it with a restriction enzyme which cuts the DNA frequently, e.g., *Sau*3A. Fractionate the DNA so that the average length of the fragments to be cloned falls within known limits. This step is essential when using λ and cosmid vectors, and very useful when applying statistical analysis to con-ventional plasmid gene banks. The approach has been discussed by Kasier and Murray in Chapter 1.

(ii)   Isolate vector DNA (see Section 4.2.2), digest it with a restriction enzyme which gives 'sticky ends' compatible with those of the chromosomal DNA, and phosphatase to prevent recircularization of the vector without a DNA insert during ligations (see Section 4.2.4).

(iii)   Ligate vector and donor DNA fragments (see Section 4.3).

These three steps are common to the preparation of any genomic DNA bank. With *Streptomyces,* DNA has to be introduced into recipients in the form of protoplasts. This step is very different from transformation of *E. coli*.

(iv)   Remove the cell wall with lysozyme to make protoplasts which must be kept in osmotically-supported medium (see Section 4.4).

(v)   Mix the DNA with the protoplasts, then add polyethylene glycol (PEG). By a process poorly understood, DNA enters the cells, i.e., they are transformed (see Section 4.5).

(vi)   Regenerate the protoplasts (transformed and untransformed) in rich medium under non-selective conditions (i.e., allow protoplasts to regenerate their cell walls). Selection for transformed cells is done subse-quently (see Section 4.6).

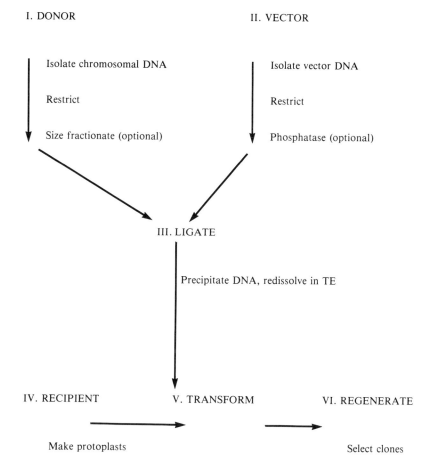

I. DONOR   II. VECTOR

Isolate chromosomal DNA   Isolate vector DNA

Restrict   Restrict

Size fractionate (optional)   Phosphatase (optional)

III. LIGATE

Precipitate DNA, redissolve in TE

IV. RECIPIENT   V. TRANSFORM   VI. REGENERATE

Make protoplasts   Select clones

Figure 6. Preparation of a genomic DNA bank of *Streptomyces*.

4.2 Preparation of DNA

The preparation of good quality, high molecular-weight DNA, which is free from RNA and protein, is essential for any cloning experiment. Restriction enzymes should digest the DNA easily. Large quantities of DNA are required to construct most Streptomycete gene banks (see Section 2.3). Therefore, if the DNA is not totally pure and needs vast excesses of restriction enzymes to digest it, the cost of construction of gene banks can become prohibitive. In addition, most restriction enzyme preparations contain trace impurities of exonucleases. If a large quantity of restriction enzyme has to be added to digest the DNA, a proportionately large amount of contaminating exonucleases will be added with it. The exonucleases will act on the sticky ends of the DNA so that it is not good for cloning. It is worth while taking a lot of care over the preparation of DNA.

Table 3. Preparation of *Streptomyces* Total DNA-Method 1.

1.	Inoculate 200 ml of TSB or YEME[a] (depending on the strain, see text) with 1 ml of a dense spore suspension. Incubate at 30°C with shaking.
2.	Harvest when the growth is dense (usually 2 days) by centrifugation in a 250 ml centrifuge tube in the Beckman JA14 rotor (or its equivalent) at 12 000 r.p.m. (22 100 g) for 10 min.
3.	Resuspend the cells in 10 ml of TE[b], add 10 mg lysozyme and incubate for 15 min at 30°C.
4.	Add 1 ml of 20% SDS, stirring gently for 15 sec with a pipette. The solution will go viscous. *Immediately* add 10 ml TE-saturated phenol and 1.5 ml 5 M NaCl. Invert the tube repeatedly and gently at room temperature for 20 min.
5.	Transfer the mixture to a 50 ml capped polypropylene tube and centrifuge at 1500 r.p.m. for 10 min in the swing-out rotor of a bench-top or medium speed centrifuge to separate the phases.
6.	Remove the viscous top layer with a wide bore pipette and transfer it to a clean tube. Do not put the tip near the interface.
7.	Add an equal volume of chloroform. Mix by inverting gently for 10 min. Centrifuge at 1500 r.p.m. for 10 min in a swing-out rotor as in step 5.
8.	Remove the viscous top layer and repeat step 7.
9.	Remove the viscous layer into a sterile 100 ml beaker. Carefully add an equal volume of isopropanol to the beaker, so that two phases are formed. Spool the DNA precipitating at the interface onto a Pasteur pipette.
10.	Redissolve the DNA in the minimum volume of TE[b]. It may be necessary to heat to 65°C to dissolve the DNA fully.
11.	Add RNase A to a final concentration of 20 μg/ml. Incubate at 50°C for 1 h.
12.	Add protease K (freshly dissolved) to a final concentration of 100 μg/ml. Add NaCl and SDS to final concentrations of 100 mM and 0.4%, respectively. Incubate at 37°C for 1 h.
13.	Add an equal volume of TE-saturated phenol and repeat steps 5 – 10.
14.	Determine the concentation and purity of the DNA by scanning absorbance from 230 to 280 nm.

[a]Media are defined in Section 5.
[b]TE is 10 mM Tris-HCl pH 8.0, 1 mM EDTA.

4.2.1 *Total Genomic DNA*

Three methods are given here for the preparation of total DNA from *Strep-tomyces* (*Tables 3 – 5*). They each have utility under different circumstances.

(i)   The first (*Table 3*) is the method of choice for most strains. It involves spooling the DNA, which is a very effective method of purification. The success of the procedure depends on efficient lysis with SDS at step 4. Stir well, but gently to prevent shearing the DNA. The phenol should be added *immediately* to prevent the action of DNA nucleases which will be released on lysis of the cells.

Usually, yields of several mg of DNA will be obtained, which provides enough DNA for many experiments. Although the procedure may appear tedious, it gives high molecular weight DNA (> 100 kb) of excellent quality and is worth the investment in time.

(ii)   The rapid procedure for preparation of total DNA (*Table 4*) involves lysis of the mycelia with SDS and purification of the DNA by centrifugation in a CsCl gradient. It has the advantage of few manipulative steps, but unfortunately does not work for all strains. Sometimes the DNA is partially degraded after centrifugation, probably because of DNase released at step

Table 4. Rapid Preparation of Streptomycete Total DNA — Method 2.

1.	Inoculate 25 ml of TSB or YEME[a] (depending on the strain, see text) with 0.1 ml of a dense spore suspension. Incubate at 30°C with shaking.
2.	Harvest when the growth is dense (usually 2 days) by centrifugation in a 50 ml centrifuge tube in the Beckman JA20 rotor (or its equivalent) at 10 000 r.p.m. (12 100 g) for 10 min.
3.	Resuspend the pellet in 25 ml TE[b], add 25 mg lysozyme and incubate for 15 min at 30°C.
4.	Add 1.25 ml of 20% SDS. The solution will go viscous. Add 26.25 g CsCl, while stirring gently and 1 ml ethidium bromide (10 mg/ml in TE). The density of this solution should be 1.58 g/ml.
5.	Add the liquid to a centrifuge tube for the Beckman VTi50 rotor (or its equivalent). Fill up tube completely with either 'top up' fluid (*Table 5*, step 8) and seal the tube following the manufacturer's instructions.
6.	Centrifuge the equilibrium. For the VTi50 rotor 45 000 r.p.m. for 16 h at 20°C is sufficient.
7.	Illuminate with long wavelength u.v. Remove the chromosomal band using a hypodermic syringe with a wide gauge needle to puncture the side of the tube. Dilute the DNA 3-fold with TE and precipitate by adding 2.2 volumes of ethanol.
8.	Chill to −20°C for an hour and pellet the DNA by centrifugation. Redissolve the pellet in minimum volume of TE. Decant the supernatant and allow the pellet to 'air-dry'.

[a]Media are defined in Section 5.
[b]See footnote[b] to *Table 3*.

4. It is possible to add phenol at step 4 (as in steps 4 − 8 of *Table 3*) to inactivate nucleases, but this increases the number of manipulations. In the event of having to add phenol, the experimenter should decide whether the procedures of *Table 3* or *Table 4* should be carried out. It is largely a matter of individual choice. Reasonable yields (about 100 μg DNA/25 ml culture) are obtained. The DNA is particularly useful for Southern Blots, where it does not matter if it is degraded slightly.

(iii) Some strains give low yields of DNA that is totally-degraded with either of the above methods. In these cases, mechanical disruption of the mycelia in liquid nitrogen followed by sarkosyl extraction has to be used (*Table 5*). After treatment with RNase and proteinase K, the lysate is loaded onto a CsCl gradient without ethidium bromide (density 1.7 g/ml) for further purification. The method gives good quality high molecular-weight DNA, but in low yield (usually < 50 μg). The problem of low yield can be overcome by growing up several litres of culture and 'scaling up' the procedure of *Table 5*, using eight tubes in the ultracentrifuge rotor. This method is a useful 'second approach' if Method 1 fails. The method also works for genera related to *Streptomyces*, e.g., proactinomycetes, for which Method 1 does not appear to work well.

4.2.2 *Plasmid DNA*

Several methods for the isolation of plasmid DNA have been given in the literature (e.g., alkaline SDS lysis, Triton lysis). In this laboratory we routinely use a method (*Table 6*) based on the early isolation of SCP2 (24). Although the isolation spans 2 days, the manipulations are simple, short, and easily reproduced. The yield of plasmid DNA is about 2 mg for high-copy vectors based on pI-J101, and so the method can be used for isolating 'reagent' quantities of cloning

Table 5. Extraction of Streptomycete Total DNA from Mycelia Disrupted with Glass Beads — Method 3.

1. Inoculate 200 ml of TSB or YEME[a] (depending on strain) with 1 ml of a dense spore suspension. Incubate at 30°C with shaking.
2. Harvest when the growth is dense (usually 2 days) by centrifugation in a 250 ml centrifuge tube in the Beckman JA14 rotor (or its equivalent) at 12 000 r.p.m. (22 100 g) for 10 min.
3. Resuspend the pellet in 0.3 M sucrose and centrifuge as in step 2.
4. Put 2 g of fine glass beads (100 mesh) into a mortar and fill it with liquid nitrogen. Scrape the mycelial pellet from the centrifuge tube using a spatula, and add it to the mortar. Grind with the pestle, adding more liquid nitrogen when necessary to keep the pellet cool.
5. Grind to a fine powder. Using a spatula, transfer the powder to 10 ml of SNET buffer[b]. Centrifuge at 10 000 r.p.m. (12 000 g) in the Beckman JA20 rotor (or its equivalent) for 20 min at 4°C.
6. Pour off the supernatant and measure its volume. Add RNase to a final concentration of 100 μg/ml and incubate at 50°C for 30 min. Add Proteinase K to a final concentration of 50 μg/ml and incubate at 37°C for 30 min.
7. Add an equal volume of TE[c]-saturated phenol. Invert the tube repeatedly and gently at room temperature for 10 min. Remove the aqueous phase after centrifugation and add it to an equal volume of chloroform. Mix by inverting gently at room temperature for 10 min.
8. Remove the aqueous phase after centrifugation and measure its volume. For each ml of aqueous phase add 1 g CsCl and stir gently to dissolve. Place in an ultracentrifuge tube and fill the tube either with CsCl solution (20 g CsCl + 20 ml TE) or paraffin oil. The density of the CsCl solution should be about 1.72 g/ml. (N.B. These tubes do not contain ethidium bromide.)
9. Centrifuge the equilibrium. For the Beckman VTi50 rotor 40 000 r.p.m. for 16 h at 20°C is sufficient.
10. Cut off the top of the tube using a hot scalpel blade. Use a 100 μl 'microcap' pipette connected to Teflon tubing to syphon the contents of the tube in 1 ml fractions into a series of microcentrifuge tubes.
11. Determine which fractions contain DNA by spotting 5 μl of each fraction onto a 1% agarose plate containing 1 μg/ml ethidium bromide.
12. Dialyse the fractions containing DNA against 100 volumes TE, with two buffer changes. Add 1/10 volume 5 M NaCl and 2.2 volumes of ethanol. Mix and chill to -20°C. Pellet the DNA by centrifugation. Decant the supernatant and allow the pellet to 'air dry.'
13. Dissolve the pellet in TE and determine DNA concentration and purity by running u.v. spectra (230–280 nm) on the DNA.

[a]Media are defined in Section 5.
[b]SNET buffer is 2% sarkosyl, 50 mM EDTA, 50 mM NaCl, 50 mM Tris-HCl pH 7.6.
[c]TE is defined in footnote[b] to *Table 3*.

vectors.

The mycelia are lysed by SDS in step 4 of *Table 6*. Good lysis at this stage is the key to obtaining high yields of plasmid. The lysates of some strains contain such large quantities of RNA that the RNA pellet on the side of the tube contaminates the plasmid band when vertical ultracentrifuge rotors are used. RNase (100 μg/ml, 50°C, 30 min) can be added to the supernatant at step 5 to prevent this.

Rapid screening of clones is best done by isolating DNA directly from the colony (29).

4.2.3 *Phage DNA*

This laboratory has no first-hand experience of isolating actinophage DNA, so it would be imprudent to include a protocol. However, we have successfully used

Table 6. Isolation of Covalently Closed Circular DNA from *Streptomyces*.

1. Inoculate 200 ml TSB or YEME[a] (depending on the strain, see text) containing the appropriate drug for continued selection of cells containing plasmid; with 1 ml of dense spore suspension. Incubate at 30°C with shaking.
2. Harvest the mycelium when growth is dense (usually after 2 days) by centrifugation in a 250 ml centrifuge tube at 12 000 r.p.m. (22 100 g) for 10 min in JA14 rotor (or its equivalent). YEME cultures will pellet better if they are diluted two-fold with sterile water before centrifugation.
3. Resuspend the mycelium in 25 ml TE[b]. Add 5 ml of 0.25 M EDTA and 75 mg lysozyme. Incubate at 30°C for 20 min.
4. Add 25 ml of TE, 15 ml of 0.25 M EDTA and 4.2 ml of 20% SDS. Immediately mix gently with a glass rod. The cells will lyse and the solution will go viscous. Add 16.8 ml of 5 M NaCl and mix gently with a glass rod. Incubate at 4°C for 4 h.
5. Centrifuge the mixture at 14 000 r.p.m. (31 000 g), 4°C, 30 min in JA14 rotor (or its equivalent). If the pellet is not 'tight' at this stage, centrifuge again. Decant the supernatant into a sterile measuring cylinder. Add half a volume of sterile 30% PEG 6000, mix by inversion, place in 250 ml centrifuge tube and stand at 4°C overnight.
6. Centrifuge at 5000 r.p.m. (3840 g) at 4°C for 5 min. Carefully decant the supernatant and discard.
7. Resuspend the pellet in 13 ml TE. Add 13 g CsCl, stirring gently to dissolve the salt. Add 500 μl ethidium bromide (10 mg/ml in TE). The density should be 1.58 g/ml.
8. For the Beckman VTi 50 rotor, make the volume up to about 38 ml with 'top-up' fluid[c], and fill ultracentrifuge tube. Seal the tube following the manufacturer's instructions.
9. Centrifuge to equilibrium. For the Beckman VTi 50 rotor at 45 000 r.p.m. for 15 h at 20°C is sufficient.
10. Remove the lower plasmid band using a wide-bore hypodermic needle.
    Steps 11 and 12 are optional, but should be used when plasmid band is still contaminated with chromosomal DNA. Otherwise go to step 13.
11. Put liquid containing the plasmid band into a tube for the Beckman VTi 65 rotor. Fill the tube with 'top-up' fluid. Centrifuge to equilibrium. For the VTi 65 rotor at 55 000 r.p.m. for 4 h at 20°C is sufficient.
12. Illuminate with long wavelength u.v. light. Remove plasmid band through the side of the tube using a wide bore hypodermic needle.
13. Remove ethidium bromide from the plasmid, by extraction with water-saturated butanol. Discard the butanol phase. Continue to re-extract until the butanol phase does not fluoresce under long-wave u.v. light.
14. Transfer the aqueous phase to a dialysis bag and dialyse against 100 volumes of TE for 16 h at 4°C. Change the TE at least twice.
15. Remove the DNA solution from the dialysis bag and precipitate with ethanol as described in steps 12 and 13 of *Table 4*. Resuspend the DNA in distilled water in a volume up to 500 μl.

[a]Media are described in Section 5.
[b]TE is defined in footnote[b] of *Table 3*.
[c]'Top-up' fluid is made up from 52 g CsCl, 52 ml TE and 2 ml of 10 mg/ml ethidium bromide.

phage DNA supplied by Chater that had been isolated using procedures described by him and his coworkers in reference 25. Actinophage DNA is more tedious to isolate than plasmid DNA, much in the same way as lambda DNA is more difficult to isolate than plasmid from *E. coli*.

4.2.4 *Phosphatasing of Vector DNA*

To maximize the number of clones containing inserts in the preparation of any gene bank, the vector should be phosphatased to prevent it ligating back onto itself without an insert during the ligation reaction. This is true of both plasmid and phage vectors. Calf intestinal phosphatase (Boehringer) is used to dephosphorylate the vector.

After digestion with a restriction enzyme, the vector DNA is precipitated with ethanol, dried under vacuum and dissolved in 0.1 M glycine (pH 10.2) at a concentration of 100 ng/μl. Alkaline phosphatase is added (0.5 units/μg DNA) and incubated at 37°C for 60 min. The reaction is terminated with an equal volume of phenol, the mixture shaken and then centrifuged in a microcentrifuge. The aqueous phase is extracted several times with chloroform and then removed to a new microcentrifuge tube for concentration by precipitation with ethanol prior to ligation.

The high pH of incubation is close to the pH optimum of the enzyme so that only very small amounts of phosphatase protein need be added. The small quantity of $(NH_4)_2SO_4$ carried into the ligation from the slurry in which the phosphatase is supplied does not affect the subsequent ligation reaction. A 'molecular biology' grade of alkaline phosphatase is available (Boehringer) which works in the restriction buffer. It does not contain $(NH_4)_2SO_4$ and also works well.

After transformation, Streptomycete protoplasts take so long to regenerate that it is worthwhile checking that the vector is phosphatased completely, *and* still capable of ligation before using it to construct gene banks. Usually only the former is checked by testing that the vector is incapable of ligating to itself. But the vector may also not ligate because its sticky ends have been 'chewed back', or because the preparation contains an inhibitor of ligation (e.g., phenol). A positive control needs to be done to show that the phosphatased vector *is* capable of ligating to an insert. In this test, the 'insert' can be another plasmid. *Figure 7* shows such a test. A 12 kb plasmid cut with *Bgl*II was tested with pUC19 cut with *Bam*HI (2.7 kb) as the 'insert'. Firstly, it should be confirmed that the cut vector can ligate to itself before phosphatasing (track 2), i.e., the restriction enzyme has left sticky ends capable of ligating. After phosphatasing, the vector should not ligate to itself (track 4), but when 'insert' is added (track 5) and then ligase (track 6) it should ligate − seen easily by the disappearance of the vector band in track 6 compared to track 5. Also present is a control of 'insert' alone before (track 7) and after (track 8) ligation. If track 8 had shown ligation and track 6 had not, this would indicate the presence of an inhibitor of ligation in the vector preparation. If, in track 6 the insert had ligated well, but the vector did not ligate well (equal intensity to the band in track 5), then the sticky ends of the vector would be suspected − it would be best to start again with fresh vector.

4.3 Ligation of DNA

To construct gene banks, the DNA is best ligated at a total DNA concentration of 20 μg/ml and an insert:vector molar ratio of 2:1. The reaction contains 66 mM Tris HCl (pH 7.6), 6 mM $MgCl_2$, 10 mM DTT and 1 mM ATP. The ATP should be neutralised with NaOH to pH 7.6 when it is made up. Heat the ligation mixture to 65°C for 10 min, then cool on ice before adding ligase. This separates the sticky ends which have undergone intramolecular hydrogen-bonding during

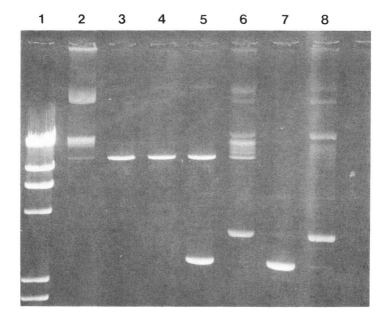

Figure 7. Test of efficiency of phosphatasing of vector. Track 1, λ *Hind*III digest; track 2, vector, (12 kb) ligated before phosphatasing; track 3, phosphatased vector, before ligation; track 4, phosphatased vector, after ligation; track 5, phosphatased vector + insert (2.7 kb), before ligation; track 6, phosphatased vector + insert, after ligation; track 7, insert alone, before ligation; track 8, insert after ligation.

preparation. Incubate the ligation reaction overnight at 14°C and precipitate the DNA with ethanol (redissolving in TE) before transformation. For example, with 5.5 μg pIJ702 (5.6 kb) and inserts of average size of 7.5 kb, 14.7 μg of insert DNA [2 x 5.5 x(7.5/5.6)] will be required to achieve a 2:1 molar ratio of insert. With 5.5 μg vector, the total DNA in the ligation is 20.2 μg. Therefore the total reaction volume will be 1.01 ml.

4.4 Preparation of Protoplasts

A generalised method is given in *Table 7*. The cells are grown in the presence of glycine, which renders the mycelium more susceptible to lysozyme treatment. It also retards mycelial growth. Protoplasts can be made from most strains when grown at a high growth rate up to late-log phase. Some strains, however, give higher transformation efficiencies when used to make protoplasts in early-mid log phase. A glycine level of 0.5% has proved optimal for most strains. The mycelium is more resistant to attack by lysozyme when the concentration is reduced below this level, whereas a higher level of glycine retards growth too much.

Care should be taken to centrifuge mycelia at low speed, because mycelial pellets collected at high centrifugal force (e.g., 10 000 *g*) are difficult to resuspend homogeneously. This results in mycelial clumps which reduce the yield of protoplasts.

Table 7. Preparation of Protoplasts of *Streptomyces*.

1.	Inoculate 25 ml TSB or YEME[a] (depending on the strain, see text) with 0.1 ml of a spore suspension. Add MgCl$_2$ and glycine to final concentrations of 5 mM and 0.5% (w/v) respectively. Incubate at 30°C with shaking for 2 days.
2.	Harvest the mycelia by centrifugation at 5000 r.p.m. (3000 g) in a Beckman JA20 rotor (or its equivalent) at 20°C for 10 min. Resuspend the pellet in 10 ml of 0.3 M sucrose and centrifuge at 5000 r.p.m. for 10 min to wash the mycelia.
3.	Resuspend the 0.3 − 1.0 g wet weight of mycelia in 4 ml 'lysozyme solution'[b] into which lysozyme has been freshly dissolved to a concentration of 1 mg/ml. Incubate at 37°C for 30 min. Keep the mycelia in suspension by stirring the solution with a pipette.
4.	Add 5 ml medium P[a] and filter the suspension through non-adsorbent cotton wool (*Figure 5*). Centrifuge at 5000 r.p.m. for 10 min.
5.	Resuspend the pellet gently in 5 ml medium P and centrifuge again.
6.	Repeat step 5.
7.	Resuspend the pellet gently in 2 ml medium P. Freeze 10 μl aliquots in a − 20°C freezer. About 5 x 10^7 protoplasts are obtained in good preparations.

[a]Media are described in Section 5.
[b]Achromopeptidase may also be added at 1 mg/ml together with the lysozyme. See the text for details.

During incubation with lysozyme, the formation of protoplasts can be followed conveniently under the microscope using a X40 objective. Using phase-contrast, protoplasts appear as small spheres which lyse when a drop of water is added to the microscope slide. Mycelia are much larger, not spherical, and of course do not lyse. Prolonged incubation with lysozyme does reduce the transformation efficiency. With uncharacterised strains, it is worthwhile removing aliquots at several incubation times, observing the number of protoplasts under the microscope and testing the transformation efficiencies of the protoplasts harvested at the different times. On the basis of microscopical analysis, a 'sixth sense' is soon developed which judges when the incubation with lysozyme should be terminated. Some strains form clumps of protoplasts very quickly. Reducing the lysozyme concentration to 0.2 − 0.3 mg/ml can prevent this.

Some strains do not form protoplasts well. In such cases, the addition of achromopeptidase can assist in protoplast formation (26). Use achromopeptidase together with lysozyme with both usually at 1 mg/ml. Varying the ratio of achromopeptidase to lysozyme has not been worthwhile.

Streptomyces protoplasts store well at − 20°C. They can be used after a year with little loss in transformation efficiency. Thawing and refreezing the protoplasts increases significantly the time for regeneration and is not advised. Times for addition of drug overlays (see Section 4.7) are difficult to judge in this situation. It is convenient to freeze protoplasts in small volumes and thaw each vial once, just before transformation. Up to 5 x 10^7 protoplasts should regenerate from each 100 μl aliquot.

4.4.1 *Troubleshooting after Obtaining Poor Yield of Protoplasts with an Unfamiliar Strain*

(i)   Check the glassware for traces of residual detergent to which protoplasts are very sensitive. If the glassware is suspect, change to polythene tubes or

Table 8. Transformation of *Streptomyces* Protoplasts.

1.	Prepare a fresh 25% (v/v) polyethylene glycol (PEG 1500) solution in 'T mix'[a] (5). It is convenient to weight out 1 g quantities of PEG 1500 into glass vials, which are stored at room temperature after autoclaving at 121°C for 10 min. When needed, a vial is heated at 60°C to thaw the PEG and 3 ml of sterile 'T mix' added to give the 25% solution.
2.	Thaw the frozen protoplast (*Table 6*) on ice. Stir the suspension gently with a micropipette tip to make it homogeneous.
3.	Add DNA in a volume of less than 10 μl. Stand for 10 sec on ice.
4.	Add 0.5 ml of 25% PEG 1500 and pipette up and down to mix. Leave for 1 min on ice.
5.	Add 2 ml of 'medium P'[a] (5) and pipette to mix. Plate the suspension on regeneration medium. If necessary, dilutions of the suspension are made in 'medium P'.

[a]'T mix' and other media are described in Section 5.

       wash the glass in nitric acid.
(ii)   Vary the incubation time for growth before harvesting the mycelia.
(iii)   Vary the level of lysozyme and the incubation time, checking protoplast formation under the microscope.
(iv)   Use achromopeptidase in addition to lysozyme.

4.5 Transformation

DNA is introduced into protoplasts (*Table 8*) using PEG. The average molecular weight of the PEG does not seem to influence transformation efficiency. PEG 1500 (BDH) has been used routinely in this laboratory, although PEG 6000 works too. It has been said that PEG samples from different suppliers, or different batches from the same supplier, contain trace impurities which can have either a positive or negative influence on transformation. Over a period of years different batches from BDH have been used without any problem.

We never add more than 1 μg of DNA to each 100 μl of protoplasts. If larger amounts of DNA are used for the construction of gene banks, several vials of protoplasts are thawed and mixed together. The transformation protocol (*Table 7*) is then 'scaled-up' accordingly, e.g., for 5 μg DNA use 5 vials of protoplasts and 2.5 ml of 25% v/v PEG with 10 ml of 'medium P'.

After dilution of the transformation mix with 'medium P' in step 5, the PEG concentration is 4 − 5%. This is acceptable for plating protoplasts. If the PEG is not diluted sufficiently, the transformation mix is too 'sticky', and spreads unevenly on the regeneration plates. If required, PEG can be removed by centrifuging the transformation mix at 3000 g for 10 min and resuspending the pellet gently in fresh 'medium P'. In the example above for 5 μg DNA, 13 ml of transformed protoplasts would be obtained (i.e., 130 plates!). Centrifugation followed by suspension in a smaller volume of 'medium P' would reduce the number of plates to be spread.

For some strains, if the Ca^{2+} content of the transformation mix ('T mix ') is reduced from 100 mM to 10 mM the transformation efficiency increases by a factor of 5 − 10.

An early report (27) that entrapment of plasmid DNA in liposomes (28) enhanced transformation frequency into protoplasts has proved difficult to repeat in this laboratory. Adding empty liposomes to the DNA prior to transformation has had no effect on transformation frequency of protoplasts whereas the use of empty liposomes to stimulate transfection by phage is well-documented. For most plasmid vectors, transformation efficiency is now sufficiently high that the technical difficulties involved in preparing and using liposomes argue against their use in plasmid transformation.

4.6 Regeneration

Streptomycete protoplasts have to be regenerated on solid medium. To date, there is no known method for regenerating the cell walls of protoplasts in liquid medium.

The medium for the regeneration of streptomycete protoplasts (R2) was originally developed for protoplast fusion experiments. It uses glucose as carbon source, is rich in nitrogen, contains high levels of Mg^{2+} and Ca^{2+}, and has a high level of sucrose, which acts as an osmotic support (see Section 5). Incompatibility of some of the ingredients during autoclaving requires that it be made up in two parts which are sterilized separately, and then combined before use, with the addition of phosphate. The temperature of the molten agar should be 50°C just before pouring. For optimal regeneration of protoplasts, the regeneration agar should be dried partially before use. The explanation for the enhanced regeneration on partially-dried plates has yet to be presented, but it does work! For *S. lividans*, a 5 – 10% weight loss is best. This may be achieved either by leaving the freshly poured plates with lids off in a flow of air from a laminar flow hood until the weight is lost, or pouring the plates 1 – 2 days before use (depending on the humidity of the atmosphere) and leaving them on the bench. With the former method the passage of air in one direction across the plates can produce uneven drying, which results in the protoplasts on one side of the plate regenerating better than on the other. Therefore the second method is used in this laboratory, but it does require forward planning! The optimal weight loss for the drying of plates is strain-dependent.

4.6.1 *Regeneration of Uncharacterised Strains*

For an uncharacterised strain, it may be necessary to change the composition of the regeneration agar to achieve optimal regeneration. The following should be tried in order of priority.

(i)   *Carbon source.* Confirm that glucose is a suitable carbon source for the strain.
(ii) *Sucrose level.* R2 contains 0.3 M sucrose. Other strains, for example *S. rimosus*, regenerates best with 0.6 M sucrose. Test a range of sucrose levels from 0.2 to 0.8 M. Other osmotic supports, e.g., KCl, sorbitol, have been tested without success.
(iii) *Divalent cations.* R2 contains 50 mM Mg^{2+} and 20 mM Ca^{2+}. Changing the ratio of the two ions in the range 0 – 100 mM can enhance transformation of some strains.

(iv)   *Phosphate level.* R2 contains 0.005% KH_2PO_4. Some strains require higher phosphate levels for optimal regeneration. Check a range up to 0.025%.

(v)   *Proline* is present in R2. For most strains it is not required. Other amino acid supplements, e.g., asparagine, may enhance regeneration or give better sporulation of uncharacterised strains.

The time taken for protoplasts of different strains to regenerate fully can vary from 3 days to several weeks. If a strain requires prolonged incubation to regenerate, it is best to wind tape round the lids of the plates to prevent the agar drying too much.

Some strains may regenerate better at a temperature which is sub-optimal for growth. Initially a temperature of 30°C should be used.

4.7 Detection of/Selection for Transformants

4.7.1 *Plasmids which Exhibit the 'Lethal Zygosis' Phenotype*

Detection of transformants of plasmids which have no selectable marker (such as a drug resistance) is usually done by 'lethal zygosis' (25). On a plate spread confluently with regenerating protoplasts, transformants can be visualised as 'pocks' as they grow up amongst the non-transformed bacteria. The presence of the plasmid-containing cells amongst a lawn of plasmid-deficient cells causes 'a pock', a translucent zone in the lawn, so called 'lethal zygosis'. The biology of the process is not understood.

At high dilutions, protoplasts will regenerate as single colonies. When the colonies have sporulated, the plate is replicated onto a lawn of a plasmid-deficient strain. Transformants are seen as 'pocks' as the replica plate grows up.

4.7.2 *Plasmids which have a Selectable Marker*

Most of the streptomycete cloning vectors developed recently have at least one drug-resistance marker which can be used to select for transformants. The drug-resistance markers are genetically dominant. In general, they code for enzymes which either (i) modify the target of the drug so that it is no longer effective, or (ii) modify the drug molecule chemically, so that it is no longer effective. An example of the first category is resistance to thiostrepton or erythromycin which is due to a specific methylation of the host's ribosomes with the result that the drugs no longer bind and inactivate ribosomal function. Examples of the second category are resistance to viomycin or to neomycin which is due to phosphotransferases that inactivate the drug such that it no longer binds to its target.

After transformation, it takes some time for the enzymes coding for drug resistance to be expressed and to build up an activity within the cell sufficient to display the resistant phenotype. Thus, transformed protoplasts would be killed if plated immediately onto medium containing the drug. Instead, they are plated on regeneration medium without the drug, and allowed to regenerate non-selectively for some time along with protoplasts which have not been transformed. The drug is then added to the regeneration agar. Non-transformed protoplasts are killed, whereas transformants survive and grow into mature colonies. The drug can be added in a soft overlay (0.6%) of regeneration agar (R2). However, it is often difficult to pick off mature colonies from plates which contain soft agar overlays, and nearly impossible to replica-plate them to carry out secondary screening because the colonies are embedded in the agar. An alternative is to add the drug in 1 ml of 0.3 M sucrose. The plate is held at an angle and rotated until the drug solution covers the entire surface of the agar. The sucrose dries into the agar in 1−2 days on further incubation at 30°C. The incubator shelves must be level, otherwise the drug solution collects at one edge of the plate which results in killing of transformants within that sector whereas on the rest of the surface there is too little drug and untransformed protoplasts survive.

The following levels of drugs are used routinely for selection of transformants using 1 ml of liquid overlay per plate: thiostrepton (200 μg/ml); erythromycin (200 μg/ml); neomycin (10 μg/ml); viomycin (200 μg/ml). The time for drug addition has to be established for regeneration of protoplasts of each strain. Drug is added to protoplasts of *S. lividans* 18−24 h after they are plated, whereas *S. rimosus* protoplasts have to be left for 24−28 h. For protoplasts of strains which take 3 weeks to regenerate fully, 3−5 days may need to elapse before drug addition. If plates are overlaid too early, the transformed protoplasts are killed because they have not acquired fully the drug-resistance. If plates are overlaid too late, mycelia which have escaped being made into protoplasts (and are present in all but the best protoplast suspensions) will outgrow the protoplasts (since they grow more quickly) and inhibit regeneration of transformed protoplasts. Therefore there is a 'narrow window' when the drug selection is best applied, and this has to be determined empirically for each strain. As a rough 'rule of thumb', when the regenerating protoplasts are just visible on the plate as a mist, the plates are ready to be overlaid.

5. MEDIA

The media used in the procedure described in this chapter are listed alphabetically below:

(i)   *Lysozyme solution*

Per 500 ml:

> 50 g   Sucrose
> 2.86 g   TES [N-tris(hydroxymethyl)methyl-2-aminoethane-sulphonic acid]
> 0.217 g   Potassium sulphate   (2.5 mM)
> 0.253 g   Magnesium chloride (2.5 mM, $MgCl_2.6H_2O$)
> 0.183 g   Calcium chloride   (2.5 mM, $CaCl_2.2H_2O$)
> 1 ml   Trace elements

The pH is adjusted to 7.2 before making up to the final volume with water. Dispense 100 ml amounts into bottles. Autoclave for 15 min at 121°C. Before use, add 0.5 ml 1% KH_2PO_4 per 100 ml of medium.

(ii) *Medium P*

Per litre:

5.73 g	TES
103 g	Sucrose
2.03 g	$MgCl_2.6H_2O$
0.5 g	K_2SO_4
3.68 g	$CaCl_2.2H_2O$
2 ml	Trace element solution

Adjust to pH 7.4 with NaOH. Autoclave for 15 min at 121°C. Just before use add 1 ml of 1% KH_2PO_4/100 ml.

(iii) *R2 agar*

Made up in two parts.

R2/A per litre:

0.5 g	K_2SO_4
20.2 g	$MgCl_2.6H_2O$
5.9 g	$CaCl_2.2H_2O$
20 g	Glucose
6 g	Proline
0.2 g	Casamino acids
4 ml	Trace element solution

Dispense in either 100 ml or 200 ml amounts, remembering that complete medium is twice the volume. Add 4.4 g agar/100 ml liquid. Autoclave for 15 min, 121°C.

R2/B per litre:

11.5 g	TES buffer. Adjust to pH 7.4 with NaOH
10 g	Yeast extract
203 g	Sucrose

Dispense 100 ml or 200 ml amounts. Autoclave for 15 min at 121°C. Before pouring plates combine R2/B into R2/A and add 1 ml 1% KH_2PO_4 per 200 ml final volume.

(iv) *Trace element solution*

Per litre

40 mg	$ZnCl_2$
200 mg	$FeCl_3.6H_2O$
10 mg	$CuCl_2.2H_2O$
10 mg	$MnCl_2.4H_2O$
10 mg	$Na_2B_4O_7.10H_2O$
10 mg	$(NH_4)_6Mo_7O_{24}.4H_2O$

(v) *Transformation medium — T mix*

Per 500 ml:

12.5 g	Sucrose

7.35 g Calcium chloride ($CaCl_2.2H_2O$: 100 mM)
0.218 g Potassium sulphate (2.5 mM)
1 ml Trace elements
2.9 g Maleic acid

Adjust the pH to 8.0 with 1 M Tris base and autoclave at 121°C for 15 min.

(vi)   *Streptomyces − YEME*

Per litre:

3 g   Yeast extract
5 g   Peptone
3 g   Malt extract
10 g   Glucose
340 g   Sucrose
1.15 g $MgCl_2.7H_2O$

Autoclave for 15 min at 121°C.

6. REFERENCES

1. Bibb,M.J., Schottel,J.L. and Cohen,S.N. (1980) *Nature,* **284**, 526.
2. Hopwood,D.A., Chater,K.F., Dowding,J.E. and Vivian,A. (1973) *Bacteriol. Rev.,* **37**, 371.
3. Hopwood,D.A. and Wright,H.M. (1983) *J. Gen. Microbiol.,* **129**, 3575.
4. Berdy,J. (1980) *Process Biochem.,* 28.
5. Fietelson,J.S. and Hopwood,D.A. (1983) *Mol. Gen. Genet.,* **190**, 394.
6. Gil,J.A. and Hopwood,D.A. (1983) *Gene,* **25**, 119.
7. Malpartida,F. and Hopwood,D.A. (1984) *Nature,* **309**, 462.
8. Bibb,M.J., Ward,J.M., Kieser,T., Cohen,S.N. and Hopwood,D.A. (1981) *Mol. Gen. Genet.,* **184**, 223.
9. Delic,V., Hopwood,D.A. and Friend,E.J. (1970) *Mutat. Res.,* **9**, 167.
10. Hunter,I.S. and Friend,E.J. (1984) *Biochem. Soc. Trans.,* **12**, 643.
11. Williams,S.T., Goodfellow,M., Alderson,G., Wellington,E.M.H., Sneath,P.H.A. and Sachin, M.J. (1983) *J. Gen. Microbiol.,* **129**, 1743.
12. Kieser,T., Hopwood,D.A., Wright,H.M. and Thompson,C.J. (1982) *Mol. Gen. Genet.,* **185**, 223.
13. Katz,E., Thompson,C.J. and Hopwood,D.A. (1983) *J. Gen. Mcirobiol.,* **129**, 2703.
14. Richardson,M.A., Mabe,J.A., Beerman,N.E., Nakatsukusa,W.M. and Fayerman,J.T. (1982) *Gene,* **20**, 451.
15. Manis,J.J. and Highlander,S.K. (1982) *Gene,* **18**, 13.
16. Thompson,C.J., Ward,J.M. and Hopwood,D.A. (1980) *Nature,* **286**, 525.
17. Bibb,M.J., Ward,J.M., Kieser,T., Cohen,S.N. and Hopwood,D.A. (1981) *Mol. Gen. Genet.,* **184**, 230.
18. Thompson,C.J., Kieser,T., Ward,J.M. and Hopwood,D.A. (1982) *Gene,* **20**, 51.
19. Schrempf,H., Hopwood,D.A. and Goebel,W. (1975) *J. Bacteriol.,* **146**, 360.
20. Lydiate,D. (1984) Ph.D. Thesis, University of East Anglia, Norwich, UK.
21. Harris,J.E., Chater,K.F., Bruton,C.J. and Piret,J.M. (1983) *Gene,* **22**, 167.
22. Chater,K.F., Bruton,C.J., Springer,W. and Suarez,J.E. (1981) *Gene,* **15**, 249.
23. Chater,K.F. and Bruton,C.J. (1984) *Gene,* in press.
24. Bibb,M.J., Freeman,R.F. and Hopwood,D.A. (1977) *Mol. Gen. Genet.,* **154**, 155.
25. Chater,K.F., Hopwood,D.A., Kieser,T. and Thompson,C.J. (1982) in *Gene Cloning in Organisms Other than E. coli,* Hofschneider,P.H. and Goebel,W. (eds.), Springer Verlag, Berlin, pp.
26. Ogawa,H., Imai,S., Satoh,A. and Kojima,M. (1983) *J. Antibiot.,* **36**, 184.
27. Rodicio,M.R. and Chater,K.F. (1982) *J. Bacteriol.,* **151**, 1078.
28. Makins,J.F. and Holt,G. (1982) *Nature,* **293**, 671.
29. Kieser,T. (1984) *Plasmid,* in press.

CHAPTER 3

Cloning in Yeast

RODNEY ROTHSTEIN

1. INTRODUCTION

This chapter is designed to serve as an introductory guide for the growing number of people who are using yeast as an experimental system. It is not intended to be a complete review of cloning methods in yeast; it will instead lead the reader through some of the concepts and techniques applicable to yeast cloning. All of the experiments describe work originally performed in *Saccharomyces cerevisiae*. Many of the techniques have been used successfully in non-*Saccharomyces* yeasts as well as some filamentous fungi. The first section will briefly describe methods for identifying cloned yeast genes. The next section includes vector development, transformation systems, and general methods for isolating clones from yeast. The final section describes techniques for manipulating cloned yeast DNA including methods for directed integration of plasmids, allele rescue, gene disruption and repeated gene replacements.

2. IDENTIFICATION OF CLONED YEAST GENES

Yeast genes have been identified by a variety of techniques. The first genes isolated from yeast were for structural RNAs, first ribosomal RNA genes (1) and later tRNA genes (2). Differential hybridisation was developed in yeast to identify inducible genes (3). Later, genes were identified by their ability to complement their cognate genetic defects in *E. coli* (4,5). The isolation of those genes directly led to the development of methods for transforming yeast (6,7). Since the development of yeast transformation, many genes have been cloned by genetic complementation in yeast using clone banks made in vectors described in Section 3.1. In the following section, several methods for identifying cloned yeast genes will be described.

2.1 Cloning Structural RNA Genes and Purified mRNA Genes

One of the easiest and most straight-forward approaches to cloning a gene is to use purified RNA as a probe. For structural RNAs such as rRNA, tRNA and snRNA, it is relatively easy to isolate a pure species. The RNA is radioactively labelled either by polynucleotide kinase and ^{32}P or is iodinated with ^{125}I and is used to screen a plasmid or phage library for the sequence of interest (8). This approach can also be extended to include mRNAs purified by virtue of their abundance or enriched by some other means, such as polysome size selection.

2.2 **Cloning by Differential Hybridisation**

Differential hybridisation screening was originally developed in yeast and used for the isolation of the galactose-inducible genes (3). This technique employs a library of yeast genomic DNA cloned in phage λ. The phage library is plated at a density of approximately 500 plaques per plate. After 24 h, duplicate nitrocellulose filters are made from the plaques, and each filter is screened with radioactively-labelled RNA or cDNA isolated from cells during two different regulatory states. For example, the galactose genes were isolated by screening for plaques that were positive only when hybridised with RNA isolated from galactose-induced cells.

2.3 **Cloning by Hybrid Selection and Antibody Screening**

A gene whose protein product has been identified can be isolated by the brute force method of hybrid selection (9). In this method, the correct clone is identified by its ability to selectively remove by hybridisation the mRNA that directs the synthesis of the protein of interest. To begin, a cDNA clone library of total mRNA is constructed. Before cloning, it is helpful to enrich for the mRNA of the gene of interest. DNA is isolated from individual clones of this library and hybridised with total mRNA from the cell. After hybridisation, unbound mRNA is translated *in vitro* and the products of translation are analysed for the presence or absence of the desired protein. A clone that, after hybridisation to total mRNA, prevents the synthesis of the protein of interest is the desired clone. Alternatively, the bound message can be translated *in vitro* and the products analysed for the presence of the desired protein (14).

A gene whose protein product has been purified can be cloned by first raising antibodies against the purified gene product. The antibodies can be used to detect production of cross-reacting material (CRM) expressed in heterologous systems, such as *E. coli*. The hexokinase gene was isolated from a clone bank of yeast DNA by first fixing total protein from the bacterial colony representing each clone to a plastic well (10). The colonies were screened by radioimmune assay for CRM using hexokinase antibody. Positive results indicated a DNA fragment in the clone directed the synthesis of the protein of interest in *E. coli*. Recently a vector, λgt11, has been devised to simplify antibody screening (11). This vector has been used to clone a number of yeast genes, including several subunits of RNA polymerase (12) and DNA topoisomerase (13). A complete discussion of this vector can be found in Chapter 2 of the first of these two volumes.

2.4 **Cloning by Cross-Hybridisation**

Genes evolutionarily conserved between species can be used to probe the yeast genome for cross-hybridising sequences. Histone genes (54), glycolytic enzyme genes (15), as well as protooncogenes (16) have been detected by this method. The technique is simple; first a genomic blot is probed at a variety of stringencies to determine whether cross-hybridisation exists. Next a phage or plasmid library is screened at the appropriate stringency (determined from the previous blots) for positive signals. Confirmation that the yeast cognate gene has been cloned requires DNA sequencing of the cloned fragment and some method to demonstrate that the sequence indeed codes for a functional protein in yeast.

2.5 Cloning by Complementation in E. coli

Yeast DNA fragments that code for enzymes in biosynthetic pathways were first isolated by complementation of *E. coli* auxotrophic mutations. For example, an *E. coli* strain containing a low reverting allele of the *leuB* gene in the leucine biosynthetic pathway was tranformed with a library of clones containing yeast DNA inserts in an appropriate bacterial vector (4). Bacterial transformants were selected on rich medium for drug resistance conferred by the vector. The transformants were next screened for their ability to grow without leucine supplement. Several clones were isolated that contained a sequence in common. This yeast sequence complemented the *E. coli* mutation and is the cognate gene for the leucine pathway in yeast, *LEU2*. This was confirmed by further studies of the cloned gene (6,17), such as genetic mapping, as well as DNA sequence analysis (18). Several other yeast genes including *HIS3* (5), *URA3* (19) and *TRP1* (20) have also been isolated by complementation in *E. coli*. These genes often serve as the basis for constructing yeast vectors.

During the construction of vectors containing these genes, it is advantageous to use a bacterial strain that can be complemented by the marker in question. This aids in the identification of the correct construction by a simple complementation test in bacteria. Some common bacterial strains containing useful markers are given in *Table 1*.

2.6 Cloning by Complementation in Yeast

Cloning a gene by complementation in yeast requires a library of wild-type yeast DNA in a vector that transforms yeast at high frequency. The construction of

Table 1. Common Bacterial and Yeast Strains used for Cloning.

	Strain	Genotype	Reference
A. *Bacteria*	C600	K12 *leuB6 thi-1 thr-1 lacY1 tonA21 supE44*	(4,21)
	hisB463	K12 *hisB463 hsdR⁺ hsdM⁺*	(5)
	trpC9830	K12 W3110 *trpC117*	(22,23)
	DB6656	K12 *pyrF::Mu trp_{am} lacZ_{am} hsdR⁻ hsdM⁺*	(19)
	BA1	K12 *leuB6 trpC1117 hisB463 Tn10::near hisB thr thi thyA str^r hsdR⁻ hsdM⁻*	Andrew Murray, personal communication
B. *Yeast*	LL20	*MATα leu2-3,112his3-11, 15*	(24)
	YNN281	*MATa his-3△ trp1-△ ura3-52 ade2-101 lys2-801*	Ron Davis, personal communication
	SR25-1A	*Mata his4-912 ura3-52*	Shirleen Roeder, personal communication
	W301-18A	*Matα ade2-1 trp1-1 leu2-3,112 his3-11,15 ura3-1 can-100*	(25)

yeast vectors and their uses will be discussed in more detail in section 3.1. The general features of a yeast vector are: a yeast genetic marker for selecting transformants, a yeast replication origin for high frequency transformation and a bacterial marker and replication origin for selection and propagation in *E. coli*. Random yeast fragments are introduced into the vector to create a library. Cloning by complementation can be accomplished in two ways. In the first, colonies that complement the gene of interest are selected directly after transformation. Alternatively, total transformants are selected by complementation of the yeast marker present in all plasmids and the pool of transformants are next screened (or selected) for complementation of the gene of interest. The first regime is chosen only when the gene of interest exhibits very low ($< 10^{-8}$) reversion frequencies. Genes that revert at higher frquencies often flood the plates with revertants, which can obscure the transformation result. In the second scheme, transformants are first selected using a low reverting allele of the selectable marker present in all the plasmids, thereby ensuring a transformation event. Next, the transformants are screened (or selected) *en masse* for complementation of the gene of interest. For auxotrophic markers, the transformants are simply selected for complementation by replica-plating to the appropriate medium. For phenotypes other than simple auxotrophy, the transformants are screened for complementation by testing the pool for those transformants exhibiting the phenotype of interest. For example, a radiation sensitive gene can be cloned by screened the pool for radiation resistant transformants (26). Depending on the library, from 1000 to 20 000 colonies need to be screened to recover a complementing fragment.

Cloning yeast genes by complementation or screening from plasmid libraries passaged through *E. coli* has been very successful. However, in two instances, certain yeast fragments were not recovered from plasmid-based yeast libraries. During a 'walk' on chromosome III, Carol Newlon and her colleagues (personal communication) were unable to find an overlapping clone between two fragments. They proceeded to clone the intervening fragment by the technique of gap repair using the adjacent sequences to promote homologous pairing (see section 4.2). Subclones containing this rescued fragment grew poorly in bacteria, suggesting that their initial failure to recover the fragment in *E. coli* was due to its under-representation in the original libraries. R.Gaber and G.Fink (personal communication) had difficulty isolating the yeast *hol1* gene from a plasmid library. They constructed a yeast library *in vitro* and isolated the clone after directly transforming an appropriately marked yeast strain. Analysis of subclones showed that a sequence adjacent to the *hol1* gene caused bacteria containing that fragment to grow poorly.

To establish that the complementation function is plasmid-borne, co-loss of both the phenotype of interest and the yeast selectable marker common to all plasmids in the pool must be demonstrated in cells that spontaneously lose the plasmid. Spontaneous loss of the plasmids can be screened by plating out $500-10\ 000$ colonies at 500 per plate and replicating to the medium originally used to select transformants. Loss of the plasmid is indicated by failure to grow on the selective medium. Colonies that have lost the plasmid are next tested for loss of complementation of the gene of interest. The number of colonies screened for loss depends on the stability of the plasmid. Some centromere-based plasmids are stable such that their frequency of loss is $< 10^{-3}$ per generation.

Finally, it is not sufficient to isolate a complementing fragment from a pool in order to be certain that the gene of interest has been cloned. Several cases are known where complementing fragments did not correspond to the gene fragment of interest (27). Methods for showing that a DNA fragment encodes the gene of interest will be discussed in Section 4.3.

3. YEAST VECTORS AND TRANSFORMATION METHODS

The design of yeast vectors and establishment of yeast transformation methods are the result of parallel developments in both areas. The first successful yeast transformation experiments utilised yeast genes that were cloned by complementation in *E. coli* (4,6). Subsequently, sequences from the endogenous plasmid (2 μm circle) were incorporated and high frequency transformation was observed (7). Chromosomal sequences were also identified that conferred high frequency transformation (20). The following selection will contain a description of vectors and their uses as well as transformation methods and related techniques for growing yeast and selecting transformants. In addition, several yeast strains commonly used for transformation experiments are included as part of *Table 1*.

3.1 **Vectors**

There are two kinds of transformation events that occur in *Saccharomyces cerevisiae*: integration and autonomous replication. These events depend on the type of vector used and on the condition of the incoming DNA, intact versus linearised (discussion in Section 4.1). In yeast, plasmid integration occurs by homologous recombination. A yeast sequence contained on a plasmid can interact with its homologous chromosomal sequence in two ways. One results in the integration of the plasmid into the genome at the homologous site (6,24). The other results in substitution of the chromosomal sequence by the yeast sequence on the plasmid without vector integration (6,24). Both of these events occur at low fequency ($\sim 1 - 10$ transformants/μg DNA/10^7 cells) using circular molecules that are unable to replicate autonomously.

High frequency transformation ($\sim 10^3 - 10^4$ transformant/μg DNA/10^7 cells) can be obtained when certain chromosomal sequences or portions of 2 μm circle DNA are included in the vector (7,20). These sequences allow autonomous replication of the vector and confer high transformation frequencies as well. These were named *ars* for *a*utonomously *r*eplicating *s*egment (20). Plasmids that contain chromosomal *ars* sequences are generally unstable. That is, after growth in non-selective conditions (10 generations), $< 5\%$ of the cells contain the plasmid. On the other hand, plasmids that contain certain portions of 2 μm circle DNA are more stable; between 60% and 95% of the cells contain the plasmid after 10 generations of non-selective growth (28). In both cases, cells that contain the plasmid have relatively high copy number ($20 - 50$ copies/cell). For 2 μm-based plasmids, the minimum amount of 2 μm circle DNA necessary to increase stabilty is the *rep3* region and the 599 base pair inverted repeat. The other 2 μm functions

Cloning in Yeast

Table 2. Yeast Vectors.

Type	Example (reference)	Description	Properties
Integrating	YIp5(23)	A pBR322 based vector that contains a selectable marker, *URA3+*.	This plasmid shows reduced recombination with certain chromosomal alleles of *ura3* (e.g., *ura3-52* or *ura3-50*), therefore yeast sequences cloned into this vector recombine by homology mainly at their chromosomal site.
Autonomous high copy-2 μm	YEp13(36) pJDB219(7)	Based on pBR322 and pmB9 respectively, these vectors contain a selectable yeast marker *LEU2+* along with the autonomous replication portion of the 2 μm circle.	These vectors transform yeast at high frequency and replicate autonomously. They are generally used for constructing libraries to clone genes by complementation. The presence of native 2 μm circle DNA in the host strain enhances the stability of these vectors.
Autonomous high copy-chromosomal	YRp7(23) YRp17 (equivalent to YRp7 with a unique *Eco*RI site)	Vectors based on pBR322 containing the *TRP1+* gene and the adjacent chromosomal *ars* sequence.	These vectors transform yeast at high frequency and replicate autonomously. The transformants are relatively unstable; about 90% of the cells lose the plasmid in the absence of selection after 10 generations. The cells that retain the vector have multiple copies. Yeast libraries have been constructed with YRp7 for cloning genes by complementation.
Autonomous high copy-telomere	YLp1(31,32)	A linear plasmid with telomeres from *Tetrahymena* macronuclear rDNA and a yeast selectable marker, *LEU2+* in pBR322.	A high copy vector that is relatively unstable; about 10% of cells maintain the plasmid non-selectively after 10 generations. Useful for cloning yeast telomere sequences and constructing mini-chromosomes.
Autonomous low copy-centromere	YCp19(37)	A circular plasmid based on pBR322 with two selectable markers *TRP1+* and *URA3+*, a chromosomal *ars* and the centromere sequence from chromosome IV.	A low copy vector that is relatively stable non-selectively; about 90% stability after 10 generations. Useful for constructing libraries and cloning genes that may be lethal when present in more than one copy per cell.
Autonomous low copy-mini-chromosome	YLp21(32)	A linear plasmid based on YLp1 with additional phage lambda sequences and the yeast centromere from chromosome III.	A low copy vector that is relatively stable. Useful for studying sequences necessary for proper chromosome function.

Table 2. continued.

Type	Example (reference)	Description	Properties
Expression-unregulated	pAAH5(33)	A circular plasmid that contains the alcohol dehydrogenase promoter with suitable sites for cloning	Useful for over-expression of a gene product in yeast cells.
Expression-regulated	YEp51(29) pYE4(34)	Galactose and phosphate regulated circular plasmids that contain regulated promoters whose expression can be manipulated by exogenous substrate concentrations or by genetic loci required for expression.	Useful for regulating a gene product whose overproduction is potentially harmful or for controlling time of expression of a gene product.
Expression-secretion	pAB112(35)	A circular plasmid that contains the signal sequence processing site for the α factor gene, MF α1.	The α factor sequence directs the secretion of small proteins that are fused to the signal portion of the MFα1 gene.

must be provided in *trans* by endogenous 2 μm plasmids (29).

Additional sequences can be added to plasmids to increase their stability. When yeast centromere sequences are added to chromosomal *ars* containing plasmids, stability increases such that 90% of the cells contain the plasmid after 10 generations (30). The centromere sequence also acts to modulate the copy number to approximately one per cell. Combining centromere sequences and 2 μm circle sequences on one plasmid results in lower stability than that described for plasmids containing both centromeres and chromosomal *ars* sequence (28).

Recently linear plasmids have been constructed by cloning telomere sequences from ciliates or yeast into plasmids to create autonomously replicating linear molecules (31). These linear plasmids are less stable than circular centromere-bearing plasmids; approximately 40% of the cells contain 20−40 copies after 10 generations. The linear molecules can be further stabilised by adding a yeast centromere sequence and also increasing the total size of the plasmid (32). The resultant molecule, a minichromosome present in one copy, is the most stable *in vitro* construction so far described, but it is still two orders of magnitude less stable than a native chromosome.

Finally, vectors have been constructed that aid in the expression and/or secretion of proteins in yeast. Some expression vectors contain strong constitutive promoters, such as one from alcohol dehydrogenase (33), while others utilise regulated promoters from genes such as UDPgalactose 4-epimerase (29) or acid phosphatase (34). The portion of the gene encoding the signal peptide of the mating pheromone α-factor has been used to create hybrid proteins for secretion (35).

Table 2 contains examples of many of the types of vectors described in the preceding section. Whenever possible, an example and reference is given, and the properties and potential uses of the vector are outlined.

3.2 **Yeast Transformation Procedures**

The first two transformation protocols developed for yeast require removal of the cell wall with a β-glucanase, exposure of the spheroplasts to Ca^{2+} and DNA, and finally polyethylene glycol (PEG 4000) treatment (6,7). The transformed cells are washed and embedded in osmotic stabilising regeneration agar. The procedure outlined in *Table 3* was derived from these initial reports. The frequencies of transformation using the spheroplast method with an autonomously replicating plasmid vary from 100 to 100 000 transformants/μg/10^7 cell depending on the yeast strain. During this procedure, it is important to treat the cells gently after they have been converted to spheroplasts. The cells should not be vortexed.

Recently, a transformation procedure that does not require cell wall digestion and spheroplast regeneration has been described. This method (*Table 4*) relies on exposure of whole cells to Li^+ salts, DNA, and finally PEG 4000 (38). Transformants can be selected directly by plating cells on the appropriate agar medium. The major disadvantage of this method is that for many strains, the

Table 3. Spheroplast Transformation Procedure.

1.	Grow yeast cells to a density of $1-2$ x 10^7/ml in 50 ml of YPD[a] liquid medium.
2.	Wash the cells twice in 10 ml of 1 M sorbitol (Aldrich) by repeated pelleting in a bench top centrifuge followed by cell resuspension.
3.	Resuspend the cells in 5 ml of 1 M sorbitol and add 5 μl of β-mercaptoethanol.
4.	Dilute an aliquot to 10^{-5} in water and plate 100 μl on YPD solid medium. This will serve as the control to calculate the percentage of killing after spheroplast formation.
5.	Add 150 μl of glusulase (Endo Laboratories, Inc.) and incubate at 30°C with gentle shaking for approximately 30 min. Determine the actual time that each strain should be incubated in glusulase by measuring the amount of time necessary to give approximately 99% spheroplast formation with $10-20$% spheroplasts capable of regeneration.
6.	Dilute an aliquot to 10^{-3} in water and plate 100 μl on YPD medium to measure the percentage survival after spheroplast formation.
7.	Pellet the spheroplasts by centrifugation 2500 r.p.m.[b] for 7 min and gently resuspend in 2 ml of 1 M sorbitol. Adjust the volume to 10 ml with 1 M sorbitol and pellet the cells. Repeat once more.
8.	Once again resuspend the spheroplasts in 2 ml of 1 M sorbitol. Adjust the volume to 9 ml with 1 M sorbitol and add 1 ml of 0.1 M Tris-HCl pH 7.4, 0.1 M $CaCl_2$. Pellet the spheroplasts and resuspend in 1.6 ml of 1 M sorbitol, 10 mM $CaCl_2$.
9.	Distribute 0.2 ml of cells for each transformation and add DNA ($1-10$ μg) keeping the total volume of DNA added less than 20 μl.
10.	Incubate the cells and DNA for 10 min at room temperature. Thoroughly resuspend the cells and add 2 ml of 50% (w/v) PEG 4000 (Fisher or Baker) mixed with 0.2 ml of 0.1 M Tris-HCl pH 7.4, 0.1 M $CaCl_2$.
11.	After 10 min at room temperature, pellet the spheroplasts by centrifugation as in Step 7 and resuspend the cells in 0.5 ml of 1 M sorbitol, 10 mM Tris-HCl pH 7.4, 10 mM $CaCl_2$ and plate in 10 ml of regeneration agar[c] (maintained at $45-50$°C) onto the appropriate omission medium.

[a]YPD medium is described in *Table 5*.
[b]Cells or spheroplasts are pelleted using the 215 rotor of the IEC Clinical Centrifuge. The equivalent centrifuge and rotor combinations from other manufacturers are equally suitable.
[c]Regeneration agar is described in *Table 5*.

Table 4. Lithium Transformation Procedure.

1.	Grow a culture of yeast in 100 ml of YPD medium[a] to OD_{600} ~0.4 (~1−2 x 10^7 cells/ml).
2.	Centrifuge the cells at 2500 r.p.m. for 7 min in a benchtop centrifuge. Resuspend the cells in TE[b] and pellet them again by centrifugation.
3.	Resuspend the cells in 1.5 ml of 0.1 M LiCl (or 0.3 M LiOAc) in TE. Incubate 3−8 h at 4°C.
4.	Remove 200 µl per transformation reaction into a capped tube and add DNA. Use 1−10 µg per tube keeping the volume of DNA added less than 1/10 the total volume. Incubate the cell-DNA mixture for 30 min at room temperature. Swirl periodically and gently.
5.	Thoroughly resuspend the cells, and then add 1.5 ml of 40% PEG 4000. Mix by inverting the tube. Incubate for 1 h at room temperature.
6.	Heat shock the cells by placing the tubes at 42°C for 5 min.
7.	Centrifuge at 1500 r.p.m. for 7 min. Wash the cell pellet twice with water by repeatedly centrifuging and gently resuspending it.
8.	Finally resuspend the cells in 100 µl water. Plate on appropriate omission medium[a].

[a]Media are defined in *Table 5*.
[b]TE is 10 mM Tris-HCl pH 7.4, 1 mM EDTA.

number of transformants obtained with autonomously replicating plasmids is 10- to 100-fold lower than with the spheroplast procedure. However, transformation using linear molecules (discussed in Section 4.1) is very efficient in Li^+ transformed cells. In addition, for some experiments, it is advantageous to collect the cells from the surface of a plate, rather than collect them from regeneration agar.

We have recently transformed cells by resuspending a fresh 2-day colony from a YPD plate in 0.1 M LiCl and incubating the cells for 4−8 h. The cells are then treated as in *Table 4* from Step 4. Transformation frequencies for linear molecules vary from 10 to 100 transformants/µg/10^7 cells.

Finally, a transformation procedure that only requires polyethylene glycol (PEG 1000) has been described (39). In that procedure, it is critical to purify the PEG to remove impurities that lower transformation frequencies. It is interesting to note that transformation frequencies for both the spheroplast and Li^+ procedures are dependent on the batch of PEG 4000 used (40). We have found that by testing different lots, one can maximise transformation frequencies.

3.3 Growth Media and Selection

Yeast cells can be grown in complex medium utilising a variety of carbon sources. Most yeast media contain glucose as the sole carbon source. *Table 5* lists some commonly used media for growing yeast. The generation time of standard yeast strains at 30°C in liquid YPD medium varies from 60 to 90 min. Cells are generally inoculated at 5 x 10^4/ml or greater for growth overnight.

Solid medium is made by adjusting the final agar concentration to 2% for both complex medium and synthetic medium. Synthetic medium without additional nutrients can support the growth of wild type yeast. Synthetic medium to grow auxotrophic yeast strains can be made by either adding only the nutrients required by the strain (SD + nutrient in question) or by adding twelve basic nutrients (complete medium). In laboratories that routinely use a single or a few yeast strains it is easier to use medium with the particular nutritional requirements of the strains. For laboratories that use many yeast strains, it is convenient to make

Table 5. Common Media for Growing Yeast.

Type of media[a]	Name	Recipe
Complex medium (rich)	YPD	1% Yeast extract 2% Peptone 2% Dextrose
Complex medium (non-fermentable carbon source)	YPG	1% Yeast extract 2% Peptone 3% Glycerol
Synthetic medium	SD	0.17% Yeast nitrogen base without amino acids and without ammonium sulphate 0.50% Ammonium sulphate 2% Dextrose
Synthetic complete	COM	SD plus 20 μg/ml each of adenine sulphate, L-arginine-HCl, L-histidine-HCl, L-methionine, L-tryptophan and uracil; 30 μg/ml each of L-isoleucine, L-lysine-HCl and L-tyrosine; 50 μg/ml of L-phenylalanine 60 μg/ml of L-leucine 150 μg/ml of L-valine
Omission medium	COM-nutrient. Generally called −nutrient (e.g., COM−URA is called −URA, read as 'minus uracil')	COM minus one (or more) nutrient(s)
Regeneration agar	RA-nutrient	SD made in 1 M sorbitol plus 0.5% YPD, the appropriate supplements at the concentrations shown for synthetic complete medium and 3% agar

[a]Liquid medium can be made solid by adding agar to 2%.

'omission' medium for the routine testing of nutritional requirements. Omission refers to the fact that one of the twelve nutrients is omitted. The medium is referred to as 'minus 'nutrient''. The phenotype of a strain can be determined by replicating a representative colony to various omission media. A nutritional requirement is indicated by failure to grow on a particular omission medium while being able to grow on complete medium.

Selection for complementation of an auxotropic marker after transformation is accomplished by using the appropriate omission medium for growth. The spheroplast method of transformaton requires regeneration of the spheroplasts in an osmotically stabilising synthetic medium (*Table 5*). The regeneration agar is made identically to the selection medium with respect to the nutrients added. For example, *LEU*⁺ transformants are selected on 'minus leu' (COM-leucine) medium by plating in 'regeneration agar minus leu' (RA-leucine). For cells transformed by the Li⁺ salts method, the cells are plated directly on the appropriate selective synthetic medium.

3.4 **Yeast Colony Hybridisation**

After selecting transformants, it is often desirable to determine whether plasmid sequences have been incorporated into the cell. This is especially important for some of the allele rescue procedures outlined in Section 4.3. The detection of plasmid (vector) sequence is easily facilitated by using the yeast colony hybridisation method (6) that is a derivative of the bacterial colony hybridisation technique (41). Radioactive probe is prepared by either nick-translation or by end-labelling. The yeast colonies are first grown for 24 h directly on an 85 mm nitrocellulose filter that is placed on a YPD plate. The filter is treated as follows:

(i)   Place the nitrocellulose filter on Whatman 3MM filter paper soaked with 2 mg/ml of Zymolyase 60 000 (or 1 mg/ml of Zymolyase 100T) dissolved in 1 M sorbitol, 20 mM EDTA pH 7.4. Next place the filter in a sealed container (usually plastic wrap over a Pyrex dish) and incubate at 37°C for 4 h or overnight to digest the cell wall.

(ii)   Check the cells for spheroplast formation by removing equivalent amounts of yeast (a toothpick full), one into a drop of 1 M sorbitol and one into a drop of 10% sodium dodecyl sulphate, and observing ghost formation under the microscope. Alternatively, assume that after overnight treatment the cell walls are sufficiently digested to allow efficient lysis.

(iii)   Next, carefully lift the nitrocellulose filter from the Whatman filter and transfer to a fresh filter that is soaked with 0.1 M NaOH. Incubate at room temperature for 7 min.

(iv)   Carefully transfer the nitrocellulose filter to a filter soaked with 1 M Tris-HCl pH 7.4, and incubate for 4 min at room temperature. Repeat this step once more.

(v)   Place the nitrocellulose filter for 2 min on a Whatman 3MM filter soaked with 2 x SSC (1 x SSC is 0.15 M NaCl, 0.015 M sodium citrate).

(vi)   Bake the filter at 80°C under vacuum for 2 h.

The filter is now ready for hybridisation. The filter is first floated on top of water in order to wet it. The filter does not always wet uniformly near the dried and fixed colonies. It is next transferred to a strong polythene bag and is sealed together with approximately 10 ml of buffer (5 x SSC and 1% sarkosyl) per filter. If more than one filter is hybridised per bag, only two are allowed to overlap, and those are placed back-to-back with the colony-containing sides outward. At some time during this pre-hybridisation step, which can last anywhere from 15 min up to 24 h, the probe ($\sim$500 000 counts/min of ^{32}P-labelled nucleic acid) is denatured by boiling in 2.5 ml of water per filter. Immediately before hybridisation an equal volume of 10 x SSC and 2% sarkosyl is added to the boiled probe. The pre-hybridisation liquid is removed from the bag and is replaced by the probe. Care is taken to remove air bubbles and the bag is sealed. Hybridisation is carried out for 12 – 18 h at 65°C. The filter is then washed twice for 30 min, with gentle shaking, at 42°C in 2 x SSC and 0.5% sarkosyl. It is then washed twice at 42°C in 2 x SSC, blotted dry and exposed to X-ray film with or without an intensifying screen (42). Using the above regimen, a single integrated copy of the plasmid can be detected in 12 – 24 h with nick-translated pBR322, labeled to 10^7 c.p.m./μg (43).

3.5 **Yeast DNA Isolation and Plasmid Rescue**

The isolation of yeast DNA is necessary to prepare clone libraries, as well as to rescue plasmids that have integrated into the genome or are replicating autonomously. Purification of yeast DNA has been greatly simplified since the first procedures worked out by Marmur and his colleagues (44). A relatively simple and rapid procedure for preparing yeast DNA was recently described by the Davis lab (45). *Table 6* describes a similar procedure that can be used to isolate DNA from 40 ml of culture that yields approximately $10-50$ μg of DNA. The entire protocol can be scaled down to 5 ml, which provides enough DNA for most procedures. DNA from these preparations can be used to re-isolate plasmids that were introduced into the cell.

Autonomously replicating plasmids can be recovered by directly transforming isolated yeast DNA into bacteria (8). Generally, $100-300$ ng of total yeast DNA will yield from 10 to 300 transformants depending on the competency of the bacterial strain and the copy number of the plasmid. It is best to transform competent bacteria (20-fold concentrated cells in 200 μl harvested at OD_{600} of 0.6) with no more than 300 ng of total DNA since the number of transformants recovered decreases in the presence of excess DNA. Integrated plasmids can be recovered by digesting total yeast DNA (~ 300 ng) with an appropriate restriction enzyme as outlined in Section 4.2. The reaction is diluted to 1.0 ml and incubated with

Table 6. Isolation of Yeast DNA.

1.	Grow the yeast culture in 40 ml of YPD medium to late log phase ($\sim 10^8$ cells/ml).
2.	Pellet the cells by centrifugation at 5000 r.p.m. for 5 min in a Sorvall RC5 using an SS34 or SA600 rotor (or their equivalent in other machines). Wash the cell pellet by resuspending in TE and pelleting again by centrifugation.
3.	Resuspend the cell pellet in 3.2 ml of 1 M sorbitol, 0.1 M EDTA and 14 mM β-mercaptoethanol, pH 7.4.
4.	Digest the cell walls with 0.1 ml of 15 mg/ml Zymolyase 60 000 or (100T) for 5 min at 37°C. Digest for longer if less concentrated Zymolyase is used (e.g., 2 mg/ml).
5.	Check visually for spheroplast formation by removing 50 μl into 200 μl of 1 M sorbitol and 200 μl of 10% sodium dodecyl sulfate (SDS). Proceed if greater than 90% of the cells form spheroplasts.
6.	Pellet the spheroplasts by centrifugation at 5000 r.p.m. for 5 min as in Step 2.
7.	Resuspend the pellet in 3.2 ml of TE. Add 0.3 ml of 0.5 M EDTA and 0.3 ml of 1 M Tris-HCl pH 7.4 and mix throughly. Add 0.16 ml of 10% SDS.
8.	Swirl the cell suspension with the SDS to mix. Place at 65°C for 30 min.
9.	Add 1 ml of 4 M potassium acetate and incubate on ice for 1 h.
10.	Centrifuge at 15 000 r.p.m. for 25 min as in Step 2. Carefully decant the supernatant into a fresh tube.
11.	Add 12 ml of 95% ethanol to the supernatant and mix. Centrifuge at 15 000 r.p.m. for 15 min as in Step 2 at room temperature. Discard supernatant and air dry the pellet.
12.	Resuspend the pellet in 3 ml of TE. If any debris is still present, centrifuge the solution at 10 000 r.p.m. for 10 min as in Step 2.
13.	Transfer the suspension to a 14 ml polypropylene tube and add 150 μl of pancreatic RNase (1 mg/ml). Incubate at 37°C for 30 min.
14.	Extract once with phenol:chloroform:isoamyl alcohol (25:24:1).
15.	Remove the top layer, adjust the salt concentration to 0.3 M, add 2 volumes of cold 95% ethanol. Mix and place at -20°C for 2 h or longer.
16.	Pellet the DNA by centrifugation at 10 000 r.p.m. for 30 min at 4°C as in Step 2.
17.	Discard the supernatant, air dry the pellet and resuspend it in 0.5 ml TE.

ATP and T4 DNA ligase at 15°C overnight (8). The ligated DNA is transformed into competent bacteria (8), and the yield is approximately as described above for autonomous plasmids.

4. MANIPULATION OF CLONED DNA

After a DNA fragment has been identified, it is often useful to re-introduce the molecule back into the genome to study gene expression. Several methods for introducing molecules into specific chormosomal locations will be described. These methods can be exploited for pre-selecting integration sites for plasmids that contain multiple yeast sequences, for allele rescue, as well as for gene disruption and gene replacement techniques. The methods all rely on homologous recombination between the incoming DNA and the yeast chromosome.

4.1 Directed Plasmid Integration

A method for effectively integrating a yeast plasmid into the genome by stimulating homologous recombination has proven useful for the manipulation of cloned DNA fragments (24). The method utilises the double-strand-break repair capabilities of yeast cells to promote recombination between an incoming linear molecule and the genome. If a plasmid contains two or more fragments homologous to the genome, a double-strand break is introduced by restriction enzyme digestion into the middle of the fragment that is homologous to the desired integration site. The linear molecules are transformed into an appropriately marked strain using either the spheroplast or Li$^+$ procedure for transformation (6,7,38). Between 1 and 10 μg of DNA is sufficient to yield from a few to thousands of transformants per 10^7 cells in a typical experiment. Colony hybridisation is performed to detect those transformants that contain the plasmid and DNA is isolated from a representative sample (6). Genomic blots are then carried out to confirm the integration site and to determine whether a multiple tandem integration of the plasmid occurred (24).

4.2 Allele Rescue Methods

Genetic studies often require the cloning of mutant alleles of the gene of interest. Once a DNA fragment containing the wild type gene has been cloned, it is relatively easy to clone mutant alleles of the gene. One strategy relies on homologous recombination between the wild type gene on the plasmid and the mutant gene in the chromosome to move a vector sequence into the proximity of the desired allele (*Figure 1A*). In the duplication created by the integration, the mutant allele is located on one side of the vector and the wild type allele is on the other. Since the precise configuration depends upon where the integration cross-over takes place with respect to the site of the mutant allele in the gene, rescue of the desired mutant sequence requires the independent cloning of each duplicated copy flanking the vector. In the example in *Figure 1A*, two independent plasmids can be isolated by separately digesting genomic DNA with the enzymes for site a and site d. Next the digested DNA is ligated and transformed into *E. coli* to recover each plasmid as described in Section 3.5.

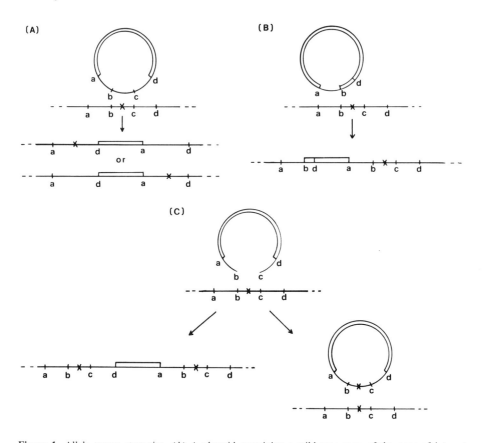

Figure 1. Allele rescue strategies. (**A**) A plasmid containing a wild type copy of the gene of interest is integrated into the genome. Recombination with the homologous chromosomal sequence containing the mutation creates a duplication that contains the mutation on one side and the wild type sequence on the other. Since either side may contain the desired mutant fragment, genomic DNA is separately digested with restriction enzymes for site a and site d. The plasmids are rescued as described in the text. (**B**) A plasmid containing a subcloned fragment that lies adjacent to the mutation of interest is integrated into the genome. This creates an integrated plasmid that lies uniquely on one side of the mutation. Digestion of genomic DNA with a restriction enzyme that cuts at site d is used to rescue the plasmid in bacteria as described in the text. The rescued plasmid must contain a bacterial origin of replication and selectable marker. (**C**) A plasmid containing an *ars* and the wild type copy of the gene of interest is linearised and gapped by digesting with a restriction enzyme(s) that removes the wild type portion of the plasmid sequence (sites b and c). The linear molecule is transformed into the mutant containing yeast and gap repair results in two types of transformants: stable integrants with the mutated site on each side of the plasmid and cells with an autonomously replicating plasmid that contains the mutated site. Rescue of these plasmids is described in the text.

Since it cannot be known *a priori* which of the duplicated genes contains the mutant allele, several methods have been developed to insure recovery of the desired allele. For one method (46), a small fragment of DNA that lies adjacent to the gene of interest is cloned into an integrating vector that contains a selectable yeast marker (e.g., YIp5). Integration into the cloned region can be directed by linearising the plasmid within the sequence as described in Section 4.1. Homologous recombination places the vector into the region adjacent to the desired allele as shown in *Figure 1B*. Transformants that contain plasmid sequences are identified by colony hybridisation. Chromosomal DNA is isolated and the integration configuration is confirmed by a genomic blot. Chromosomal DNA is digested with the appropriate restriction enzyme for plasmid rescue (site d in *Figure 1B*). The enzyme should digest the vector near the end upstream of the gene of interest and digest the adjacent chromosomal DNA downstream from the gene. Selection of the site should insure that the final plasmid for rescue in *E. coli* contains a selectable gene (e.g., ampicillin resistance) and the plasmid origin of replication.

Another method that insures recovery of the mutant allele takes advantage of plasmid gap repair (*Figure 1C*, 40). A plasmid is constructed that contains a selectable marker and the wild type copy of the entire genetic region of interest. The plasmid is digested with the appropriate restriction enzyme(s) to delete the wild type portion of the gene (sites b and c in *Figure 1C*), leaving homology to the chromosomal region on both ends of the plasmid. The linear molecule is transformed into the mutant strain and transformants expressing the selectable marker are identified. During the integration event, the gap in the plasmid is repaired using the mutant chromosomal sequences as template. The resultant integrant contains the plasmid sequences flanked by two identical mutant copies of the chromosomal allele. This is confirmed by colony hybridisation to detect vector sequences and a genomic blot to demonstrate the desired integration configuration. Plasmid rescue of either copy adjacent to the plasmid sequences results in the cloning of the mutant chromosomal allele.

The gap repair technique described in the preceding section can be modified to produce an autonomously replicating plasmid that contains the desired mutant chromosomal allele (*Figure 1C*, 40,47). The plasmid can be isolated directly from a yeast DNA preparation by transforming competent *E. coli* cells, eliminating the need for recovering the plasmid from yeast chromosomal DNA by a restriction enzyme digestion and subsequent ligation. This allele rescue technique requires an autonomous replication segment on the rescuing plasmid. As in the preceding section, a linear molecule that contains a selectable yeast marker as well as an *ars* sequence is gapped to delete the wild type complementing portion of the desired chromosomal mutation. Transformants are selected and two types of recombination events are recovered with equal frequency after gap repair of the deleted segment. One is the event described in the previous paragraph, an integrated plasmid flanked by direct repeats of the chromosomal mutant allele. The other event results in an autonomously replicating plasmid that contains the mutant allele after repairing the gap from the mutant chromosome. Colonies containing the autonomously replicating plasmid can easily be identified by growing the cells overnight on non-selective medium. Unlike colonies containing the chromosomally intregrated plasmid, they will be unstable for the selectable marker. Yeast DNA is isolated from the transformant and used to transform *E. coli* and rescue the plasmid containing the allele of interest.

4.3 Gene Disruption Strategies

It is often desirable to disrupt a gene either to examine the null phenotype or to show gentically that a cloned fragment encodes a given phenotype. In the latter case, it is not sufficient to demonstrate tight genetic linkage of the gene to the plasmid after it undergoes integration by homologous recombination with its chromosomal site. It is a formal possibility that a tightly linked suppressor mutation had been cloned, and therefore, direct evidence is required to demonstrate that a cloned fragment encodes a given phenotype. This is accomplished by disrupting a region within the fragment, substituting the disrupted allele for the intact gene at its chromosomal location and demonstrating loss of the wild type phenotype. A number of strategies have been developed to address these problems; they have been termed 'gene transplacements', 'gene disruptions' or 'gene replacements' by various researchers. In each case, the goal is the same: introduce a non-functional copy of the gene in question into its normal chromosomal position.

The first step in any gene replacement strategy is localisation of all genetic regions on the clone. This can be accomplished by constructing a transcript map or analysing the DNA sequence for potentially functional genes (8). These regions serve as the targets for *in vitro* manipulations such as insertion or deletion mutagenesis on the cloned fragment. In the case where a fragment has been identified by complementation, the mutagenised clone is reintroduced into the appropriate genetic background, and once again tested for complementation. Failure to complement indicates that the gene of interest on the clone was disrupted. The types of mutations introduced can be simple deletions of restriction fragments or random insertions of linkers (48) or even gene-sized restriction fragments (25). The next step is to introduce the mutation into the genome.

Once its suspected location has been identified there are several strategies available to disrupt the gene. One method requires the subcloning of an internal fragment of the gene into an integrating plasmid containing a selectable marker (49). Homologous recombination between the plasmid fragment and the chromosome results in a disruption of the gene by physically separating the intact 5' end from the intact 3' end (*Figure 2A*). Since the cloned internal fragment does not contain an intact 5' or 3' end, both duplicated genetic regions flanking the plasmid fail to constitute an intact gene. This disruption strategy offers the advantage of both using a single integration step to create the disruption and linking a genetic marker, namely the selectable gene on the plasmid, with the disruption phenotype. The gentic linkage aids in genetic studies by providing a phenotype that can be scored independently of the disruption. The limitations of the method are that:

(i)   Only a single type of defect can be generated.
(ii)   Many genes do not contain restriction fragments suitable for such a disruption experiment.
(iii)   The size of the gene to be disrupted is limited since fragments less than 250 base pairs in length inefficiently recombine with the genome.
(iv)   The disruptions are unstable due to breakdown of the duplication by genetic recombination, an event that can occur at a frequency as great as 10^{-2} for some disruptions.

Another method of gene disruption involves the 'transplacement' of a mutated sequence on a plasmid into the genome to replace the normal chromosomal copy (50). This method is depicted in *Figure 2B*. First, a mutation is introduced into

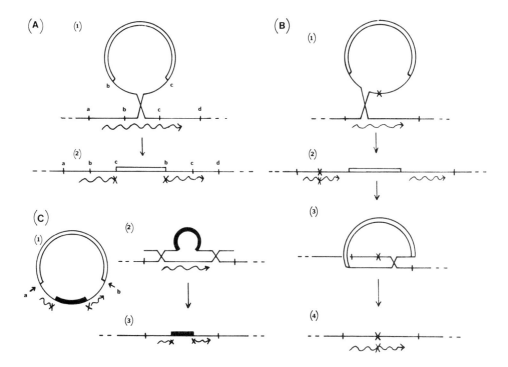

Figure 2. Gene disruption methods. (**A1**) A plasmid containing an internal fragment of the gene of interest is integrated by homologous recombination into the chromosomal site. The integration disrupts the gene with plasmid sequences (**A2**). (**B1**) A plasmid containing the desired alteration of the gene of interest is integrated by homology into the genome. (**B2**) The integration creates a duplication that contains the mutation on one side and wild type sequence on the other. (**B3**) Excisions of the plasmid are selected or screened for. (**B4**) In some portion of the cells the desired alteration is left in the genome after excision. (**C1**) The desired region is disrupted by cloning a selectable marker (solid bar) into the appropriate plasmid. The plasmid is digested with restriction enzymes (sites a and b) to liberate the vector sequences from the plasmid. (**C2**) The linear DNA is used to transform yeast selecting for the inserted marker. (**C3**) Homologous recombination results in the replacement of the chromosomal region by the disrupted one.

the cloned gene of interest on a plasmid containing a selectable marker (usually *URA3*+). Next, the plasmid is forced to undergo homologous recombination with the wild type sequence in the genome. This recombination event can occur by means of a crossover on either side of the mutated site, thus creating a duplication of the target region with one copy of the mutant and one copy of the wild type gene. The next step requires a crossover to excise the plasmid and leave the mutated copy in the genome. The success of this method requires that the second crossover, resulting in loss of the vector, occur in sequences on the side of the mutation opposite the side of the first crossover. In the case of *URA3*+ containing plasmids, plasmid loss can be facilitated by 5-fluoro-orotic acid selection (described in Section 4.4). Recombinants are screened for the absence of vector sequence by colony hybridisation and for the presence of the mutated fragment, either by the resulting phenotype or by a genomic blot of the chromosomal DNA. The major advantage of the transplacement method is that the introduction of virtually any alteration of a gene fragment is possible as long as sufficient

DNA homology exists for the crossovers to occur. The limitation of this method is that two steps are required to generate the desired mutation. This is especially cumbersome when dealing with repeated gene replacements in a particular region of interest, since each disruption has to be introduced by the two-step process just described.

The third gene replacement strategy relies on the ability of free DNA ends to stimulate gentic recombination (25). A selectable genetic marker is used to disrupt a cloned copy of the gene of interest (*Figure 2 C1*). In the next step, all (or most) of the plasmid vector sequence is removed, resulting in a linear molecule whose free ends can pair with the chromosome. Homologous recombination at the ends results in the replacement of the chromosomal copy by the *in vitro* disrupted gene. The gene disruptions generated by this method can be simple insertions resulting from cloning the selectable marker into a single restriction site. Deletions can also be generated by replacing a restriction fragment on the plasmid with the selectable marker. The advantages of this method are that:

(i)   It requires only one transformation to create the disruption.

(ii)   A scorable genetic marker is linked to the disrupted region.

(iii) The disruption is not revertable.

One limitation is that restriction sites are not always available for simple insertions.

4.4 Gene Replacement Techniques

The power of yeast molecular genetics lies in the fact that a gene can be altered *in vitro* and reintroduced into its precise genomic position to assess the genetic consequence of the alteration. Several methods have been developed to facilitate gene replacements where a series of mutated fragments are independently introduced into the same chromosomal site. One method, transplacement, was described in the previous section (50). Another method that utilises a recessive drug resistance marker and its corresponding wild type (Sensitive) fragment to facilitate transplacement, has recently been described (51). First, the strain of interest is made cyclohex-imide resistant at the *cyh2* locus, a gene coding for a ribosomal protein. Next, the wild type cyclohexamide sensitive gene, *CYH2*, is inserted into the sequence of the desired gene cloned in a plasmid containing a selectable marker. By transplacement methods, the wild type genomic copy of the gene of interest is substituted by the *CYH2* disrupted gene (*Figure 3A*). This creates a strain that is cyclohexamide sensitive due to expression of the *CYH2* fragment. Next, a mutated copy of the gene of interest is transformed into the *CYH2* disrupted strain either on a circular plasmid or as a linear fragment. After allowing for phenotypic lag, recombinants are selected on cycloheximide containing medium. These results from substitution of the mutated DNA for the *CYH2* disrupted chromosomal region since loss of the *CYH2* DNA allows expression of the recessive cycloheximide resistance gene, *cyh2^r*. A similar strategy using canavaline resistance and its cognate wild type gene has also been successful for repeated gene replacements. The major limitation of this method is that all strains containing the altered fragments are in a cycloheximide resistant genetic background. This background could adverse-

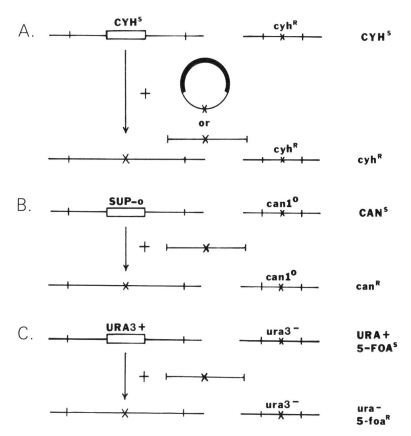

Figure 3. Gene replacement strategies. (**A**) First, a cyclohexamide resistant strain (*cyh2ʳ*) is used as a recipient for a disruption of the gene of interest by a fragment containing the wild type *CYH2* gene. This creates a cycloheximide sensitive strain. Next, either circular plasmid molecules or linear molecules (without plasmid sequence) containing the desired mutation are transformed into the strain. The desired replacement results in a cycloheximide resistant strain. (**B**) A strain containing an ochre suppressible canavanine resistance allele (*can1ᵒ*) is used to disrupt the gene of interest with an ochre suppressor containing fragment (*SUP-o*). Transformation with linear molecules that contain the desired mutation are used to generate the replacement – a canavanine resistant strain. (**C**) A *ura3⁻* strain is used as a recipient for a disruption of the gene of interest by a *URA3+* fragment. This strain is 5-fluoro-orotic acid sensitive. Linear molecules containing the desired mutation are transformed into the strain. Successful replacements have lost the *URA3+* fragment and are 5-fluoro-orotic acid resistant.

ly affect the study of the gene in question (especially translational studies).

Another method recently developed for multiple gene replacements takes advantage of yeast suppressor genetics (Kim Nasmyth, personal communication). Strains containing the *can1-100* gene are resistant to canavanine, an arginine analogue and cell poison. Strains containing an efficient ochre suppressor and the *can1-100* allele are sensitive to canavanine since *can1-100* is ochre suppressible. In the method, an ochre suppressor is introduced into the gene of interest and substituted for the wild-type gene in a genetic background containing *can1-100*, thus creating a canavanine sensitive strain (*Figure 3B*). Linear DNA fragments that contain alterations of the gene of interest are transformed into the suppressor bearing strains. Loss of suppressor activity can occur by many routes, therefore the linear molecule is co-transformed with a circular autonomously replicating molecule to assure a population of canavanine resistant cells enriched for transformants. The resulting canavanine resistant colonies are screened for the successful gene replacement by genetic phenotype and/or genomic blots.

Strategies that rely on co-transformation have been applied even when no selection exists for the successful replacement (J.Strathern and A.Klar, personal communication; H.Rufold and A.Hinnen, personal communication, 52). In these cases, a linear fragment containing the desired alteration is co-transformed with a circular autonomously replicating plasmid, enriching for transformants. Approximately 1% of the transformants contain the desired replacement. This method is useful only when a recognisable phenotype arises from the replacement. To insure a successful replacement using this method the starting strain should contain a genetically marked disruption of the gene of interest (e.g., a *HIS3* insertion into the gene). Transformants are subsequently screened for loss of the original insertion (in this example *his3⁻*) by replica-plating to the appropriate medium.

The final gene replacement method to be discussed uses the drug 5-fluoro-orotic acid to select for replacement transformants. Wild type yeast are sensitive to 5-fluoro-orotic acid at concentrations of 750 μg/ml (53). Resistant colonies occur by loss of *URA3+* function (in addition to other ill-defined pathways). The resistance of *ura3⁻* cells to 5-fluoro-orotic acid has been exploited in gene replacement studies (R.Rothstein *et al.*, unpublished observations). Starting with a *ura3⁻* yeast strain, the gene of interest is disrupted using the *URA3+* gene fragment in a one-step gene disruption creating a uracil independent cell (*Figure 3C*). Next, the strain is transformed with a linearised DNA fragment containing the *in vitro* altered gene and transformants are selected for 5-fluoro-orotic acid resistance. A successful replacement is confirmed by a genomic blot.

Several of the gene replacement strategies just described use a dominant insert to disrupt the gene of interest. Next, a fragment mediated event is used to replace the insert simultaneously uncovering a recessive chromosomal allele. However, the fragment mediated event is not the only event that can occur. We found among the 5-fluoro-orotic resistant colonies some that arose from gene conversion of the *URA3+* insert by the homologous chromosomal *ura3⁻* mutation. In addition, we found new *ura3⁻* mutations in the *URA3+* insert. Both of these events can be a significant portion of the colonies that arise in selective medium if the transformation frequency is low. It is advisable to distinguish the successful replacement from these 'background' events by a genomic blot.

5. ACKNOWLEDGEMENTS

The author thanks Brian Gallay, Barbara Thomas and John Wallis for critical reading of the manuscript, Irwin Gelman for photography and Louis Purcell for typing the manuscript. The author also thanks Carol Newlon, Kim Nasymth and Gerry Fink for permission to quote unpublished results. This work was supported by grants from NIH and NSF.

6. REFERENCES

1. Kramer,R.A., Cameron,J.F. and Davis,R.W. (1976) *Cell,* **8**, 227.
2. Beckmann,J., Johnson,P. and Abelson,J. (1977) *Science,* **196**, 205.
3. St. John,T.P. and Davis,R.W. (1979) *Cell,* **16**, 443.
4. Ratzkin,B. and Carbon,J. (1977) *Proc. Natl. Acad. Sci. USA,* **74**, 487.
5. Struhl,K., Cameron,J.R. and Davis,R.W. (1976) *Proc. Natl. Acad. Sci. USA,* **73**, 1471.
6. Hinnen,A., Hicks,J. and Fink,G.R. (1978) *Proc. Natl. Acad. Sci. USA,* **75**, 1929.
7. Beggs,J.D. (1978) *Nature,* **275**, 104.
8. Maniatis,T., Fritsch,E.F. and Sambrook,J. (1982) *Molecular Cloning,* published by Cold Spring Harbor Laboratory, New York.
9. Paterson,B.M., Roberts,B.E. and Kuff,E.L. (1977) *Proc. Natl. Acad. Sci. USA,* **74**, 4370.
10. Clarke,L., Hitzeman,R. and Carbon,J. (1979) in *Methods in Enzymology,* Vol. **68**, Wu,R. (ed.), Academic Press Inc., New York, p. 436.
11. Young,R.A. and Davis,R.W. (1983) *Proc. Natl. Acad. Sci. USA,* **80**, 1194.
12. Young,R.A. and Davis,R.W. (1983) *Science,* **222**, 778.
13. Goto,T. and Wang,J.C. (1984) *Cell,* **36**, 1073.
14. Hereford,L., Fahvner,K., Woolford,J., Rosbash,M. and Kaback,D.B. (1979) *Cell,* **18**, 1261.
15. Holland,M.J., Holland,J.P. and Jackson,K.A. (1979) in *Methods in Enzymology,* Vol. **68**, Wu,R. (ed.), Academic Press Inc., New York, p. 408.
16. DeFeo-Jones,D., Scolnick,E., Koller,R. and Dhar,R. (1983) *Nature,* **306**, 707.
17. Hicks,J. and Fink,G.R. (1977) *Nature,* **269**, 265.
18. Andreadis,A., Hsu,Y.-P., Kohlhaw,G.B. and Schimmel,P. (1982) *Cell,* **31**, 319.
19. Bach,M.-L., Lacroute,F. and Botstein,D. (1979) *Proc. Natl. Acad. Sci. USA,* **76**, 386.
20. Stinchcomb,D.T., Struhl,K. and Davis,R.W. (1979) *Nature,* **282**, 39.
21. Appleyard,R.K. (1954) *Genetics,* **39**, 440.
22. Yanofsky,C., Horn,V., Bonner,M. and Stasiowski,S. (1971) *Genetics,* **69**, 409.
23. Struhl,K., Stinchcomb,D.T., Scherer,S. and Davis,R.W. (1979) *Proc. Natl. Acad. Sci. USA,* **76**, 1035.
24. Orr-Weaver,T.L., Szostak,J.W. and Rothstein,R.J. (1981) *Proc. Natl. Acad. Sci. USA,* **78**, 6358.
25. Rothstein,R. (1983) in *Methods in Enzymology,* Vol. **101**, Wu,R., Grossman,L. and Moldave,K. (eds.), p. 202.
26. Schild,D., Konforti,B., Perez,C., Gish,W. and Mortimer,R.K. (1983) *Curr. Genet.,* **7**, 85.
27. McKay,V.L. (1983) in *Methods in Enzymology,* Vol. **101**, Wu,R., Grossman,L. and Moldave,K. (eds.), p. 325.
28. Tschumper,G. and Carbon,J. (1983) *Gene,* **23**, 221.
29. Broach,J.R., Li,Y., Wu,L.C. and Jayaram,M. (1983) in *Experimental Manipulation of Gene Expression,* Inouye,M. (ed.), Academic Press, Inc., New York, p. 83.
30. Clarke,L. and Carbon,J. (1980) *Nature,* **287**, 504.
31. Szostak,J.W. and Balckburn,E.H. (1982) *Cell,* **29**, 245.
32. Murray,A.W. and Szostak,J.W. (1983) *Nature,* **305**, 189.
33. Ammerer,G. (1983) in *Methods in Enzymology,* Vol. **101**, Wu,R., Grossman,L. and Moldave,K. (eds.), Academic Press Inc., New York, p. 192.
34. Kramer,R.A., DeChiara,T.M., Schaber,M. and Hilliker,S. (1984) *Proc. Natl. Acad. Sci. USA,* **81**, 367.
35. Brake,A.J., Merryweather,J.P., Coit,D.G., Heberlein,U.A., Masiarz,F.R., Mullenbach,G.T., Urdea,M.S., Valenzuela,P. and Barr,P.J. (1984) *Proc. Natl. Acad. Sci. USA,* **81**, 4642.
36. Broach,J.R., Strathern,J.N. and Hicks,J.B. (1979) *Gene* **8**, 121.
37. Stinchcomb,D.T., Mann,C. and Davis,R.W. (1982) *J. Mol. Biol.,* **158**, 157.
38. Ito,H., Fukada,Y., Murata,K. and Kimura,A. (1983) *J. Bacteriol.,* **15**, 163.
39. Klebe,R.J., Harriss,J.V., Sharp,Z.D. and Douglas,M.G. (1983) *Gene,* **25**, 333.
40. Orr-Weaver,T.L., Szostak,J.S. and Rothstein,R.J. (1983) in *Methods in Enzymology,* Vol. **101**, Wu,R., Grossman,L. and Moldave,K. (eds.), Academic Press Inc., New York, p. 228.
41. Grunstein,M. and Hogness,D.S. (1975) *Proc. Natl. Acad. Sci. USA,* **72**, 3961.
42. Laskey,R.A. and Mills,A.D. (1977) *FEBS Lett.,* **82**, 314.
43. Rigby,P.W., Dieckmann,R., Rhodes,C. and Berg,P. (1977) *J. Mol. Biol.,* **113**, 237.
44. Cryer,D.R., Eccleshall,R. and Marmur,J. (1975) *Methods Cell Biol.,* **12**, 39.
45. Davis,R.W., Thomas,M., Cameron,J.R., St. John,T.P., Scherer,S. and Padgett,R.A. (1980) in *Methods in Enzymol.,* Vol. **65**, Gross,L. and Moldave,K. (eds.), Acacemic Press Inc., New York, p. 404.
46. Stiles,J.I., Szostak,J.W., Young,A.T., Wu,R., Consul,S. and Sherman,F. (1981) *Cell,* **25**, 277.
47. Orr-Weaver,T.L. and Szostak,J.W. (1983) *Proc. Natl. Acad. Sci. USA,* **80**, 4417.

48. Heffron,F., So,M. and McCarthy,B.J. (1978) *Cold Spring Harbor Symp. Quant. Biol.*, **43**, 1279.
49. Shortle,D., Haber,J.E. and Botstein,D. (1982) *Science*, **217**, 371.
50. Scherer,S. and Davis,R.W. (1979) *Proc. Natl. Acad. Sci. USA*, **76**, 3912.
51. Struhl,K. (1983) *Gene*, **26**, 231.
52. Siliciano,P.G. and Tatchell,K. (1984) *Cell*, **37**, 969.
53. Boeke,J.D., Lacroute,F. and Fink,G.R. (1984) *Mol. Gen. Genet.*, in press.
54. Smith,M.M. and Murray,K. (1983) *J. Mol. Biol.*, **169**, 641.

CHAPTER 4

Genetic Engineering of Plants

CONRAD LICHTENSTEIN and JOHN DRAPER

1. INTRODUCTION

It is only recently that plants have become respectable eukaryotic organisms for study by molecular biologists. One reason why they have not been studied before is that they lack the advantages of *Drosophila*, for example, as a genetic system. The wealth of *Drosophila* genetics has attracted molecular biologists to study the genes involved in regulating the organism's development. Work in animal systems has also attracted the funding, and consequently research workers, owing to its obvious applications to human genetic diseases and cancer. This has led to the development of animal virus-based vectors to transform mammalian tissue culture. Plants do, however, have one major advantage over animals; namely the ease with which undifferentiated somatic tissue can be redifferentiated into mature fertile plants *in vitro*. This feature opens up the possibility for molecular biologists to study the expression of genes newly introduced into the whole organism. It also supplies the plant breeder with new techniques to engineer genetic improvements of agriculturally important crops and so complement the more classical methods of plant breeding.

Thus, owing to the recent developments of methods to transform plants, a new revolution is now taking place. In Section 2 of this chapter we review the latest vectors that have been constructed to introduce foreign cloned genes to plants. Whilst various vectors based on plant viruses have been suggested, none have been extensively developed to date. Possible candidates include vectors based upon the single-stranded DNA viruses in the Gemini group, the double-stranded DNA viruses of the Caulimovirus group (that infect Brassicae) and Tobacco Mosaic Virus, the single-stranded RNA virus. These candidates, however, suffer from the disadvantages of restricted host range; packaging constraints on the amount of foreign DNA that can be included; and stability of inheritance. The latter point is made since these viruses replicate autonomously and do not integrate into the plant nuclear genome. For this reason we have chosen to focus upon vectors based upon the tumour inducing (Ti) plasmids of *Agrobacterium tumefaciens*. This soil microbe is a natural genetic engineer since during infection of a plant it transfers a portion of the Ti plasmid DNA (known as T-DNA) to plant cells. The T-DNA is integrated into the plant nuclear genome and expressed. Additionally, Ti plasmid-based vectors have the advantage of a broad host range. They will infect most dicotyledenous plants and apparently have no packaging constraints.

In Section 3 we describe the ways in which Ti plasmid based vectors can be used to transform plants. This is discussed with respect to transformation of plant tissue *in planta* and to *in vitro* transformation methods (transformation of protoplasts). We have decided to focus upon transformtion of tobacco and other *Nicotiana* species as model systems. This is because these plants have proved the most responsive in tissue culture with respect to the generation of healthy protoplasts that can be regenerated into plants. The development of good tissue culture systems remains the major obstacle for the genetic engineering of other agriculturally important crop species. Since Ti plasmid based vectors cannot be used to transform monocotyledenous plants we have included alternative protocols to introduce foreign genes in plants.

In Section 4 we discuss recent methods to prepare nucleic acids from plant tissue. These methods are needed in order to analyse the organisation and expression of foreign genes that have been transformed into plant tissue. It is obviously also required to prepare genomic DNA and cDNA libraries in order to clone and characterise specific plant genes. Such genes are often identified using heterologous probes and by exploiting mRNA abundance of active genes. As many of these techniques are well established and generally applicable we will not discuss them in detail in this chapter. Those plant genes that have been studied include multi-gene families encoding major storage proteins expressed in seeds (for example the zein gene of maize), plant transposable elements and light inducible genes involved in photosynthesis (see reference 1 for recent developments in the field).

Finally in Section 5 we discuss the various approaches to analyse the organisation and transcription of foreign DNA integrated into the plant chromosome. We also provide protocols for enzyme assays of enzymes encoded by genes that have commonly been used to transform plant tissue.

2. PLASMID VECTORS TO DELIVER FOREIGN DNA TO PLANTS

As discussed in the previous section many techniques are available for the transfer of DNA to plants. At the moment the most successful and simplest method makes use of the natural gene transfer system of *A. tumefaciens*. For recent reviews on this phenomenon see reference 1 (pages 143 – 176) and reference 2.

A. tumefaciens is a soil bacterium that is capable of infecting a wide range of gymnosperms and dicotyledenous angiosperms. Subsequent to infection the wounded tissue proliferates to produce a mass of tissue known as a Crown Gall tumour. The induced tumour is able to continue growing without the inciting bacterium. The transformed plant tissue is able to grow axenically in tissue culture without the hormone supplements, auxins and cytokinins, normally required by plant cells in culture. Until recently phytohormone independent growth has been the only selectable marker for transformed tissue. In addition to this attribute, transformed plant tissue synthesizes one or more of a variety of compounds, not found in non-transformed tissue, known collectively as opines (for example, octopine and nopaline). These opines can be metabolised specifically by the bacteria reponsible for inciting the tumour. All these functions are encoded

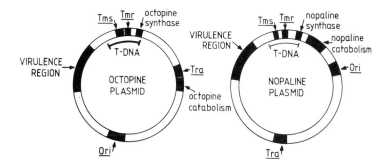

Figure 1. Genetic maps of an octopine Ti plasmid (left) and a nopaline one (right). The T-DNA has been defined. Within it gene loci governing synthesis of octopine or nopaline have been mapped, as have genes affecting tumour morphology, which are revealed by the 'rooty' (*tmr*) and 'shooty' (*tms*) mutations. In both plasmids a region to the left of the T-DNA governs the virulence functions that enable the plasmid to induce a tumour. Regions outside the T-DNA also govern the catabolism of octopine or nopaline and other opines.

on large Ti plasmids of *A. tumefaciens*, the most commonly studied of which are the octopine type and the nopaline type Ti plasmids. A genetic map of each of these Ti plasmids is presented in *Figure 1*. A specific region of these plasmids, the T-DNA, is transferred from the plasmid to the nucleus of a susceptible plant cell. Southern blotting experiments show that the T-DNA is integrated into the plant nuclear genome. Northern blotting experiments demonstrate that the integrated T-DNA is transcribed; for the octopine type Ti plasmid, eight polyadenylated transcripts have been identified representing 0.001% of the total polyadenylated mRNA in the plant tissue. As T-DNA transcription is inhibited by α-amanitin it is thought to be RNA polymerase II dependent. DNA sequence analysis indicates open reading frames that correlate with these transcripts which are preceded by TATA boxes and in some cases CAAT boxes and succeeded by AATAAA boxes. The former two sequence boxes are thought to be involved in initiation of transcription and the latter one in polyadenylation of the transcript in a wide variety of eukaryotes. S1 mapping experiments confirm that these sequences are correctly positioned with respect to the beginning and end of T-DNA transcripts. Genetic analysis using transposon and deletion mutagenesis has identified four genetic loci in the T-DNA that map within open reading frames. The *ocs* locus encodes the octopine synthase enzyme. The *tmr* locus encodes an enzyme involved in cytokinin biosynthesis (3). Mutations in this locus (rooty mutants) result in massive root proliferation from the incited tumour. The *tms1* and *tms2* loci encode functions involved in auxin biosynthesis (4). Mutations in either of these loci result in shoot proliferation from the incited tumours (shooty mutants). The developmental effects of altered levels of auxins and cytokinins seen in tumour tissue transformed by wild-type, rooty and shooty mutant T-DNA are also reflected in *in vitro* culture of untransformed tissue under various auxin and cytokinin concentrations.

The mechanism of transfer of T-DNA is not yet fully understood. What is known is that the T-DNA is flanked by 25 base-pair near perfect repeats. These sequences are thought to be involved in transfer of the T-DNA to the plant genome since the end points of the integrated T-DNA are close to these sequences. Removal of the right border by deletion mutagenesis of the Ti plasmid abolishes transfer of T-DNA. When a 25 base-pair synthetic oligomer identical to the wild-type sequence is cloned into the right border site to replace the large deletion that removed it, virulence and transfer of T-DNA are recovered. However, this is only true if the sequence is cloned in the same orientation as the wild-type sequence (5).

Other loci are also involved in transfer of T-DNA to the plant genome. A large locus (50 kb), the *vir* region, containing a cluster of many genes maps in a region of the Ti plasmid that is not adjacent to the T-DNA. Mutations in any of these genes abolishes T-DNA transfer and hence virulence. Recently it has been shown that some of these genes are not expressed unless the bacterial cells are either mixed with plant tissue or separated from growing plant tissue by a dialysis membrane (6). The inducer factor produced by the plant tissue has not yet been identified. Other mutations that map in the *A. tumefaciens* chromosome also abolish virulence. Some of these genes are thought to be involved in the attachment of *A. tumefaciens* to plant tissue during the process of infection.

Other functions encoded by the Ti plasmid are genes encoding functions responsible for conjugal transfer of the Ti plasmid to avirulent strains of *Agrobacterium*. These genes are normally not expressed. Expression results from exposure to the opine produced in the plant tissue. The opine acts as a molecular 'aphrodisiac' to derepress both the genes involved in conjugal transfer and the genes involved in opine catabolism. Thus the Ti plasmid has devised an excellent evolutionary strategy for survival. It infects plant tissue, the resultant callus proliferates and produces opines which *Agrobacteria* harbouring the Ti plasmid can metabolise as a carbon and nitrogen source and also results in the transfer of this plasmid to other avirulent *Agrobacteria* that do not possess a Ti plasmid.

Given the efficiency with which T-DNA is transferred to plant tissue subsequent to infection and the fact that foreign DNA cloned into the T-DNA is also transferred, the Ti plasmid is an excellent choice as a vector to deliver DNA to plant tissue. Up to 50 kb of foreign DNA has thus far been transferred (7). The upper limit has not yet been determined.

However, the main disadvantage of the Ti plasmid is that it causes tumours in plants. The genetic engineer would prefer to transform plant tissue with foreign DNA in such a way that tumours did not develop since tumour tissue is difficult to regenerate into normal plants. For this reason Ti plasmid based vectors have been developed that have deletions of the *onc* region (the genes *tms1, tms2* and *tmr*) yet retain the border sequences that allow T-DNA transfer. As phytohormone independence is now lost these vectors contain genes encoding resistance to antibiotics in plants to serve as selectable markers. For example, kanamycin and methotrexate are both very toxic to plants and genes encoding resistance to these drugs have proved to be powerful dominant selectable markers. Other bacterial antibiotics are not as toxic to plants and cannot be used for selection. As these genes are bacterial in origin they lack the eukaryotic features required for transcription in plant tissue. For this reason chimaeric genes have been constructed that contain promoters, known to function in plants, fused to the coding region of the drug resistance genes and followed by the signal sequence AATAAA required for polyadenylation. A popular choice for such a promoter has been that of the nopaline synthase gene of the Ti plasmid. This gene is the most actively transcribed of all T-DNA genes and is known to be constituitively expressed in a wide variety of plants. Such chimaeric constructs allow expression of these antibiotic resistance genes in plants and thus enable selection of transformed tissue.

The construction of vectors based on Ti plasmids and that contain selectable markers is currently the focus of intense research in a number of laboratories around the world. Consequently many of the vectors to be discussed in this section may soon become obsolete as new improved vectors become available.

Those vectors that we shall discuss serve to illustrate the variety of techniques that are needed to introduce foreign DNA to plants. These techniques involve exploiting the wealth of expertise in recombinant DNA technology and the ease and efficiency of transforming *E. coli* with newly constructed plasmids. Having verified these constructs they must then be transferred to *A. tumefaciens via* conjugation or transformation for eventual transfer to plant tissue. Owing to the large size of the Ti plasmids it is clearly not possible to clone directly into them. Two alternative strategies have therefore been developed to overcome this problem. The first one uses a disarmed (i.e., non-oncogenic) Ti plasmid in which the border sequences flank a pBR322 copy. Any DNA sequence suitably cloned into pBR322 can then be cointegrated into this Ti plasmid between the border sequences by homologous recombination between the pBR322 sequences. Upon plant cell infection all the DNA between the border sequences is transferred to the plant chromosome.

The second system uses two plasmids. One is the vector and is able to replicate in *E. coli* and *A. tumefaciens*. It contains the T-DNA border sequences flanking a selectable marker and suitable cloning sites. This plasmid is transferred to *Agrobacterium* containing a second helper plasmid, a Ti plasmid, that retains the *vir* region and yet suffers a large deletion comprising the T-DNA. Complementation *in trans* then allows transfer of the DNA between the border sequences of the vector plasmid to the plant cell nuclear genome.

As *A. tumefaciens* has not yet been shown to infect monocotyledenous plants such as cereals, Ti plasmid based vectors cannot be used. We have therefore included protocols for alternative methods to transform such plants. A recent report provides evidence that lily and narcissus can be infected by Ti plasmids based upon expression of opines in the infected plant tissue (8). However, stable inheritance of foreign genes has not yet been demonstrated in these plants. It may therefore be appropriate to develop vectors based on plant viruses to transform monocots. Such vectors would be analogous to SV40 based vectors used to transform animal cells, and would include an origin of replication and a dominant selectable marker.

2.1 **Shuttle vector and Ti plasmid derived acceptor vector for the transfer of foreign DNA to plants**

2.1.1 *Onc⁻ vectors*

As outlined in the introduction to this section this system uses a non-oncogenic Ti plasmid derived vector to transfer DNA to plants. The acceptor vector, pGV3850, (9) is shown in *Figure 2*. It is derived from the nopaline type Ti plasmid pTi C58 and has the following features:

(i)   It contains the T-DNA border regions and all contiguous Ti plasmid sequences outside the T-DNA region.

(ii)   The T-DNA sequences near to the right border encoding the nopaline synthase enzyme are present. Activity of this enzyme serves as a marker to identify transformed cells.

(iii)   The internal T-DNA genes encoding the tumourigenic functions that prevent the normal differentiation of transformed plant cells are deleted.

(iv)   This internal T-DNA deletion is replaced with the cloning vehicle pBR322.

The presence of the pBR322 DNA makes pGV3850 a versatile acceptor plasmid since any gene cloned into a pBR322-like plasmid (the shuttle vector) can easily be inserted between the T-DNA borders by homologous recombination to form a plasmid cointegrate. The shuttle vector can be mobilised from *E. coli* to *A. tumefaciens* by conjugation (Section 2.2). As pBR322 derived plasmids cannot replicate in *A. tumefaciens*, selection for an antibiotic resistance gene encoded on the shuttle vector selects for this single crossover event. Section 2.3 describes a rapid method to verify that the desired cointegrate has been obtained.

The type of shuttle vector depends upon which gene it is desired to be transferred to plants. Those that have already been constructed include pBR322 based plasmids containing the nopaline synthase (*nos*) promoter fused to the coding region of one or other of the following genes: *ocs* (octopine synthase from an octopine type Ti plasmid), *npt* II (neomycin phosphotransferase II, the kanamycin resistance gene from transposon Tn5), *dhfr* (the dihydrofolate reductase gene), the methotrexate resistance gene from the plasmid R67 and *cat* (the gene for chloramphenicol acetyltransferase, that determines chloramphenicol resistance and which was originally taken from transposon Tn9). These constructs also contain DNA sequences from the non-coding 3′ ends of either the *nos* or *ocs* genes positioned at the 3′ end of the coding region of the chimaeric gene construct to allow accurate splicing and polyadenylation of the transcript (7,10,11 – 13). The bacterial coding regions allow selection for antibiotic resistance. Recently, Herrera-Estrella *et al.* (13) have fused the promoter of the nuclear gene for the small subunit ribulose bisphosphate carboxylase (*ss Rubisco*) of pea to the coding region of the *cat* gene. They demonstrate light-inducible expression of this gene upon introduction to *Nicotiania tabacum*.

If one wishes to clone a characterised gene with no selectable phenotype into a particular crop plant to confer some desirable new trait that increases the value of the crop, then one requires a shuttle vector with the following properties:

(i)   It should contain pBR322 or vector sequences derived from pBR322 to allow cointegration of the construct into pGV3850 as described previously.

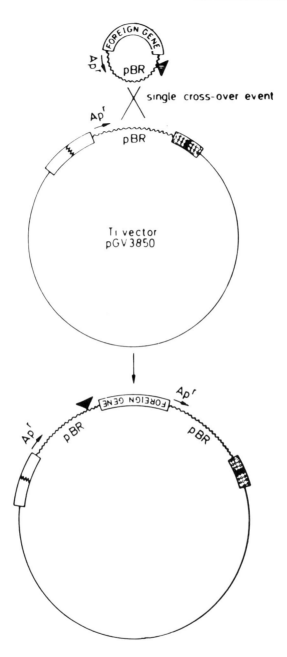

Figure 2. Use of the Ti plasmid vector pGV3850 as a recipient for any foreign gene of interest whose expression is to be monitored in plants. The gene of interest is first cloned into a pBR322-type vector in *E. coli*. This plasmid is then mobilised to *Agrobacterium*. A single recombination event will allow cointegration of the recombinant plasmid containing the foreign gene of interest into pGV3850. The solid triangle represents an antibiotic resistance gene other than the carbenicillin resistance gene. The pBR322 DNA (wavy line) in pGV3850 is flanked by T-DNA which contains the left and right border sequences and the *nos* gene. Taken from Figure 5 reference 9 and reproduced with kind permission.

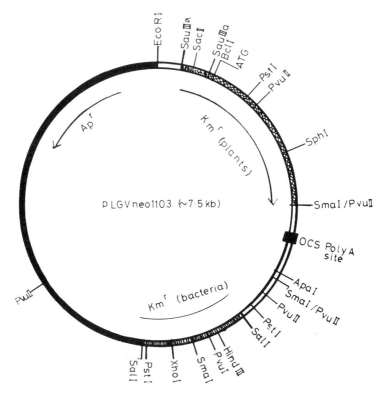

■■■■ pBR322
▨▨▨▨ NPT I gene (Tn 903)
▨▨▨▨ NPT II gene (Tn 5)
▨▨▨▨ NOS Promoter

EcoRI
SauIIIa
SacI
SauIIIa
BclI
ATG
PstI
PvuII
SphI
SmaI/PvuII
OCS PolyA site
ApaI
SmaI/PvuII
PvuII
PstI
SalI
HindIII
PvuI
SmaI
XhoI
PstI
SalI
PvuII
Ap^r
Km^r (plants)
Km^r (bacteria)
pLGVneo1103 (~7·5 kb)

Figure 3. Restriction map of pLGVneo1103, a shuttle vector that can be cointegrated into pGV3850 to allow selection of kanamycin resistant plant cell transformants. From Luis Herrera-Estrella and published with his kind permission.

(ii) It should have *npt* II gene coding region fused to the *nos* promoter to allow selection of transformants by virtue of kanamycin resistance. This fusion lacks the Shine-Dalgarno sequence and is not expressed in prokaryotes.

(iii) The vector will therefore need an alternative selectable marker that is expressed in bacteria to allow selection of transformants in *E. coli* and selection of co-integrates in exconjugant *A. tumefaciens*. A suitable gene is the Kn^r gene of Tn903 (*npt* I) this gene has no significant DNA sequence homology with *npt* II and will not therefore lead to deletion of plasmid DNA *via* homologous recombination.

A vector that fits the above criteria has recently been constructed by Herrera-Estrella *et al.* (14). This plasmid, pLVG neo 1103, is illustrated in *Figure 3*. It contains a unique *Eco*RI site suitable for cloning the gene of interest.

Alternatively one may wish to study a particular plant gene to monitor the regulation of its expression with respect to tissue specificity, stage of plant development, light inducibility, response to wounding, or infection of the plant by a pathogen. This gene may have been isolated under specific environmental conditions or from a specific tissue by cDNA cloning and may have no known gene product. It may then be appropriate to replace the coding region of this gene with a gene that can be readily assayed. A suitable gene is chloramphenicol acetyltransferase (CAT) from the bacterial transposon Tn9. Whilst CAT activity cannot reliably be used as a selectable marker in plants a rapid assay has been developed for this enzyme and has already been widely used in animal systems (see Chapter 6 of this volume). The coding region of this gene can be isolated as a *Sau*IIIa fragment from pBR325. Measurement of CAT activity provides a rough guide to the level of transcription from the plant promoter. Deletion mutagenesis of the 5' and 3' non-coding regions of this gene fusion can then be performed to establish which domains of the DNA sequence are responsible for modulation of the type of transcriptional regulation under investigation.

Construction of plasmid recombinants have been described in detail elsewhere (15) and will not be discussed here. Ligation mixtures can be used to transform *E. coli* strain HB101 with selection for appropriate antibiotic resistance. Verification of plasmid constructs can be achieved by restriction mapping of small scale plasmid preparations of a number of independent transformants. Having done this, the construct is now ready for cointegration into pGV3850 in *A. tumefaciens*. This is discussed in Section 2.2.

2.1.2. Onc⁺ vectors

The pGV3850 vector system described above is an *onc⁻* system and is therefore most appropriate to use where one wishes to study either the expression of the foreign gene in intact plants that have been regenerated from callus tissue, or for plant genetic engineering for crop improvement. In some instances one may wish to study the expression of foreign genes in crown gall tissue and prefer an *onc⁺* vector. The advantage of this system is that given the rapid proliferation of crown gall tissue, transformants are more rapidly and easily obtained owing to the hormone independent phenotype (see Section 3 for details).

Two *onc⁺* acceptor vectors similar to pGV3850 could be used. One vector pGV3851 (16) contains a less extensive deletion of internal T-DNA than pGV3850; the left internal region has been deleted and replaced by pBR322, the right T-DNA region including the *tmr* gene is retained. This plasmid generates phytohormone independent shooty mutants. The other vector pMPG1 (17) contains an insertion of the cosmid pHC79 (sharing homology with pBR322) in the nopaline synthase gene of pTiC58. This plasmid generates wild-type *nop⁻* tumours in plants.

2.1.3 Split End Vectors

The Monsanto group (11) have constructed an *onc⁺/onc⁻* vector. This vector, based upon the octopine type Ti plasmid pTiB6S3, is shown in *Figure 4*. It contains two right border sequences. After co-cultivation and selection for kana-

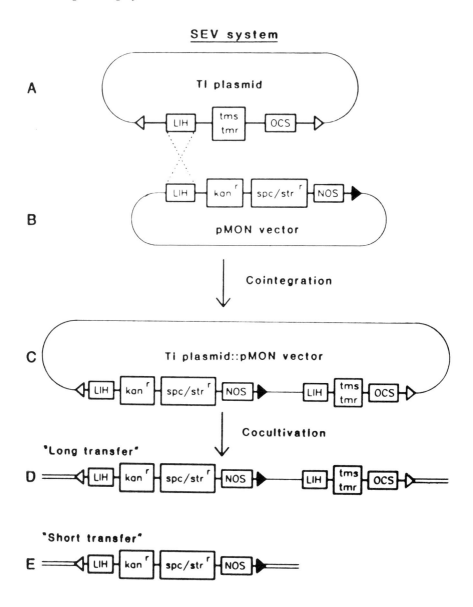

Figure 4. Steps in the SEV system for plant cell transformation. The arrows represent the T-DNA border sequences. *LIH* is a region of homologous DNA for recombination. The tumour genes are represented by *tms* and *tmr* (11); *OCS* and *NOS* are octopine and nopaline synthase genes, respectively. The chimeric kanamycin resistance gene is designated as *kan*[r]. The bacterial spectinomycin-streptomycin resistance determinant for selection of cointegrates is designated *spc/str*[r]. Reciprocal recombination of (**A**) a resident Ti plasmid (pTiB6S3) and (**B**) pMON120 derivative (pMON128) yields (**C**) the cointegrate, pTiB6S3::pMON128. After co-cultivation and selection for kanamycin-resistant plant cells either (**D**) the entire hybrid T-DNA or (**E**) a truncated T-DNA without tumour genes is transferred into the plant genome. Taken from reference 12 and reproduced with kind permission.

mycin resistant plant cells (Section 3) either the entire hybrid T-DNA is transferred to plant genome (from the right-most border sequence) or a truncated T-DNA without tumour genes (from the internal right border). The short transfer transformation can be regenerated to give intact plants.

2.2 Transfer of Plasmids to A. tumefaciens by Conjugation

Having constructed the desired recombinant by cloning gene X into pBR322 or a pBR322 derivative as described in Section 2.1, it is now necessary to introduce this construction into *A. tumefaciens*. As pBR322 is unable to replicate in *A. tumefaciens*, it can only be maintained by integration into the Ti plasmid pGV3850 in this strain. Transfer of a new construct from *E. coli* to *A. tumefaciens* can be achieved by conjugation.

ColE1 type plasmids (such as pBR322) are not self-transmissible and must be supplied with transfer functions *in trans* by a self-transmissible Iα type plasmid such as R64drd11. Also required for conjugation are the *mob* functions (ColE1 encoded) that recognise a specific sequence, the *bom* site. Transfer is initiated from the origin of transfer at or near the *bom* site. The plasmid pBR322 retains the *bom* site but the *mob* functions must be supplied *in trans* by a helper plasmid. The plasmid pGJ28 serves this purpose; it retains both *mob* functions, its own *bom* site and is compatible with pBR322. Compatibility means that both pGJ28 and pBR322 can replicate in the same *E. coli* host. Thus, an *E. coli* donor harbouring R64drd11, pGJ28 and the pBR322 derived construct will mobilise all three plasmids into *A. tumefaciens*. R64drd11 is unstable in *A. tumefaciens* and neither of the other two plasmids can replicate at all. Using this system Van Haute *et al.* (18) have shown that the pBR322 derivative is transferred to the recipient with a frequency of 4.5×10^{-3}. As pBR322 cannot replicate after transfer this was shown using a plasmid construct containing the pBR322 derivative cloned into a non-transmissible non-mobilisable plasmid vector able to replicate in *A. tumefaciens*. Using a pBR322 plasmid containing 3.3 kb of Ti plasmid DNA the frequency of stable introduction was found to be 6.7×10^{-6}. Assuming a transmission frequency of 4.5×10^{-3} this implies a recombination frequency of 1.5×10^{-3} or 0.5×10^{-6} per base pair of sequence homology.

The pBR322 derived plasmid construct containing the foreign gene X(pBR::X) is first introduced into HB101 by transformation (*Table 2* gives a list of bacterial strains used in this Chapter). The absence of the other plasmids aids the analysis of transformants as discussed in Section 2.1. To allow conjugal transfer to *A. tumefaciens*, the plasmids pGJ28 and R64drd11 must then be introduced to HB101 (pBR322::X). This can be done by conjugation between the donor strain GJ23 (harbouring pGJ28 and R64drd11) and the recipient HB101 (pBR322::X). The *recA* phenotype of HB101 prevents homologous recombination between pGJ28 and pBR322::X. Exconjugants are selected on LB plates containing ampicillin, kanamycin and streptomycin. The concentrations of the drugs that are to be used are given in *Table 3*. An exconjugant from this cross can then be picked and crossed with the *A. tumefaciens* recipient GV3101 (pGV3850). Exconjugants may be selected by plating on AB medium (*Table 1*) with counter selection for the

Table 1. Bacterial Media.

All quantities are given for 1 l of medium. Agar media are made by addition of 15 g/l of Bactoagar (Difco). Unless otherwise stated, the pH is adjusted by addition of 1 M NaOH to pH 7.0.

YEB medium

Yeast extract (Oxoid)	1 g
Beef extract (Lab Lemco, Oxoid)	5 g
Peptone (Difco)	5 g
Sucrose	5 g
$MgSO_4.7H_2O$	0.5 g

Luria broth (LB)

Bactotryptone (Difco)	10 g
Bacto yeast extract (Difco)	5 g
NaCl	10 g

Nutrient broth (NB)

Difco nutrient broth	8 g

AB medium

	20x stock	1x
Solution I		
K_2HPO_4	60 g	3 g
NaH_2PO_4	20 g	1 g
Solution II		
NH_4Cl	20 g	1 g
$MgSO_4.7H_2O$	6 g	0.3 g
KCl	3 g	0.15 g
$CaCl_2$	3 g	0.01 g
$FeSO_4.7H_2O$	50 mg	2.5 mg

Autoclave 20x stocks of solutions I and II separately. Combine to give a 1x solution, adding glucose to 0.5% from sterile 20x stock solution (10% w/v)

SM medium

K_2HPO_4	2.05 g
KH_2PO_4	1.45 g
$MgSO_4.7H_2O$	0.5 g
NaCl	0.15 g
$CaCl_2$	0.02 g
$(NH_4)_2SO_4$	3.0 g
Glucose	4.0 g

SM liquid mating medium

This is identical to SM medium but contains in addition:

$FeSO_4$	0.001 g
$MgSO_4$	0.001 g

SMNO medium (SM medium with no nitrogen)[a]

K_2HPO_4	10.25 g
KH_2PO_4	7.25 g
$MgSO_4.7H_2O$	0.50 g
NaCl	0.15 g
$CaCl_2$	0.02 g
Octopine (or nopaline)	0.10 g
Glucose	2.00 g
NaOH	0.014 g
Bromophenol blue	0.15 g

Table 1. *continued.*

Mannitol glutamate (MG) medium	
Mannitol	10 g
Sodium glutamate	2.32 g
KH$_2$PO$_4$	0.5 g
NaCl	0.2 g
MgSO$_4$.7H$_2$O	0.2 g
Biotin	2 g

LBMG medium
This is made by mixing equal proportions of LB and MG media.

λ *dil buffer*	
10 mM	Tris-HCl pH 7.4
100 mM	NaCl
10 mM	MgCl$_2$

[a]To make SMNO agar, it is necessary to wash the agar in double distilled water before use to remove nitrogenous contaminants. Wash 15 g of agar with 1 l of water.

donor and supplemented with kanamycin or any other antibiotic for which resistance is encoded by the pBR322::X construct. Bacterial conjugations are performed following the protocol given in *Table 4*.

2.2.1 Triparental Matings

An alternative way to introduce the new construct into *A. tumefaciens* is to carry out a triparental mating. This method is quicker and relies upon the fact that the conjugal transfer of R64drd11 and pGJ28 from one *E. coli* strain to another is extremely efficient (~50% of recipients acquire these plasmids). Grow up cultures of GJ23, HB101 (pBR322::X) and GV3101 (pGV3850) as described in *Table 4*. Then mix and incubate to allow conjugal transfer as described in *Table 4* and select on AB plates (*Table 1*) selecting for GV3101 (pGV3850) supplemented with kanamycin to select for pBR322::X.

A. tumefaciens exconjugants are now ready for analysis to show that cointegration between pBR322::X and pGV3850 has occurred. This is described in Section 2.3. Strains containing stable cointegrates must be maintained by appropriate drug selection.

2.2.2 Conjugal Transfer of Ti Plasmids

It is unlikely that the need will arise for this technique since many of the Ti plasmid based vectors are already available in appropriate *A. tumefaciens* strains. However, the occasion may arise when one wishes to mobilise a particular Ti plasmid construct to a new chromosomal background. Since the *tra* genes required for conjugal transfer of Ti plasmids are negatively regulated, they must be derepressed by addition of inducer. The inducer is octopine for octopine type Ti plasmids and nopaline for nopaline type Ti plasmids. The inducer also derepresses the operons involved in catabolism of either of these opines. Thus transconjugants may be selected where these opines are supplied as a sole nitrogen source. The method for conjugal transfer (19) is given in *Table 5*.

Table 2. Bacterial Strains and Plasmids

Strain	Features	Reference
E. coli strains		
HB101	F^-, r_B^-, m_B^-, RecA, ara, proA, lacY, galK, str, xyl5, mtl, SupE	15
GJ23	RecA derivative of AB115 harbouring pGJ28 and R64drd11	18
JM83	K12 with lacZ ΔM15 on a φ80 integrated into the chromosome (ara, Δlac-pro, strA, thi, φ80dlacZ ΔM15)	38
A. tumefaciens strains		
A136	C58 rif^r cured of Ti plasmid	20
LA4404	A136 harbouring pAL4404	22
GV3102	Equivalent to A136	9
Plasmids		
pBR322	Ap^r, Tc^r	15
pGJ28	Kn^r, Nm^r, Cda$^+$, Ida$^+$ colD replicon carrying ColE1 mob and bom	18
R64drd11	Tc^r, Sm^r Iα type plasmid, transfer derepressed derivative of R64	18
pGV3850	Ap^r onc$^-$ derivative of nopaline pGV3839 plasmid carries pBR322 sequences	9
pAL4404	pTiAch5 (octopine plasmid) derivative with deletion of T-DNA region	22
pLGVneo1103	Ap^r Kn^r pBR322 derivative with nos-neo gene fusion	14
pRK2013	Kn^r ColE1 derivative with tra genes of RK2	23
pGV2215	Kn^r tms mutant of octopine type Ti plasmid pTiB653	16
pGV3851	Ap^r tms$^-$ derivative of nopaline Ti plasmid C48 that carries pBR322 sequences	16
pMP61	Derived from pGV3105 Tra^c noc^c occ$^+$ with insertion of pHC79 in the nos gene	17
pHC79	Cosmid Ap^r Tc^r	17

Table 3. Antibiotic Concentrations for use with Bacteria

Antibiotic	Abbreviation	Concentration (μg/ml) for E. coli cells	Concentration (μg/ml) for A. tumefaciens cells
Carbenicillin	Cb	200	100
Chloramphenicol	Cm	30	–
Kanamycin	Kn	50	50
Neomycin	Nm	50	50
Rifampicin	Rif	100	50
Spectinomycin	Sp	100	100
Streptomycin	Sm	100	100
Tetracycline	Tc	10	2.5

2.3 Rapid Mapping of Cointegrates of the Shuttle Vector into the Ti-Plasmid Based Vector in A. tumefaciens

Having mobilised the pBR322 derived shuttle vector from *E. coli* to *A. tumefaciens*, it is now necessary to show that the vector has integrated into the pBR322 DNA present within the T-DNA of the Ti plasmid mutant pGV3850. Given the large size of Ti plasmids (150–250 kb), restriction endonuclease digestion will generate too many fragments for easy analysis on a gel. Thus Southern blotting is

Table 4. Bacterial Conjugations.

1.	Using a sterile toothpick, pick a single colony of the donor and recipient bacteria and grow them separately overnight with shaking to stationary phase in 5 ml LBMG[a]. *E. coli* strains may be grown up in LB at 37°C for conjugation. The lower salt of LBMG is preferred for *A. tumefaciens* which is grown at 30°C since at higher growth temperatures Ti plasmids are unstable.
2.	Mix 0.1 ml each of donor and recipient bacterial cultures and concentrate the cells on a small 0.22 μM Millipore filter by filtration.
3.	Place the filter on NA[b] plates. For crosses of *E. coli* x *E. coli* incubate the plates for 4 – 6 h at 37°C. For *E. coli* x *A. tumefaciens* crosses incubate overnight at 30°C. (LB plates may be used for *E. coli* x *E. coli* crosses.)
4.	Resuspend the cells from the filters in 1 ml λ dil and plate out dilutions on appropriate selective media (e.g., AB minimal plates containing antibiotics to select for exconjugants and to determine the titre of donor and recipient cells. The transfer efficiency is expressed as the ratio of exconjugant over recipient titre. Incubate the plates overnight at 37°C for *E. coli* x *E. coli* crosses or 2 – 3 days at 30°C for *E. coli* x *A. tumefaciens* crosses.
5.	Pick and streak several exconjugant colonies on selective media to isolate purified colonies. Incubate the plates overnight at 37°C or 2 – 3 days at 30°C as appropriate.

[a]See *Table 1* for the composition of media.
[b]NA is NB (*Table 1*) plus 1.5% agar.

Table 5. Conjugal Transfer of Ti Plasmids.

1.	Pick a colony of the donor and recipient bacteria and grow shaking slowly overnight at 30°C to stationary phase in 5 ml of SM medium supplemented with $FeSO_4$ and $MnSO_4$.
2.	Dilute 1:20 and incubate for a further 6 h. The optical density at 600 nm (O.D.$_{600}$) should be about 0.8 (2 x 10^9 cells/ml).
3.	Gently mix equal proportions of donor and recipient cells and add 0.2 ml of the mixture to Millipore filters as described in *Table 2*. Place the filters on SM plates supplemented with 1 mg/l $MnSO_4$ and 2 g/l octopine or nopaline (as appropriate). Incubate the plates for 2 – 4 days at 30°C.
4.	Resuspend the cells from the filters and plate out on selective plates as described in *Table 2*[a]. To select for opine utilisation, plate on SMNO plates (*Table 1*).
5.	Streak purify exconjugants as described in *Table 2*.

[a]Selection of exconjugants is simplified if the Ti plasmid to be transferred contains a drug resistance marker. pGV3850, for example, encodes carbenicillin resistance.

preferred. Since large scale preparation of a Ti plasmid DNA is both expensive and time consuming it is simpler to prepare total *A. tumefaciens* DNA from 10 – 20 different colonies following the protocol in *Table 6*. This protocol, which was developed by Garfinkel *et al.* (20) to map Ti plasmid mutants generated by insertions of transposons, should give a yield of 3 – 6 μg of DNA. The DNA can be digested with the appropriate restriction endonuclease using 15 μl of DNA in a digest volume of 20 μl. DNA should also be prepared from the *A. tumefaciens* strain GV3101 which lacks a Ti plasmid, and from the strain with the parental plasmid pGV3850.pBR322::X DNA should also be digested with the restriction endonuclease. Having digested the DNA samples mix 1 ng of the pBR322::X plasmid with DNA prepared from GV3101 to reconstruct the copy number of the plasmid DNA expected in the strains to be screened. Remember to include suitable molecular weight markers such as phage λ digested with *Hind*III. The

Table 6. Preparation of DNA from Single Colonies of *A. tumefaciens* for Southern Blotting.

1.	Pick a single colony into 5 ml of LBMG[a] and grow overnight at 30°C on a shaking incubator to late logarithmic phase.
2.	Pipette 1.5 ml of the culture into microcentrifuge tubes and harvest the cells by centrifugation for 5 min.
3.	Resuspend the pellet in 380 μl TE[b]. Add 20 μl of 25% SDS and 100 μl of 5 mg/ml pronase (previously predigested at 27°C for 2 h), and incubate for 30 min at 37°C.
4.	Add 50 μl of 5 M NaCl and incubate for 30 min at 68°C.
5.	Phenol extract with 1 volume of phenol (equilibrated with 100 mM Tris-HCl pH 8.0, 0.5 M NaCl).
6.	Re-extract the aqueous (upper) phase with an equal volume of a 12:12:1 mixture of phenol:chloroform:isoamylalcohol.
7.	Re-extract the aqueous phase with an equal volume of chloroform:isoamyl alcohol (24:1).
8.	Add two volumes of ethanol to the aqueous phase to precipitate the DNA. It is not necessary to add more salt. Chill to -70°C in a dry ice/ethanol bath and pellet the DNA by centrifugation in a microcentrifuge.
9.	Rinse the DNA pellet with 70% ethanol. Dry the pellet under vacuum and resuspend it in 200 μl of DNA buffer[c].

[a]See *Table 1* for description of media.
[b]TE is 50 mM Tris-HCl pH 8.0, 20 mM EDTA.
[c]DNA buffer is 10 mM Tris-HCl pH 8.0, 5 mM NaCl, 1 mM EDTA.

samples are now ready for agarose gel electrophoresis and Southern transfer using previously published methods (16).

An alternative approach to the Southern transfer is to dry the gel down. This has the advantage of being quicker and it saves the cost of nitrocellulose or 'Genescreen'. The DNA in the gel is denatured and subsequently reneutralised as if for a Southern transfer. The gel is then placed on two sheets of 3 mm filter paper, covered with a sheet of 'clingfilm' or 'Saranwrap' and dried down on a gel dryer under vacuum. Do not heat the gel with the gel drying apparatus since this can melt the gel. However, it is a good idea to provide some heat by illumination with an infra red lamp. After the gel is dry it can be peeled gently off the filter paper backing, rinsed in $3-5$ x SSC (1 x SSC is 0.3 M NaCl, 0.03 M sodium citrate) to remove adhering shreds of paper and then treated just like a normal blot with respect to pre-hybridisation and hybridisation with probe. With this technique it is best to use a high gelling temperature (HGT) agarose to cast the gel. This produces a more durable gel once it has been dried down.

The blot should be probed with nick translated shuttle vector DNA and the data interpreted after autoradiography. The pattern of bands that results should confirm that integration of the shuttle plasmid into pGV3850 has occurred. The strain showing the correct configuration of bands can now be used to transform plant tissue as described in Section 3.

2.3.1 *Preparation of A. tumefaciens Ti Plasmid DNA*

On occasion it may be desirable to prepare Ti plasmid DNA. Large scale Ti plasmid DNA preparations have been described previously (21). The protocol in *Table 7* is a rapid miniprep that can be used to isolate up to 2 μg of plasmid from

Table 7. Mini-preparation of Ti Plasmid DNA.

1.	Pick a single colony into 10 ml of YEB[a] and grow at 28°C in a fast orbital shaker for 20 h.
2.	Add 1.65 ml of the culture to each of two microcentrifuge tubes, and pellet the cells by centrifugation for 2 min. Pour off the supernatant and add a further 1.65 ml of the culture to each tube. Repeat until the cells from about 5 ml of culture have been pelleted in each tube.
3.	Resuspend the pellets by repeated pumping with a Gilson pipetteman or its equivalent in 300 μl each of STET buffer[b].
4.	Add 25 μl of lysozyme solution (freshly made up at 10 mg/ml), mix by vortexing briefly and incubate at room temperature for 1 min.
5.	Place the tubes through a thin piece of polystyrene tile and float in a boiling water bath for 1 min.
6.	Plunge the tubes immediately into ice water and leave for 5 min.
7.	Centrifuge the tubes for 10 min and remove the supernatant to new microcentrifuge tubes using a Gilson pipetteman or its equivalent. Use a pipette tip from which the first 1 mm has been cut to increase the bore size, thus reducing shearing of DNA. From this step onwards, all manipulations should be carried out using wide bore tips with the minimum of agitation.
8.	Extract the supernatant three times with 400 μl of chloroform:isoamyl alcohol (24:1), centrifuging between the extractions to separate phases and re-extracting the aqueous (uppermost) phase.
9.	Precipitate the plasmid from the supernatants by adding 300 μl of cold (− 20°C) isopropyl alcohol (Propan-2-ol). Place tubes at − 20°C for 30 min. If a DNA precipitate is not visible after this time, leave overnight.
10.	Pellet the DNA by centrifugation for 2 min in a microcentrifuge. Decant the supernatant, dry the pellet under vacuum and re-dissolve it in sterile distilled weater. Heat to 56°C for 5 − 10 min to aid dissolution if required.
11.	Pool the samples and extract with one volume of phenol equilibrated with 100 mM Tris-HCl pH 8.0, 0.5 M NaCl. Centrifuge and remove aqueous (upper) phase.
12.	Repeat the extraction with one volume chloroform and once with diethylether.
13.	Heat to 68°C for 10 min to drive off the remaining ether[c].
14.	If required, measure the DNA concentration using the diphenylamine reaction (see Section 4.1.1) and dilute to the required concentration.

[a]See *Table 1* for a description of media.
[b]STET buffer is 50 mM Tris-HCl pH 8.0, 50 mM EDTA, 8% (w/v) sucrose, 5% (v/v) Triton X-100.
[c]Do not handle large quantities of ether near naked flames or electrical switches.

a 10 ml culture. The Ti plasmid DNA is suitable for restriction with most common endonucleases but will still contain small amounts of contaminating RNA.

2.4 Binary Vectors for the Transfer of Foreign DNA to Plants

Binary vectors exploit the observation that when the *vir* region and the T-DNA region of the Ti plasmid are on two separate plasmid replicons the *vir* region on one plasmid can complement *in trans* to effect transfer of T-DNA to plants. The advantage of this system over that described in Section 2.2 is that it is not necessary to construct plasmid cointegrates. The disadvantage, if any, is that the efficiency of transfer to plant tissue in co-cultivation experiments (see Section 3) may not be as high (16). The binary vector system was first reported by Hoekema *et al.* (22), using the plasmid pLA4404 as helper to donate the virulence functions. This plasmid is derived from octopine type Ti plasmid pTiAch5 and is shown in *Figure 5*.

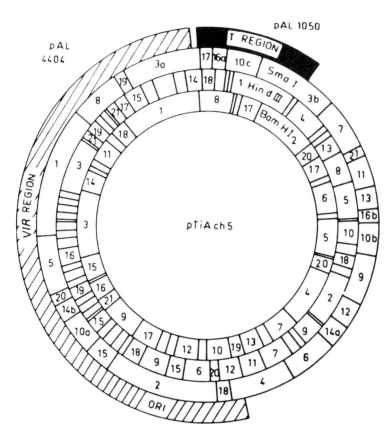

Figure 5. The physical map of pTiAch5 is shown for the restriction endonucleases *Sma*I, *Bam*HI and *Hind*III. In the hatched and shaded areas the Ti plasmid segments are indicated that are present separately on pAL4404 and pAL1050. Taken from reference 23 and reproduced with kind permission.

The binary vector system was initially demonstrated using pLA4404 as the helper plasmid and the T-DNA region cloned into a wide host range plasmid vector. The latter construct is however not suitable for transfer of foreign genes to plants. To date no suitable DNA transfer vectors have been published. However, we are currently constructing such a vector with the following features:

(i) Plasmid RK2 replication functions to allow replication in *E. coli* and *A. tumefaciens*. It is also compatible with the Ti plasmid.

(ii) The kanamycin resistance gene (*npt* I) from Tn903 to allow selection in bacteria.

(iii) The left and right octopine Ti plasmid border sequences to allow transfer of DNA to plants.

(iv) These border sequences flank a nos-neo fusion to allow selection of plant transformants, as well as the *Hae*II fragment from puc18 which includes the polylinker sequence in a portion of the β-galactosidase gene. This polylinker sequence offers multiple restriction sites to clone the gene of in-

Table 8. Direct Transformation of *A. tumefaciens* with Plasmid DNA.

1.	Pick a colony of LBA4404[a] into 5 ml LBMG or YEB and grow overnight at 30°C on a shaking platform.
2.	Inoculate 200 ml of fresh medium with the overnight culture and grow in 1 l flasks at 30°C for 5−6 h.
3.	Centrifuge at 4000 r.p.m. in the Sorvall GSA rotor or its equivalent for 5 min. Resuspend the cell pellet in 100 ml of 10 mM Tris-HCl pH 8.0.
4.	Pellet the cells by centrifugation as in step 3 and resuspend them in 2 ml of YEB (or LBMG)[b].
5.	Mix 200 µl of cells with 100 µl of plasmid DNA (0.1−1.0 µg) and freeze to −70°C in a dry ice/ ethanol bath for 5 min.
6.	Thaw in a water bath at 37°C for 25 min.
7.	Add 2 ml of YEB (or LBMG) and incubate for 1 h at 30°C, shaking.
8.	Plate 100 µl of cells on LBMG plates plus kanamycin or any other antibiotic for which resistance is encoded by the plasmid DNA used in the transformation and incubate at 30°C for 1−2 days.
9.	Pick and streak for single colonies on the same selective plates. Incubate at 30°C.

[a]See *Table 2* for a description of *A. tumefaciens* LBA4404.
[b]Cells from this step can be divided into 200 µl aliquots and stored long term at −70°C. To use, add DNA and proceed from step 6.

terest. Thus, plasmid recombinants give white colonies when transformants are plated on media containing isopropyl-1-thio-β-D-galactoside (IPTG) and 5-bromo-4-chloro-3-indolyl-β-D-galactoside (X-gal) (as with the pEMBL vectors, see Chapter 5 of the first volume of this book).

When the DNA transfer vector containing gene X of interest has been constructed, it is necessary to mobilise the construct from *E. coli* to *A. tumefaciens* by conjugation. Since the vector lacks transfer functions these can be supplied *in trans* by the plasmid pRK2013 (23) in a tri-parental cross as described in Section 2.2.1. In this experiment the two *E. coli* donors are HB101 (pRK2013) and JM83 (DNA transfer vector) (*Table 2*). The recipient is LA4404 harbouring pLA4404. Exconjugants are selected on minimal AB (to counterselect the donors) supplemented with kanamycin to select for the DNA transfer vector. Although pRK2013 also encodes kanamycin resistance it is unable to replicate in *A. tumefaciens*. The new construct is now ready for transformation of plant tissue.

2.4.1 *Transformation of A. tumefaciens with Plasmid DNA*

An alternative way to introduce the DNA transfer vector to the *A. tumefaciens* strain harbouring pLA4404 is by direct transformation as detailed in *Table 8*.

3. THE DELIVERY OF DNA TO PLANT CELLS AND THE REGENERATION OF TRANSFORMED PLANTS

Transformation procedures for higher plants using DNA vectors have been developed over the past few years following the discovery of the Ti plasmid of *A. tumefaciens* and the elucidation of its involvement in the aetiology of crown gall tumour induction (see Section 2). Virulent agrobacteria harbouring Ti plasmids are able to promote the transfer and integration of *in vitro* engineered genes into the genome of dicotyledenous plants where they are subsequently expressed.

Table 9. Plant Growth Media.

A.   *Media based on Murashige and Skoog (M S) salts* (39)

1.   *MSO*

Component	Concentration per litre
$CaCl_2.2H_2O$	0.44 g
NH_4NO_3	1.65 g
KNO_3	1.90 g
KI	0.83 mg
$CoCl_2.6H_2O$	0.025 mg
KH_2PO_4	0.17 g
H_3BO_3	6.2 mg
$Na_2MoO_4.2H_2O$	0.25 mg
$MgSO_4.7H_2O$	0.37 g
$MnSO_4.4H_2O$	22 mg
$CuSO_4.5H_2O$	0.025mg
$ZnSO_4.7H_2O$	8.6 mg
Fe NaEDTA	36.7 mg
Glycine	2 mg
Inositol	0.1 g
Nicotinic acid	0.5 mg
Pyridoxine HCl	0.5 mg
Thiamine HCl	0.1 mg
Sucrose	30 g

Adjust the pH to 5.6 with 0.1 M HCl. The above components with the exception of sucrose are marketed in preweighted packets by Flow Labs.

2.   *MSO IC*

Add carbenicillin to molten MSO held at 45°C to give a final concentration of 1 mg/ml. A filter sterilised stock solution of carbenicillin at 100 mg/ml is convenient to use. Add 1 ml to 100 ml of molten MSO before pouring plates or 0.4 ml to 40 ml of MSO in 6 oz Powder Rounds using an automatic pipette (e.g., Gilson P1000) fitted with a sterile tip.

3.   *MSPl*

This is identical to MSO but contains in addition:

NAA (naphaleneacetic acid)   2.0 mg/l

6BAP (6 benzylaminopurine)   0.5 mg/l

Dissolve 500 mg of NAA in 20 ml of 50% ethanol. Add 80 μl to 1 l of MSO. Dissolve 126 mg of 6BAP in 1.0 ml of 1 M HCl and then add to 99 ml of distilled water. Add 400 μl to 1 l of MSO.

4.   *MSPl 9M, 7M, 3M*

Add mannitol in the required amount (9% w/v, 7% w/v or 3% w/v) to MSPl before adjusting to the final volume. For MSPl 9M, add 90 g/l; for MSPl 7M, add 70 g/l; and for MSPl 3M, add 30 g/l.

The above media may be solidified by the addition of 0.8% agar (8 g/l). The agar is usually dissolved by steaming the medium before autoclaving.

Table 9. continued

	Component	Concentration per 100 ml of 100x stock	Final concentration (mg/l)
B. *Caboche Medium (40)*			
100x Stock Solutions			
100x CA stock 1	NH_4NO_3	4 g	400.0
	$CaCl_2.2H_2O$	2.93 g	293.0
100x CA stock 2	$MgSO_4.7H_2O$	2.46 g	246.0
100x CA stock 3	$FeSO_4.7H_2O$	0.27 g	27.0
	Na_2EDTA	0.37 g	37.0
100x CA stock 4	KH_2PO_4	0.68 g	68.0
100x CA stock 5	H_3BO_3	62 mg	6.2
	$MnSO_4.H_2O$	1.7 mg	0.17
	$ZnSO_4.7H_2O$	2.8 mg	0.28
	$CoCl_2.6H_2O$	0.25 mg	0.024
	$CuSO_4.5H_2O$	0.25 mg	0.025
	$Na_2MoO_4.2H_2O$	0.24 mg	0.024
Components added Directly			
	Inositol		180.0
	Pyridoxine		0.5
	NAA		0.1
	6BAP		1.0
	Sucrose		20000.0
	pH 5.6		

This medium requires the addition of glutamine to a concentration of 2 mM before use. Glutamine may be dissolved at a concentration of 23.36 mg/ml in distilled water, sterilised by filtration and stored at $-20°C$ until required. Add 1.2 ml to liquid medium or molten agar-solidified medium held at 45°C. To make a pyridoxine stock, dissolve at a concentration of 5 mg/ml in distilled water and add 100 μl/l. Agar may be added if required at a concentration of 8 g/l.

Various vector systems based on the Ti plasmid have been developed (Section 2). Some vector systems (*onc*[+]) utilise the natural oncogenicity of the Ti plasmid T-DNA to induce tumours whilst in other vector systems (*onc*[−]) the *onc* gene functions have been removed from the T-DNA and replaced by dominant selectable markers to aid the recovery of transformed, but non-tumourous cells. A major disadvantage of the Ti plasmid system is the apparent inability of virulent *A. tumefaciens* strains to transform monocotyledons. It is proposed in this section to describe the main successful strategies for transforming plant cells with *onc*[+] and *onc*[−] vectors using virulent agrobacteria to deliver the DNA to plant cells. Secondly, a brief outline will be given of several systems which have been used to deliver naked DNA vectors to plant protoplasts in an attempt to extend transformation studies into the monocotyledons. Finally, the combinations of vector and delivery systems will be discussed with relation to the regeneration of transformed plants and the stability of the transferred genes through meiosis.

Table 10. Antibiotics for Inclusion in Plant Growth Media.

Carbenicillin (Sigma)	Dissolve in distilled water at a concentration of 100 mg/ml and sterilise by filtration. Store at 4°C for up to 1 month or freeze at −20°C.
Cefotaxime (Hoechst)	Dissolve in distilled water at 50 mg/ml and sterilise by filtration. Store at −20°C. Cefotaxime will degrade in the light in culture medium after approximately 3 weeks.
Kanamycin (Sigma)	Dissolve in distilled water at 10 mg/ml and sterilise by filtration. Store at 4°C for up to 1 month or freeze at −20°C.
Rifampicin (Sigma)	Dissolve in methanol at 20 mg/ml. Place a weighed amount of rifampicin in a sterile Universal bottle, add methanol to dissolve, cap the bottle and then shake. Store at −20°C for up to 3 months.
Tetracycline (Sigma)	Dissolve in distilled water at 10 mg/ml and sterilise by filtration. Store at 4°C for up to 1 month or freeze at −20°C.

3.1 Tumour Induction on Sterile Seedlings or Explants with Virulent and Oncogenic Agrobacteria

With most plant species the most suitable material for tumour induction are sterile seedlings or well established plants that have been propagated *in vitro*. The use of sterile plant material is very important to avoid the contamination of the transformed plant tissue with microorganisms, other than *Agrobacterium*. Such contaminants could otherwise infect the culture medium and possibly kill the plant cells. In addition they would contaminate any biochemical or molecular analysis performed on the transformed material. If seeds or propagated material are not available, then in some cases it is possible to inoculate plants *in vivo*. Alternatively it is possible to take explants from greenhouse-grown material and achieve successful tumour induction *in vitro*, provided the material is checked rigorously for sterility. To maintain sterility all plant work must be performed in a laminar flow culture hood using autoclaved materials. Media for plant culture are described in *Table 9* and antibiotics for inclusion in plant growth media are given in *Table 10*. The preparation of explants and their inoculation with *A. tumefaciens* is described in *Tables 11* and *12*. A scheme showing inoculation methods is given in *Figure 6*.

3.2 Transformation of Explants with *onc*⁻ Vectors Containing Dominant Selectable Markers

At present the most widely used dominant selectable transformation marker is that of resistance to the aminoglycoside antibiotic kanamycin conferred by the enzyme neomycin phosphotransferase (NPTII). The structural gene for NPTII, driven by the *nos* promoter (see Section 2), can give resistance for up to 500 μg/ml (or more) of kanamycin in cultured cells, shoots or seedlings (9). A protocol for this dominant selection is given in *Table 13*. Strategies for the direct regeneration of transformed shoots expressing dominant markers are presented in Section 3.5.

Table 11. Preparation of Plants for Inoculation.

A.	*From sterile seedlings*

1. Surface sterilise seeds in a 10% (v/v) solution of commercial bleach (sodium hypochlorite) having a final concentration of about 1% active chlorine. The sterilisation time will depend on the species but somewhere in the order of 10 − 30 min is usually sufficient. If the seeds have a very waxy seed coat or are very hairy or irregular, sterilisation may be aided by a prerinse in 70% ethanol for 5 − 10 sec).
2. Wash the seeds with five changes of sterile tap water. If the seeds are very small then it may be convenient to contain them in a sterile nylon gauze or in a nylon or stainless steel sieve (e.g., a tea strainer) during the sterilisation and washing manipulations.
3. Germinate the seeds on MSO medium[a] solidified with 0.8% agar in a suitable culture vessel[b]. Incubate at 25°C at a medium light intensity (1000 lux).

B. *From explants*

1. Excise young shoots from greenhouse-grown plants and surface sterilise them with 10% (v/v) commercial bleach for 20 − 30 min. This can be done conveniently in a Pyrex casserole dish (Corning).
2. Rinse the shoots with five changes of sterile tap water.
3. Working on a sterile glass plate, white glazed tile or in a large Petri dish, remove any bleach-damaged areas from the shoots, using a sterile scalpel. Trim off the larger basal leaves and cut the shoots into 4 cm lengths (or the maximum size that will stand up in the culture vessel).
4. Insert the explants, basal end downwards, into agar-solidified MSO medium.

[a]See *Table 9* for a description of all plant culture media.
[b]Suitable culture vessels are 'Universal' bottles or 'Powder Round Jars' supplied by Beatson Clark and Co. Ltd., Rotherham, Yorkshire, UK. Use 6 ml of medium per Universal or 40 ml of medium per Powder Round Jar.

Table 12. Inoculation with *Agrobacterium tumefaciens*.

1. Cut the tops off *in vitro* propagated plants or germinated sterile seedlings with a sterile scalpel or use freshly initiated stem explants.
2. Inoculate the cut surface with a loop of an overnight culture of an oncogenic strain of *A. tumefaciens*, as illustrated in *Figure 6A*. Alternatively, wound the plant stem by pricking with a hypodermic needle (25 gauge) previously dipped in an overnight culture of *A. tumefaciens*.
3. Incubate the inoculated plants for 3 − 6 weeks at 25°C with a low light intensity (600 − 1000 lux).
4. Remove the tumours when they are 3 − 10 mm in diameter and place on MSO IC medium[a] (5 ml of medium in a 5 cm plastic Petri dish). MSO IC contains 1 mg/ml of the antibiotic carbenicillin and is only suitable for killing agrobacteria which do not carry a beta-lactamase (Ap[r]) gene. Alternative antibiotic treatments[b] to kill bacteria in the tumours include:
    Rifampicin (50 μg/ml) plus tetracyline (10 μg/ml)
    Cefotaxime (500 μg/ml)
    Vancomycin (100 μg/ml)
5. Subculture after 3 − 6 weeks (depending on the state of growth of the excised tumour) on to MSO containing 0.5 mg/ml of carbenicillin and from then on maintain on MSO containing 0.25 mg/ ml of carbenicillin. Alternatively, use reduced levels of the other antibiotics if carbenicillin is inappropriate.
6. Periodically check the tumour for sterility by incubating a small amount for two days in nutrient broth at 25°C with rapid shaking. When the tumour shows no sign of persisting bacteria, grow it on MSO without antibiotics.

[a]See *Table 9* for a description of plant growth media.
[b]See *Table 10* for details of antibiotic solutions.

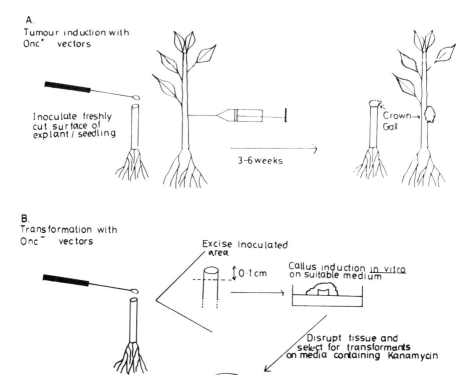

A.
Tumour induction with
Onc⁺ vectors

Inoculate freshly
cut surface of
explant / seedling

3–6 weeks

Crown→
Gall

B.
Transformation with
Onc⁻ vectors

Excise inoculated
area

↕ 0·1 cm

Callus induction in vitro
on suitable medium

Disrupt tissue and
select for transformants
on media containing Kanamycin

Figure 6. Scheme showing the methods to inoculate plant tissue with *onc⁺* and *onc⁻* Ti plasmid-based vectors.

Table 13. Dominant Selection of Transformed Callus.

1.	Prepare stem explants or topped seedlings and inoculate as described in *Table 12*.
2.	Incubate for 2–3 weeks at 25°C with a light intensity of 600–1000 lux. Remove the top of the stem, making the cut 1–2 mm below the cut surface (see *Figure 6B*).
3.	Place the inoculated explant on a suitable medium for callus induction which also contains antibiotics to kill any contaminating agrobacteria as described in *Table 12*. The composition of the medium will depend on the plant species. For tobacco use MSP1.
4.	After callus induction tease apart the tissue, or chop it up with a scalpel (depending on friability) and place the tissue portions on selective medium, e.g., MSP1 containing 100 μg/ml of kanamycin.

3.3 Transformation of Plant Cells By Co-cultivation with A. tumefaciens

During crown gall tumourigenesis *Agrobacterium* is able to transform cells adjacent to wounded tissue which are undergoing cell division in response to wounding (wound response). Cells in an intact plant show maximal tumourigenic response to infection with *A. tumefaciens* between 1.5 and 2 days after wounding. In several species this competent phase has been correlated with the onset of wound response cell divisions. It has been shown (24) that the wound response can be mimicked in tissue culture by converting quiescent leaf mesophyll cells or actively dividing suspension cultured cells into protoplasts and then allowing cell wall regeneration and cell division to occur in appropriate media. At a specific phase in the development of the protoplast culture the cells are competent to be transformed by co-culture with *A. tumefaciens*. The co-cultivation technique allows the generation of large numbers of transformants under controlled conditions. By careful manipulation of the culture conditions, plating density and selection procedure clonally derived transformed colonies can be obtained. It is important to realise that crown galls are a mixture of non-transformed and several types of independently transformed cells and consequently the phenotype of the gall is the sum of all the phenotypes of the cell types which are present in the tumour. However, in transformants derived by co-cultivation there is normally only one type of transformed cell and so phenotypic masking does not occur, making selection and analysis much less prone to artifactual results. Transformed colonies can be selected on the basis of hormone independent growth in the case of *onc*⁺ vectors or by selection on medium containing antibiotics in *onc*⁻ vectors having dominant selectable markers such as kanamycin resistance.

3.3.1 *Protoplast Isolation from Higher Plants*

Protoplast isolation procedures vary from species to species and also depend on the starting material (e.g., leaf mesophyll or suspension cultured cells). Protoplast isolation and culture protocols have been presented in the literature for many plant species (25,26). However, it must be pointed out that to survive co-cultivation with *Agrobacterium* only fairly vigorous protoplast systems can be used. For the purpose of this chapter the isolation, culture and transformation of leaf mesophyll protoplasts from *N. tabacum* var. Xanthi will be used as a model system. *Figure 7* shows stages in cell wall synthesis and early cell division, while *Figure 8* shows stages of mitotic activity in *N. tabacum* cv. Xanthi protoplasts. A protocol for protoplast isolation is given in *Table 14*. The solutions required for this procedure are in *Table 15* and staining procedures to visualise protoplast development are given in *Table 16*.

3.3.2 *Co-cultivation of Protoplasts with Agrobacterium and the Selection of Transformed Colonies*

The protoplast culture should be regularly monitored using an inverted microscope. Note when the first cell divisions appear. Observation may be aided by staining small samples of the preparation with fluorescent cell wall stains (e.g., Tinopal) which bind to cellulosic material and can be used to visualise newly-formed cell walls and cell plates by u.v. microscopy (see *Table 16A*). Alternatively, the development of the protoplast culture may be monitored by examining nuclear divisions using acridine orange staining and u.v. microscopy (*Table 16B*). When 20 – 30% of the cell population are in their first division, the protoplasts

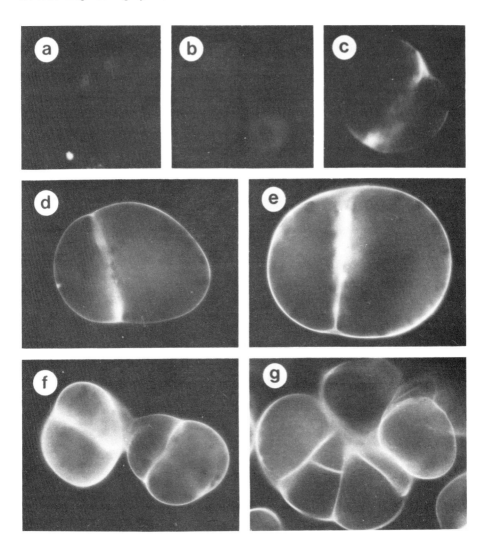

Figure 7. Progressive stages in cell wall synthesis and early cell divisions of *N. tabacum* cv. Xanthi leaf mesophyll protoplasts as viewed under u.v. illumination following tinopal staining. **(a)** Freshly isolated protoplasts (x 400). **(b)** Protoplasts after one day in culture exhibiting a weak fluorescence resulting from a limited amount of cell wall regeneration (x 325). **(c)** Protoplasts after approximately five days in culture with cell plate formation visible (x 583). **(d)** Protoplasts undergoing first cell division after seven days of culture (x 1142). **(e)** Cell expansion following first division after nine days in culture (x 909). **(f)** Cell plate formation prior to second cell division in a ten day old protoplast culture (x 440). **(g)** Small colony after 14 days in culture (x 600).

are co-cultivated with *Agrobacterium* following the protocol in *Table 17*.

Transformed plant cell colonies may be selected either by plating on hormone free medium in the case of *onc*$^+$ vectors or by plating on media containing high levels of normally toxic antibiotics if dominant selectable marker genes (e.g., resistance of kanamycin) are incorporated into *onc*$^-$ vectors (see *Figure 9*). There

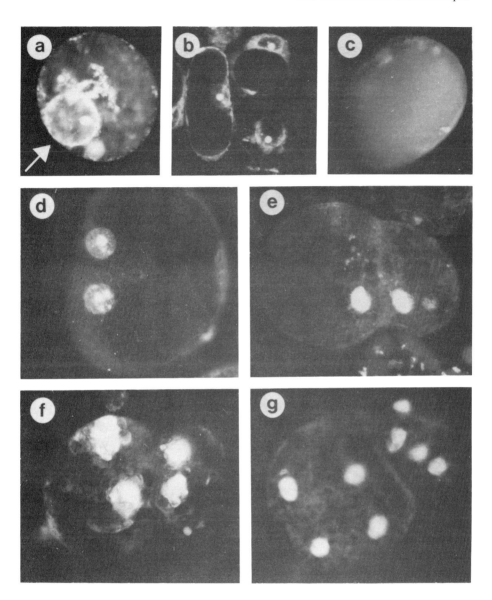

Figure 8. Progressive stages of mitotic activity in *N. tabacum* cv. Xanthi leaf mesophyll protoplasts during the early stages of culture as viewed under u.v. illumination following acridine orange staining. **(a)** Freshly isolated protoplasts with one nucleus clearly visible (arrowed) (x 600). **(b)** Protoplasts after three days in culture have reformed cell walls, but each protoplast still has only one nucleus visible (x 370). **(c)** Viable protoplast after four days in culture with no nucleus visible, presumed to be undergoing mitosis (x 633). **(d)** Protoplasts following first mitotic division after approximately six days in culture (x 1222). **(e)** Eight day old protoplast after first cell division showing two nuclei (x 1000). **(f)** Protoplast after approximately ten days in culture with four nuclei visible following the second mitotic division (x 1333). **(g)** Small colony following the third mitosis after 12 days in culture (x 1028).

93

Table 14. Isolation of Protoplasts from Tobacco Plants.

1.	Excise almost fully expanded leaves from 2 months old tobacco plants.
2.	Sterilise the leaves by soaking them for 20 min in 10% (v/v) commercial bleach and then rinse them five times with sterile tap water.
3.	Allow the leaves to wilt in the airstream of a laminar air flow cabinet before removing the lower epidermis with fine forceps.
4.	Place the peeled leaf pieces with the exposed mesophyll in contact with the surface of 30 ml of CPW 13M medium[a] in a 14 cm plastic Petri dish to plasmolyse the cells.
5.	When the surface of the liquid is completely covered with peeled leaf pieces replace the CPW 13M with 20 ml of enzyme mixture[b]. Incubate in the dark at 25°C overnight (16−18 h).
6.	Release the protoplasts by gently agitating the digested leaf pieces in the Petri dish. Tilt the dish and move the large debris out of the protoplast suspension using sterile forceps. Pass the suspension through a sterile sieve (64−100 micron mesh).
7.	Remove the protoplasts to a screwcapped Pyrex centrifuge tube (16 x 120 mm, Corning). Centrifuge the tube at 100 g for 5 min to pellet the protoplasts.
8.	Using a Pasteur pipette gently resuspend the protoplast pellet in CPW 13M. Re-pellet the protoplasts by centrifuging at 100 g for 5 min. At this stage it may be beneficial to separate the intact from the disrupted protoplasts and cells. This may be carried out by resuspending the protoplasts in CPW 21S[a] and centrifuging at 150 g for 10 min. After centrifugation the intact protoplasts will have formed a tight band at the top of the CWP 21S whilst the dense disrupted protoplasts and debris will have pelleted. The protoplasts can be removed with a Pasteur pipette and diluted with MSP1 9M before proceeding with the isolation protocol. However, many protoplast systems are sensitive to incubation in high concentrations of sucrose and other protoplast types are simply too dense to float and so this further purification may not be possible.
9.	Wash the protoplasts twice by pelleting and resuspending in MSP1 9M.
10.	On the final wash resuspend the protoplasts in 10 ml (or any other known volume) of MSP1 9M and count the number of viable protoplasts using a haemocytometer. Viable protoplasts will be completely spherical and will have their chloroplasts intact (oval in shape) and distributed evenly around the cytoplasm. Non-viable protoplasts are often non-spherical and have misshapen chloroplasts all clumped at one side of the cell.
11.	Plate the protoplasts at a density of about 2 x 10^5 live cells per ml in 5 ml of liquid MSP1 9M on top of 12 ml of MSP1 9M solidified with 0.8% agar in a 9 cm plastic Petri dish. This gives a final plating density of 5 x 10^4 protoplasts per ml.
12.	Incubate at a low light intensity (600 lux) at 25°C.

[a]See *Table 15* for a description of solutions for the isolation of protoplasts.
[b]The enzyme mixture consists of 0.2% cellulose and 0.05% pectinase in CPW 13M. See *Table 15*.

are several points worth mentioning concerning selection conditions. Firstly, the microcolonies must be of sufficient size to allow their plating at low densities. Plating at too high a density allows colonies to grow into each other. This makes it difficult to recover clonal material, and also causes problems resulting from crossfeeding effects. Secondly, the selection conditions must be such that massive necrosis of non-transformed material does not inhibit the grow-through of transformed colonies. Thus a primary selection on media containing borderline inhibitory concentrations of antibiotics may be desirable, especially if the protoplast system is not particularly vigorous. Any putative transformants can then be picked off onto media containing optional levels of selective agent. Thirdly, it should be noted that the inherent resistance of a plant colony to a toxic antibiotic is often greater with increased colony size (age!). A protocol for the selection of *onc*$^+$ transformants on hormone free medium (MSO) is given in *Table 18*. The selec-

Table 15. Solutions for the Isolation of Plant Protoplasts.

A.	*CPW salts*			
		Component	Concentration for 1 l of 100x stock	Concentration (mg/l)
	CPW stock 1	KH_2PO_4	2.72 g	27.2
		KNO_3	10.1 g	101.0
		$MgSO_4.7H_2O$	24.6 g	246.0
		KI	16 mg	0.16
		$CuSO_4.5H_2O$	2.5 mg	0.025
	CPW stock 2	$CaCl_2.2H_2O$	148 g	1480.0

CWP 13M consists of CPW salts plus 13% w/v mannitol (130 g/l). CPW 9M is CPW salts plus 9% w/v mannitol (90 g/l), and CPW 21S is CPW salts plus 21% w/v sucrose (210 g/l). The media are adjusted to pH 5.8 and sterilised by autoclaving.

B.   *Enzyme solution for the isolation of tobacco leaf mesophyll protoplasts*

Cellulase 'onozuka' R-10	0.2% w/v
Macerozyme R-10	0.05% w/v

These enzymes may be obtained from the Yakult Pharmaceutical Industry Co. Ltd., Japan. Dissolve both enzymes in CPW 13M, sterilise by filtration through a 0.45 μ membrane and then store at $-20°C$ until required.

Table 16. Fluorescent Staining Techniques for Examination of Plant Protoplast Development.

A.   *Visualisation of cell wall regeneration using Tinopal*

1.   Make a saturated solution of Tinopal (Ciba Geigy UK Ltd.) in distilled water and store it in the dark at 4°C.
2.   Without disturbing the undissolved material add 0.1 ml of the saturated Tinopal solution to 20 ml of CPW 9M and mix. This solution may be sterilised by filtration and stored in the refrigerator without degradation for a couple of months.
3.   Mix one drop of Tinopal solution with one drop of protoplast suspension on a clean glass microscope slide and cover with a cover slip.
4.   Leave at room temperature for 5 min.
5.   Examine on a microscope fitted with an epifluorescence facility (e.g., Nikon Optiphot, Model XF-EF[a]) having an excitation filter in the u.v. range 330 − 380 nm (e.g., Nikon excitation filter cassette U; 365 nm in conjunction with an auxillary excitation filter u.v.; 330 − 380) and a barrier filter cutting out fluorescence at around 420 nm (e.g., Nikon eyepiece-side absorbtion filter 420 K). An example of Tinopal stained protoplasts is shown in *Figure 7*.

B.   *Visualisation of nuclei using acridine orange*

1.   Prepare the acridine orange stain by dissolving 10 mg of acridine orange (Sigma) in 4 ml of 50 mM glycine-NaOH buffer pH 8.5 and mix with 6 ml of CPW 13M. Store in the dark at 4°C.
2.   Add one drop of the acridine orange solution to 3 ml of the protoplast suspension and incubate in the dark for 15 min at room temperature.
3.   To wash the protoplasts, centrifuge at 100 *g* for 5 min and resuspend the pellet in CPW 9M. Do this twice.
4.   View by u.v. microscopy. Use a blue excitation filter[a] (e.g., Nikon excitation filter cassette code B; 495 nm in conjunction with an auxillary excitation filter; 460) and a barrier filter cutting out fluorescence around 515 nm (e.g., Nikon eyepiece-side absorbtion filter 515 W). An example of acridine orange stained protoplasts is shown in *Figure 8*.

[a]These examples are of Nikon equipment used by the authors. Equivalent equipment from other manufacturers is also suitable.

Table 17. Co-cultivation of Protoplasts with *Agrobacterium*.

1.	When 20 – 30% of the cell population are in their first division (*Figure 7* stage d) wash the protoplasts in MSP1 7M and replate over MSP1 7M. This will normally be 3 – 10 days after isolation, depending on the protoplast preparation. At this stage most of the cell population will have just finished the first nuclear division and will be in the G1 phase of the 2nd division, preparing to initiate DNA replication (*Figure 8* stage d). The normal development of *N. tabaccum* protoplasts is shown in *Figures 7* and *8*.
2.	Using a Pasteur pipette, add three drops of an overnight culture of *Agrobacterium* to the protoplasts to give approximately 100 bacteria per plant cell.
3.	Incubate at a low light intensity (600 lux) at 25°C for 40 h. At the end of this time the plant cells and bacteria will have gummed together to form a mat of plant cells linked by bacterial aggregates.
4.	Separate the clumped plant cells by gentle repeated pipetting with a Pasteur pipette. Pellet the cells by centrifugation at 100 g for 5 min. Resuspend the cell pellet in MSP1 7M containing 1 mg/ml of carbenicillin or other suitable antibiotics and re-centrifuge.
5.	Finally resuspend cells for plating in 6 ml of liquid MSP1 7M containing suitable antibiotics (e.g., 1 mg/ml carbenicillin). Plate the cells over 12 ml of the same medium solidified with 0.8% agar. The same antibiotic regimes as those used for cleaning up crown galls (Section 3.1.2) may also be used to inhibit the growth of agrobacteria in co-cultivation experiments.
6.	After a further 7 – 10 days growth, transfer the plant cells to MSP1 3M containing antibiotics and incubate for a further week.

tion of *onc*⁻ transformants expressing kanamycin resistance is given in *Table 19*, for a vector that determines resistance to 100 μg/ml of kanamycin when expressed in tobacco colonies (Section 2). Normal cells show little growth at kanamycin concentrations much over 50 μg/ml.

3.4 Delivery of DNA to Plant Cells by Isolated DNA Vectors

Transformation of intact plants or plant cells by interaction with *onc*⁺ or *onc*⁻ *A. tumefaciens* strains is very efficient but does have some disadvantages. The major problem is that monocotyledons are not amenable to transformation by *Agrobacterium*. Secondly, many crop species, especially legumes, show a poor response to inoculation with *Agrobacterium* and protoplasts from these species are often not hardy enough to withstand the rigours of co-cultivation. The agrobacteria can also contaminate extracts of transformed tissues used for transcriptional or translation studies of transformed genes and lead to artifactual results. Finally, the present methods for the engineering of Ti plasmids are tedious and time consuming. For these several reasons there have been many attempts to try and develop a DNA delivery system for higher plant cells which is independent of virulent *A. tumefaciens*.

Several techniques have been developed to introduce biologically active nucleic acids into eukaryotic cells, most of which have been tried with plant protoplasts with rather limited success.

(i)   The chemically stimulated uptake of naked DNA or DNA precipitates by endocytosis/fusion enhancers such as polyethylene glycol (PEG).

(ii)   The fusion or endocytosis of plasmid-containing liposomes with plant protoplasts.

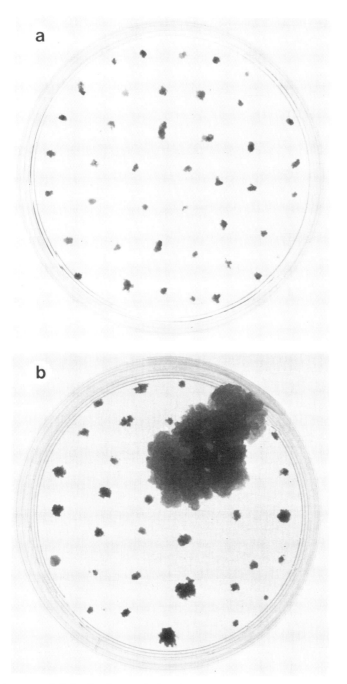

Figure 9. Colony growth following four weeks in culture on hormone-free medium. Fifty colonies of approximately 1.5 – 2.0 mm diameter were placed on 30 ml of MSO medium containing 1 mg/ml of carbenicillin in 9 cm plastic Petri dishes. **(a)** Colonies from non-*Agrobacterium*-treated *P. parodii* protoplasts. **(b)** Colonies from seven day old *P. parodii* which were infected with *A. tumefaciens* pTiAch5 (10^7/ml) and incubated for 36 h.

Table 18. Selection of *onc*⁺ Transformants on Hormone Free Medium.

1.	Transfer the 16 – 30 days old colonies (depending on the growth rate of protoplasts) from step 6 of *Table 17* to centrifuge tubes and wash them with 12 ml of MSO[a] containing 0.5 mg/ml of carbenicillin. Determine the number of colonies per ml.
2.	Embed the colonies at a density of 500 colonies per ml in an 8 ml layer of molten MSO-agar containing 0.5 ml of carbenicillin, pre-cooled to 40°C, and layered onto 10 – 15 ml of the same medium in a 9 cm dish. In systems less vigorous than tobacco it might be advisable to use low gelling temperature (LGT) agarose to solidify the MSO.
3.	Pick off any colonies growing independently of hormones onto MSO containing 0.25 mg/ml of carbenicillin or suitable levels of other antibiotics. Place 50 colonies of approximately 2 – 5 mm in diameter on 25 ml of medium in a 9 cm Petri dish (see *Figure 9*).

[a]See *Table 9* for description of media.

Table 19. Selection for *onc*⁻ Transformants Expressing Kanamycin Resistance.

1.	Transfer the colonies from step 6 of *Table 17* to centrifuge tubes and wash them with 12 ml of Caboche medium containing 100 μg/ml of kanamycin together with 0.5 mg/ml of carbenicillin (or other suitable antibiotic) to inhibit growth of contaminating agrobacteria.
2.	Determine the number of colonies per ml and embed the colonies at a density of 500 colonies per ml in an 8 ml layer of agar-solidified Caboche medium (containing suitable antibiotics) with 100 μl/ml of kanamycin over a 10 ml layer of the same medium in a 9 cm Petri dish.
3.	Pick off any kanamycin resistant colonies which develop and transfer to MSP1 medium containing reduced levels of the antibiotics which kill agrobacteria and optimal levels of kanamycin (e.g., 200 μg – 1000 μl/ml) to select the kanamycin resistant transformed plant cell colonies.

[a]See *Table 9* for description of media.

(iii)   The fusion/endocytosis of plasmid-containing bacterial spheroplasts with plant protoplasts.

(iv)   The direct introduction of plasmid vectors into plant protoplasts and their nuclei *via* microinjection techniques.

These techniques have been reviewed recently (26) and to date the introduction of Ti plasmid DNA into plant protoplasts by PEG has proved the most successful (27,28) with transformed cells recovered at an extremely low frequency ($\sim 10^{-5}$). However, it is certain that all these methods do deliver DNA in an intact form at least to the protoplast cytoplasm and so the lack of transformation may be due to the vector design or the competence of the protoplast system.

A protocol for the PEG stimulated uptake of plasmid DNA into protoplasts is given in *Table 20*. This same protocol may be used to stimulate the uptake of liposomes or *E. coli* spheroplasts into plant protoplasts. In the case of spheroplasts, the use of a higher PEG concentration (25% w/v) may be beneficial.

3.5 Regeneration of Transformed Plants

Research over the past few years has indicated that it is impossible to regenerate transformed plants from tissues transformed by oncogenic strains of *A. tumefaciens*. In some cases, however, transformed shoots are formed that will rarely produce roots. This is true of crown gall tumours (especially those induced by

Table 20. Transformation with Isolated Vector Plasmid using PEG Stimulated Uptake into Plant Protoplasts.

1.	Prepare protoplasts as outlined in *Table 14*. Pellet $1-3$ x 10^6 protoplasts in MSP1 9M in a 16 x 125 mm Pyrex centrifuge tube.
2.	Add the plasmid DNA (in sterile distilled water) in as small a volume as possible (e.g., 25 μg in 20 μl) and mix by gentle pipetting with a Pasteur pipette for 10 sec.
3.	Add 2 ml of sterile 15% (w/v) PEG 6000 (Koch-Light Ltd.) dissolved in MSP1 9M and mix by rolling the incubation tube between the hands.
4.	Incubate at 27°C for 30 min and then slowly add 10 ml of MSP1 9M.
5.	Pellet protoplasts by centrifugation at 100 g for 10 min and then resuspend them in 1.0 ml of MSP1 9M. Leave at room temperature for 30 min.
6.	If required, intact protoplasts may be separated from disrupted protoplasts by banding on CPW 21S as outlined in *Table 14*.
7.	Wash, by resuspending protoplasts pelleted by centrifugation, in MSP1 9M. Count the protoplasts and then plate as in *Table 14*.
8.	Select the transformants by plating microcolonies on suitable selective media as outlined in *Table 19*.

'shooty' mutant agrobacteria), and tissues derived by co-cultivation with *onc*$^+$ strains. Where intact transformed plants have been recovered (29) transformation markers have only been shown to pass through meiosis in a stable manner if the T-DNA had incurred deletions/rearrangements that abolished the functions of the oncogenic loci (*tms1, tms2, tmr*). Consequently most effort has been directed towards the development of *onc*$^-$ vectors (see Section 2) which, in the presence of *Vir* gene functions, and the T-DNA border direct repeats, are still capable of transformation where the resulting tissue is non-tumourous. Transformant selection in this case is normally possible by the insertion of a dominant selectable marker (e.g., Km^r) between the T-DNA ends (30).

3.5.1 *Regeneration on Shoot-inducing Medium from Transformants Incited by Completely Non-oncogenic Strains of A. tumefaciens*

Transformation is induced as outlined in Section 3.2 and transformed cells enriched by growth on suitable selection media. Callus derived from such transformed tissues are placed on media previously shown to induce regeneration from normal plant callus of the same species. For example *N. tabacum* can be induced to regenerate shoots when cultured on agar solidified MSO medium containing $1-2$ mg/l of 6 benzylaminopurine and 3% sucrose when grown at 25°C at a light intensity of 1000 lux). In general, a high cytokinin to auxin ratio will induce regeneration in many species. If the tissues are still contaminated with live agrobacteria then suitable antibiotics must be included in the medium. Regenerated shoots can then be excised from the callus and rooted on MSO medium. When the shoots are $3-5$ cm in height, they may be transferred to compost, and grown in a greenhouse or growth room, taking care that the plantlets are protected from desiccation for $1-2$ weeks. This can be done simply by covering the plant pot with a clear plastic bag for the first few days after removing from culture. The bag can subsequently be slit along one side to reduce humidity and then finally

Figure 10. Crown gall tumour on *N. tabacum* resulting from infection by a *tms* mutant octopine type Ti plasmid. Note that the initial response is callus proliferation followed by shoot development. When young sterile plantlets are innoculated with *tms* mutant Ti plasmids more shoots proliferate; thus *tms* mutants are suitable to use in coinfection experiments with *onc⁻* vectors.

dispensed with as the plant hardens off. Alternatively, the shoots may be hardened off in a mist propagator.

3.5.2 *Regeneration of Transformed Shoots by Co-infection of Shooty Mutant Strains of A. tumefaciens with onc⁻ Vectors*

Ti plasmid mutants in either the *tms1* or *tms2* gene generates tumours with the shooty phenotype (*Figure 10*). The expression of these two loci normally functions to suppress shoot formation in crown galls by causing the overproduction of auxins in the tumour. The expression of the *tmr* gene functions to suppress root formation by causing an overproduction of cytokinins. In the absence of or at reduced levels of expression of *tms1* or *tms2*, cytokinin overproduction caused by the *tmr* gene product thus drives spontaneous shoot regeneration in such tumours. Shoots derived from octopine type shooty mutant tumours are rarely found to be transformed and thus it is very probable that the transformed cells in shooty tumours are stimulating regeneration in surrounding non-neoplastic cells. This property can be exploited to drive regeneration from cells transformed with *onc⁻ A. tumefaciens* vectors by co-infecting wound sites with both a shooty mutant (e.g., pGV2215) and an *onc⁻* strain of *Agrobacterium* (16). Transformation is performed as outlined for *onc⁺* vectors in Section 3.2 using a mixture of cells in a ratio of 5:1 *onc⁻*:*tms* mutant. Regenerated shoots can be excised from the transformed tissue and screened for the expression of transformation markers as outlined in Section 5. In cases where kanamycin resistance is used as a marker, transformed shoots around 1 cm in height can be selected from non-transformed shoots by transfer to MSO medium containing appropriate levels of kanamycin (usually 100 μg/ml) where only transformed shoots will survive.

4. ISOLATION OF PLANT DNA AND RNA

The isolation of DNA and RNA is of absolutely fundamental importance to the study of gene organisation and gene expression at the molecular level, as well as in the genetic engineering of plants using recombinant DNA techniques. In the first instance, isolated DNA is required for the production of clone banks of genomic DNA. The isolation of RNA, in particular mRNA, is required for the production of cDNA libraries in order to identify useful genes and to provide probes for the screening of genomic DNA libraries for clones containing such genes. Once a particular gene has been cloned and its structure and organisation in the genome elucidated then it is possible to introduce the cloned gene back into plants (in a modified form if required) *via* Ti plasmid based vectors (see Section 2). Of great importance is the identification and manipulation of the promoter regions of engineered genes to allow expression in the transformed plant at sufficient levels at the correct developmental stage of the life cycle. The isolation, by cDNA cloning techniques, of genes which are highly expressed in particular organs of a plant species at a specific developmental stage in the life cycle should provide a source of such promoters as well as a wealth of fundamental knowledge about gene expression during development. Besides their use in cloning, DNA and RNA isolation procedures are valuable tools for the analysis of the organisation of expression of the exogenous DNA in both transformed plant tissues and whole regenerated plants.

4.1 Preparation of Total Plant DNA

The isolation of DNA has been reported from different organs of a wide range of plant species using a variety of techniques. In some cases subcellular fractionation of plant tissues has enabled the specific isolation and cloning of nuclear, mitochondrial or chloroplast genomic DNA. In this section it is proposed to outline a general procedure which has proved to be widely applicable for the isolation of total DNA from a wide range of plant organs from a variety of species. Subsequently, a brief outline will be presented of procedures for the isolation of nuclear, mitochondrial and chloroplast DNA.

A major problem in the isolation of plant DNA is the efficient disruption of the plant cell wall. Unfortunately, the same techniques used to break open cells also shear DNA and thus a compromise between DNA length and quantitative recovery has to be reached. The use of freeze dried material does offer some advantages with regard to the problem (Section 4.1.1). However, the isolation of highly polymerised DNA is only one part of the problem as plant extracts are loaded with polysaccharides, tannins and pigments which, in some instances, are very difficult to separate from the DNA. To make matters worse, some polysaccharide-like contaminants are almost impossible to detect by normal non-degradative analytical techniques. These contaminants interfere with the quantification of nucleic acids by spectrophotometric methods and also cause anomalous hybridisation kinetics. Much more importantly, they also inhibit the activity of most restriction enzymes and other DNA modifying enzymes, thus making the analysis of DNA by Southern blotting and the cloning of DNA very difficult.

4.1.1 *Large Scale Preparation of Plant DNA*

Freeze drying (lyophilisation) is a convenient way of storing plant material before DNA isolation. This is especially important if DNA is to be prepared from many samples of plant tissue which can be harvested when in prime condition, washed, freeze dried and then all processed simultaneously at a later date. Dry plant materials can be stored desiccated for several years with little loss in DNA quality. The dry tissue can be disrupted very efficiently by dry milling or grinding, and when the DNA is in a non-hydrated state, it is less susceptible to damage by shearing. Nucleolytic degradation is also minimised as the DNA is hydrated immediately in the presence of chelating agents and detergents which inhibit nuclease activity.

The technique outlined in *Table 21* relies on the fact that nucleic acids will form stable soluble complexes with the detergent cetyl triethylammonium bromide (CTAB) under high salt conditions (0.7 M NaCl). When the salt concentration is reduced below 0.4 M NaCl the CTAB/nucleic acid complex will precipitate leaving the majority of polysaccharides in solution (31). DNA can be separated from RNA and CTAB by the usual CsCl/ethidium bromide (EtBr) density gradient ultracentrifugation methods. The CTAB/nucleic acid pellet is redissolved in salt

Table 21. Preparation of DNA from Freeze Dried Plant Material.

1.	Mix 1.0 g of freeze dried plant tissue with 4 g of alumina (Sigma, type 305) and grind to a very fine powder with a pestle and mortar. This is probably the most important step in efficient disruption of the plant cell wall and is the key to good DNA recovery.
2.	Tip the powder into a 50 ml polypropylene centrifuge tube (e.g., Nalgene type 3100-0500 or its equivalent). Add 15.0 ml of extraction buffer (EB)[a] and mix gently with a disposable plastic stirring rod (Sarstedt) until the powder is homogeneously dispersed.
3.	Cap the tube and incubate in a water bath at 56°C for 20 min, occasionally agitating the tube gently to keep the extract mixed.
4.	Allow the incubation mixture to cool to room temperature. It is important that the temperature is not allowed to fall below 15°C as precipitation of CTAB will occur.
5.	Add 15 ml of chloroform:octanol (24:1 v/v), cap the tube, and then mix by gentle inversion until a one-phase emulsion is formed[b].
6.	Pellet the debris and separate out the organic and aqueous phases by centrifugation at 8000 *g* (for example at 10 000 r.p.m. in a 8 x 50 ml Sorval SS34 rotor) for 10 min at 20°C.
7.	Remove the aqueous phase (top layer) into a clean 50 ml tube. Avoid taking any denatured protein present at the interface.
8.	Add 1/10 volume (~1.5 ml) of 10% (w/v) CTAB in 0.7 M NaCl. Mix gently and repeat the chloroform:octanol extraction.
9.	Avoiding the debris, remove the aqueous phase to suitable sterile glass centrifuge tubes (e.g., Corex tubes or good quality Pyrex low speed centrifugation tubes such as Corning 16 x 125 mm Pyrex tubes).
10.	Add an equal volume of precipitation buffer[c], mix gently and leave to stand at room temperature for 30 min while the precipitate forms.
11.	Pellet the precipitate at 1500 *g* for 10 − 15 min at room temperature. It is very important not to pellet the precipitate too hard as a compact pellet is very difficult to redissolve. In a good preparation the pellet should be white or only slightly discoloured.
12.	Drain the pellet by inverting the tube, held in a rack, onto a wad of Kleenex. At this stage the preparation may be left overnight at 4°C if required, or proceed to the banding protocol in *Table 22*.

[a]Extraction buffer (EB) is 50 mM Tris-HCl pH 8.0, 0.7 M NaCl, 10 mM EDTA, 1% (w/v) CTAB (Sigma H-5882), 1% (v/v) 2-mercaptoethanol (2ME). The solution should be made up without 2ME on a heated stirrer avoiding foaming. It should then be autoclaved and 2ME added when it has cooled to room temperature.
[b]Work in a fume hood and occasionally release the cap of the tube to avoid pressure build up.
[c]Precipitation buffer is 50 mM Tris-HCl pH 8.0, 10 mM EDTA, 1% (w/v) CTAB.

solutions (~1 M) and the DNA then purified on a CsCl/EtBr gradient. The procedure given in *Table 22* is designed for a Beckman VTi 65 tube (5.2 ml) but the volumes may be adjusted to suit any type of high performance vertical or fixed angle rotor, provided the final density is the same (1.55 g/ml).

Large scale preparations of DNA can also be carried out using fresh plant material.

(i)   Swab down the plant material briefly with 70% ethanol to sterilise its surface. Rinse with distilled water. Rapidly freeze 10 g of tissue in liquid nitrogen.

(ii) Transfer the frozen tissue to a pre-chilled mortar and break up the large pieces with a pestle. Add 4 g of alumina and grind quickly to a fine powder before the tissue thaws.

(iii) Remove the homogenate to a 50 ml centrifuge tube and add 0.2 ml of

Table 22. Banding Plant DNA.

1.	Redissolve the precipitate from step 12 of *Table 21* in 2.4 ml of 1 M CsCl/EtBr[a] by heating in a 56°C water bath. The ethidium bromide will allow the visualisation of any nucleic acids remaining adhered to the centrifuge tube. It is important to note that although on addition of the 1 M CsCl/EtBr much of the opaque CTAB pellet will dissolve quickly, the nucleic acids take longer to dissolve.
2.	Remove the redissolved nucleic acids to a Beckman VTi 65 ultracentrifuge tube or its equivalent. Add 2.9 ml of 7 M CsCl/ EtBr[b], seal the tubes and mix the contents by gentle inversion.
3.	Centrifuge at 58 000 r.p.m. for 4 h or 40 000 r.p.m. overnight (16 h).
4.	Visualise the DNA band under longwave (320 nm) u.v. illumination and remove it by side puncture of the tube using a 1 ml syringe fitted with a 19 gauge needle. The RNA will have pelleted on the side of the tube and care must be taken not to dislodge it.
5.	Remove the ethidium bromide by partitioning 5 times against NaCl-saturated isopropanol (propan-2-ol). The ethidium bromide will be taken up into the organic phase which can be discarded into a waste container.
6.	Remove the CsCl by dialysis against 2 l of distilled water at 4°C, changing the water at least three times over a 24 h period and keeping it mixed on a magnetic stirrer.
7.	Measure the concentration of the DNA solution at 260 nm using silica cells (1.0 O.D. unit = 50 μg/ml).
8.	Add 1/10 volume of 3 M sodium acetate followed by two volumes of cold ethanol and precipitate the DNA at -20°C overnight.
9.	Pellet the precipitated DNA by centrifugation for 3 min in a microcentrifuge or for 15 min at 8000 g in a siliconised Corex tube in a Sorvall or the equivalent centrifuge. Dry the pellet under vacuum and redissolve it in sterile distilled water to the required concentration. The DNA can be stored at -20°C and will remain stable for several years. Alternatively, the DNA may be stored as a washed ethanol precipitate under 80% ethanol.

[a]1 M CsCl/EtBr is 1 M CsCl, 50 mM Tris-HCl pH 8.0, 10 mM EDTA, 0.2 mg/ml ethidium bromide.
[b]7 M CsCl/EtBr is 7 M CsCl, 0.1% Sarkosyl (Ciba Geigy NL35), 0.2 mg/ml ethidium bromide.

2-mercaptoethanol followed by 10 ml (or more if required) of 2x extraction buffer (see footnote a of *Table 21*), previously heated to 95°C. Mix immediately with a plastic stirring rod.

(iv)   Incubate the capped tube in a water bath at 56°C for 20 min.

The procedure is then the same as for freeze dried plant material from step 4 of *Table 21*.

4.1.2 *Rapid DNA Mini-preps for Southern Blot Analysis of Transformed Plant Tissues*

The CTAB DNA isolation procedure can be scaled down to cope with a situation when nucleic acid isolations are required from a large number of plant samples. DNA prepared by the method in *Table 23* can be analysed by Southern blotting techniques (see Section 5) without purification on CsCl gradients thus giving great savings on plant material, time, expensive ultracentrifuge equipment and running costs.

4.1.4 *Determination of DNA Concentration in Nucleic Acid Preparations using the Diphenylamine Reagent*

The diphenylamine reagent can be used in a quantitative colourometric assay

Table 23. Mini Preparations of Plant DNA for Southern Transfer Hybridisation.

1. Weigh out either 0.1 g of freeze dried callus or 0.07 g of freeze dried leaf material with the mid-ribs removed. Grind the tissue with 0.3 g of alumina in a small mortar and pestle (9 cm diameter). If more tissue is available it is often convenient to grind a larger sample and then to weigh out the required amount for DNA extraction procedure.

2. Remove the powdered plant tissue to a microcentrifuge tube. Add 600 μl of extraction buffer (EB)[a] and gently disperse the powder with a plastic stirring rod. Incubate for 10 min at 56°C.

3. Add 600 μl of chloroform:octanol (24:1) and emulsify by shaking. Centrifuge for 2 min in a microcentrifuge and then remove the aqueous phase (top layer) to a clean microcentrifuge tube.

4. Wash the denatured protein interface in the original tube with 100 μl of EB. Recentrifuge and pool the aqueous phases.

5. Add 1/10 volume of 10% CTAB ($\sim$ 70 μl) to the pooled aqueous phases and mix. Repeat the chloroform:octanol extraction.

6. Avoiding the debris, remove the aqueous phase to a clean siliconised Pyrex centrifuge tube such as a 16 x 125 mm Corning tube. Add an equal volume of precipitation buffer[b] (600 μl), mix and leave at room temperature for 20 min.

7. Spin in a bench centrifuge (2000 g, 15 min) to pellet the precipitate. Pour off the supernatant and drain the pellet. If the precipitate does not pellet then spin it down in a siliconised Corex tube for 15 min at 8000 g at 20°C.

8. Redissolve the nucleic acid/CTAB pellet with 400 μl of 1 M NaCl, heating to 56°C if required. Remove the solution to a sterile microcentrifuge tube.

9. When the nucleic acid/CTAB pellet is fully dissolved add two volumes (800 μl) of ethanol, mix and place at −20°C overnight.

10. Centrifuge down the precipitated nucleic acids in a microcentrifuge for 2 min and wash the pellet with 1 ml of 65% ethanol for 1 min. Pour off the ethanol and repeat the wash twice. Wash in a similar fashion with 85% ethanol. If at any time the pellet should become dislodged then repellet in the microcentrifuge. The ethanol washes will remove any CTAB from the nucleic acid preparation and allow the DNA to be cut with restriction enzymes.

11. Dry the pellet in a vacuum desiccator and then redissolve it in 85 μl of sterile water.

12. Use 10 μl of this nucleic acid sample in a small scale diphenylamine test to accurately determine the amount of DNA in the solution (see *Table 24*).

13. Dilute or concentrate the sample accordingly to give the required DNA concentration (e.g., 200−500 μg/ml). This DNA preparation is now suitable for analysis using restriction enzymes and Southern blotting techniques (see Section 5). Any RNA present in the sample will not interfere with the analysis as in most cases it will run off the gel during electrophoresis.

[a]See footnote a of *Table 21* for details of EB.
[b]See footnote c of *Table 21* for details of precipitation buffer.

specific for thymidine residues in crude extracts of plant tissues (*Table 24*). The specificity of the reaction for thymidine bases means that the assay can discriminate easily between RNA and DNA, and is much more accurate than reading optical density of the nucleic acid solution at 260 nm. The difference in the nucleic acid content determined by O.D. 260 and the diphenylamine test will obviously give an indication of the amount of RNA in the preparation.

4.2. The Isolation of Nuclear, Mitochondrial and Chloroplast DNA

Methods for the isolation of nuclear and organelle specific DNA have been reported for use with callus, leaf, protoplasts and single cells. The techniques vary from one laboratory to the next but generally involve the disruption of fresh

Table 24. The Diphenylamine Assay to Quantitate DNA.

1. Make up the reagent solutions as follows.

    A. *3N Perchloric acid*. Make 49.2 ml of perchloric acid (Analar grade; BDH) up to 200 ml by the addition of distilled water. Mix the solution and store in a stoppered bottle in the fume hood. Treat with caution as 3N perchloric acid is extremely corrosive. This solution is stable for months at room temperature.

    B. *Diphenylamine solution*. Make a solution of 4% diphenylamine with 0.01% paraldehyde in glacial acetic acid. The ingredients are all dangerous, especially the diphenalamine, and so all work must be performed in a fume hood or extraction hood.
    (i) Add 20 μl of paraldehyde (Analar grade; BDH) to 198 ml of glacial acetic acid (Analar) in a 250 ml Pyrex Erlenmeyer flask.
    (ii) Weigh out carefully (wear gloves) 8 g of diphenylamine (Analar). Add to the paraldehyde/glacial acetic acid solution and quickly cover with tin foil to exclude any light. Shake until all the diphenylamine has dissolved. Store in the dark at room temperature in a fume hood for up to 6 weeks in a stoppered Pyrex container.

2. Working in a fume hood add the components to the reaction mixture in the order listed. The reaction may be conveniently performed in a microcentrifuge tube.
    140 μl of sterile distilled water
    10 μl of DNA solution
    150 μl of 3N perchloric acid
    180 μl of diphenylamine solution.

3. Cap the tubes and mix by inversion and incubate in the dark for 16 – 20 h at 30°C. It is important to stick rigidly to the same time and temperature of incubation, otherwise a calibration curve will have to be constructed using standard DNA samples each time the assay is performed.

4. Read the optical density at 600 nm using micro cuvettes and compare the O.D. reading with a calibration curve made from measurements on a standard DNA dilution series. Each time the assay is performed it is important to have a blank control (distilled water; no DNA) in order to zero the spectrophotometer.
    There is a linear relationship between $O.D._{600}$ and DNA concentration between the limits of 5 – 50 μg/ml of DNA. Above this level the linearity breaks down and the sample will have to be diluted for an accurate estimation of DNA concentration.

plant material, fractionation of subcellular components by differential centrifugation, DNase treatment of intact organelles, lysis, deproteinisation and then purification of nucleic acids on CsCl/EtBr gradients. It is proposed in this section to outline protocols which may be used as starting points in the development of nuclear and organelle DNA isolation procedures from any particular type of plant material.

The isolation of nuclear DNA from leaf or callus tissue is described in *Table 25*. It may be necessary to modify the force and duration of the homogenisation procedure in step 4 of this protocol to suit the plant material under investigation. Similarly, step 6 may have to be modified to avoid excessive contamination of nuclei with chloroplasts. Throughout the procedure it is wise to check the purity of the nuclei preparation microscopically by staining small aliquots with acridine orange (*Table 16B*).

Good quality, transcriptionally active nuclei from plant protoplasts can often be isolated by gentle lysis of the protoplasts in hypertonic buffers (32). However,

Table 25. Isolation of Nuclear DNA from Fresh Leaf or Callus Tissue.

1.	Weigh out 200 g of tissue and wash with distilled water at 2°C in a large Pyrex beaker. Carry out all subsequent manipulations on ice (2 − 4°C), preferably in a cold room.
2.	Remove the midveins (if using leaves) and slice the tissues into small pieces with a single-sided razor blade. This may be conveniently done in a large Petri dish (14 cm, Sterilin).
3.	Place the tissue in a 1 l Pyrex beaker, cover with diethyl ether and stir for 2 min.
4.	Pour off the ether, rinse with cold sterile distilled water and then homogenise the plant material in 400 ml of nuclei isolation buffer A[a] using a 'Polytron' set at medium speed for 1 min. Alternatively, disrupt the tissue in a pre-chilled Waring blender with two 5 sec bursts at medium speed[b].
5.	Filter the homogenate through four layers of sterile muslin and one layer of Miracloth (Calbiochem) into a sterile 1 l beaker.
6.	Transfer the homogenate to sterile, large volume centrifuge bottles and collect a crude nuclear pellet by centrifugation at 2000 g for 10 min (4°C)[b].
7.	Resuspend the pellets in 25 ml of nuclei wash buffer B[c] using the sterile Potter-Elvehjem homogeniser (Jencons) and centrifuge in a siliconised sterile Corex tube (Corning) using a swing-out rotor at 2000 g for 10 min at 4°C. Continue with washing until the nuclei are free from cellular debris (3 − 4 washes).
8.	If ultra pure preparations are required, the nuclei can be banded on a discontinuous Percol gradient. Resuspend the nuclei in 7 ml of nuclei isolation buffer A and layer over a discontinuous Percol density gradient made up from a 7 ml layer of 35% Percol and a 7 ml layer of 80% Percol (Pharmacia) in nuclei isolation buffer A and set up in a sterile, siliconised Corex tube. Centrifuge in a swing-out rotor at 8000 g for 30 min at 4°C. The zone between the two layers should contain nuclei freed from cellular debris. This band is removed and diluted with 25 ml of nuclei isolation buffer A and the nuclei pelleted and then washed once in the same buffer (2000 g, swing-out rotor). Purified nuclei can be stored in nuclei storage buffer[d] at −70°C.
9.	Resuspend the nuclei on ice in 5 ml of nuclei lysis buffer C[e] (which does not contain an osmoticum) and disrupt the nuclei by ten 5 sec burst of vortexing over a 5 min period, cooling on ice between bursts[f].
10.	Add 0.1 ml of 2-mercaptoethanol followed by 5 ml of 2x extraction buffer[g] preheated to 95°C and mix immediately with a plastic stirring rod. Proceed as outlined for the isolation of DNA from freeze dried plant material (*Table 21*, step 4 onwards).

[a]Nuclei isolation buffer A is 1 M sucrose, 10 mM Tris-HCl pH 7.2, 5 mM $MgCl_2$, 2 mM 2-mercaptoethanol (2ME). With the exception of the 2ME, the buffer components should be autoclaved. 2ME is then added immediately before use.
[b]See text for additional comments on this step.
[c]Nuclei wash buffer B is identical to nuclei isolation buffer A (footnote a), but in addition contains 0.5% Triton X-100.
[d]Nuclei storage buffer is 250 mM sucrose, 20 mM Hepes pH 7.8, 5 mM $MgCl_2$, 1 mM dithiothreitol, 50% glucose.
[e]Nuclei lysis buffer is 10 mM Tris-HCl pH 7.2, 10 mM EDTA, 5 mM 2ME, 0.5% Triton X-100.
[f]DNA extraction may be improved by freeze-thawing the nuclear pellet and protease treating the lysed nuclei for 30 min at 37°C in 5 ml of nuclei lysis buffer C containing 500 μg/ml of pre-digested pronase.
[g]See footnote a of *Table 21* for extraction buffer.

this process is limited to plant species where protoplasts can be obtained in reasonable numbers.

(i)   Isolate protoplasts as outlined in *Table 14*.
(ii)   Resuspend the protoplasts in protoplast nuclei isolation buffer (100 mM glycine, 1% hexylene glycol, 2% sucrose adjusted to pH 8.5 with saturated

Table 26. Isolation of Chloroplast and Mitochondrial DNA.

1.	Homogenise 200 g of plant material as described in *Table 25* steps 1 – 5 using organelle isolation buffer F[a] instead of nuclei isolation buffer A.
2.	Spin out the nuclei at 100 g for 10 min at 4°C and discard. Centrifuge at 1800 g (4°C, 10 min) to collect a chloroplast pellet. Pour off filtrate into a clean tube and place chloroplast pellet on ice.
3.	Centrifuge at 10 000 g for 15 min at 4°C to collect the mitochondrial fraction.
4.	Resuspend the organelle pellets separately in 25 ml of DNase buffer G[b] and incubate on ice (0°C) for 1 h with occasional swirling. The DNase concentration in this incubation may have to be determined empirically to avoid contamination of organelle DNA with fragments of nuclear DNA. Concentrations of between 150 and 500 μg/ml have been used successfully (33).
5.	Terminate the DNase treatment by the addition of EDTA (pH 8.0) to a final concentration of 10 mM.
6.	Collect chloroplast and mitochondrial pellets by centrifugation at 1800 g and 10 000 g, respectively, for 15 min at 4°C.
7.	Resuspend the organelle pellets separately (using a Potter-Elvehjem homogeniser if required) in 7 ml of organelle isolation buffer F and purify the organelle preparation by centrifugation on a discontinuous Percol gradient as described in *Table 25*, step 6).
8.	Dilute the chloroplast and mitochondrial fraction separately with 25 ml of organelle isolation buffer F and centrifuge the fractions at 1800 g and 10 000 g, respectively, for 15 min at 4°C.
9.	Wash the pelleted organelles by re-suspending and re-pelleting in 25 ml of organelle wash buffer H[c].
10.	Freeze the pellets separately in a dry ice/ethanol bath and thaw them out in a 22°C water bath. Repeat the freeze/thawing three times and then resuspend the pellets in 5 ml of the same buffer used to isolate nuclear DNA (nuclei lysis buffer F[d]).
11.	Proceed with DNA isolation as outlined for nuclei in *Table 25*, steps 9 and 10.

[a]Organelle isolation buffer F is 300 mM mannitol, 50 mM Tris-HCl pH 8.0, 3 mM EDTA, 0.1% bovine serum albumin (BSA), 1 mM 2-mercaptoethanol (2ME). The solution is made up and autoclaved without the BSA and 2ME, which are added immediately before use.
[b]DNase buffer G is identical to organelle isolation buffer F, but in addition contains 13 mM MgCl$_2$ and 150 – 500 μg/ml of DNase added immediately before use.
[c]Organelle wash buffer H is 150 mM NaCl, 100 mM EDTA pH 8.0.
[d]Nuclei lysis buffer C is described in footnote e of *Table 25*.

Ca(OH)$_2$ solution) at 4°C using 1 ml per million protoplasts. Remove the suspension to a siliconised Corex tube (30 ml).

(iii) Add Triton X-100 to a final concentration of 0.1% and incubate on ice for 10 min whilst gently lysing the protoplasts by repeated pipetting with a plastic disposable Pasteur pipette.

(iv) Check microscopically to see if all the protoplasts have burst then pellet the nuclei by centrifugation at 200 g for 10 min at 4°C.

(v) Proceed as indicated for the isolation of nuclear DNA from leaf/callus tissue from step 5 of *Table 25*, scaling down the volumes as appropriate.

The isolation of chloroplast and mitochondrial DNA is described in *Table 26*. The methods used for the isolation of organelle DNA are very similar to those used for nuclear DNA preparation and usually involve the pelleting of the heavier nuclei after disrupting the plant cells followed by a medium speed spin to harvest chloroplasts and a subsequent higher speed centrifugation of the same extract to collect a mitochondrial pellet (33). After washing and DNase treatment, the DNA is isolated from the organelle pellets in much the same way as the procedure described for nuclear DNA isolation.

4.3 **Preparation of Plant RNA**

A major objective in the isolation of RNA is the quantitative recovery of the pure nucleic acid with a minimum of degradation. The extraction technique relies heavily on the successful inhibition of ribonuclease activity and the efficient disruption of RNA-protein complexes (ribonucleoprotein or RNP). There are many published methods for the isolation of RNA from various plant sources using either;

(i)   Phenol as a protein denaturant;
(ii)   Strong salt solutions (e.g., 4 M guanidinium chloride) to dissolve the RNA and precipitate denatured protein at the same time;
(iii)   Protease solutions to degrade the protein components of RNP (34).

Of these methods the techniques involving phenol are perhaps the most widely used and so in this section a general protocol is presented that can be used with a wide range of plant material. Ribonuclease inhibition is the major concern in RNA isolation and a few basic precautions should be undertaken to avoid degradation of the RNA by extracellular enzymes.

(i)   Glassware must be pretreated by baking in an oven (150°C for 3 h), autoclaving or soaking in 0.5% diethylpyrocarbonate followed by incubation at 100°C for 5 min.
(ii)   Solutions to be used in the extraction procedure should be autoclaved if possible.
(iii)   Extreme cleanliness should be observed in the work area and the researcher should wear disposable rubber gloves to guard against contamination by ribonuclease present in perspiration.
(iv)   RNA should be stored in the presence of a detergent or under ethanol at low temperature ($-20°$ to $-80°C$).
(v)   Only chemicals of the highest purity should be used, e.g., ribonuclease free sucrose is available.

A protocol for making total plant RNA is given in *Table 27*. In this procedure, the volume of homogenisation buffer and weight of plant material is designed for use with fresh leaf material from broad leaved plants (36). It will be necessary to adjust this protocol for use on other plant material, e.g., tissue cultured cells, fruits, etc., to ensure efficient RNA extraction and handling of the extract.

Eukaryotic messenger RNA (mRNA) can be separated from other RNA species in a total RNA preparation by affinity chromatography by virtue of the presence of a polyadenlyic acid 'tail', $20-250$ bases long, at the $3'$ end of the molecule. Both oligo(dT) cellulose or poly(U) Sepharose are used routinely and are available commercially. These consist of polymers of $10-20$ nucleotides (T or U) and will hybridise and bind poly(A)$^+$ containing RNA under high salt conditions providing the 'tail' is at least $15-20$ bases in length. Ribosomal RNA and tRNA will not bind to the poly(A) cellulose or poly(U) Sepharose columns and can be washed through. Bound mRNA can then be eluted with low salt buffers. A protocol to enrich an RNA preparation for poly(A)$^+$ RNA is given in *Table 28*.

A good guide to the purity of an RNA preparation is to test its translation ef-

Table 27. Preparation of Total Plant RNA.

1.	Weigh out fresh plant material into convenient amounts (5 g) and freeze as soon as possible after excision in liquid nitrogen. Store under liquid nitrogen until required.
2.	Take 5 – 10 g of frozen leaf material and grind in a liquid nitrogen cooled mortar to a fine powder. Add liquid nitrogen periodically to keep the tissue frozen.
3.	Add 50 ml of homogenisation buffer[a]. Grind briefly as the buffer thaws and then immediately squeeze the slurry through four layers of sterile muslin (autoclaved) into a 250 ml conical flask.
4.	Transfer the homogenate to sterile 30 ml Corex tubes and centrifuge at 8000 *g* for 10 min at 4°C (8 x 50 ml rotor).
5.	Filter the supernatant through four layers of autoclaved muslin into a sterile 250 ml conical flask.
6.	Add 1/20 volume of 10% SDS to give a final concentration of 0.5%.
7.	Add an equal volume of water saturated phenol. Add an equal volume of chloroform: isosamyl alcohol (24:1) and mix for 20 min on a magnetic stirrer at room temperature using a stirring speed just sufficient to keep the phases emulsified.
8.	Transfer the mixture to sterile 50 ml Pyrex centrifuge tubes and separate the phases by centrifugation in a bench centrifuge at 2500 *g* for 10 min. Denatured protein will collect at the interface between the organic and aqueous phase.
9.	Remove the aqueous phase (top layer) and interface to a clean 250 ml flask and add an equal volume of chloroform. Mix for 10 min and separate the phases by centrifugation and remove only the aqueous phase.
10.	Repeat the chloroform extraction until only a thin disc of protein is visible (usually once is sufficient). Avoiding the denatured protein, remove the aqueous phase to sterile 30 ml Corex tubes. Add 1/20 volume of 3 M sodium acetate followed by two volumes of ethanol and then precipitate out the nucleic acids at $-20°C$ overnight.
11.	Pellet the precipitated nucleic acids by centrifugation at 8000 rp.m. for 20 min at 4°C in a 8 x 50 ml Sorval SS34 rotor (or its equivalent).
12.	Wash the pellet twice in 1 – 2 ml of 3 M sodium acetate pH 6.0 at 4°C (on ice) to remove the ethanol and DNA.
13.	Wash the pellet twice in 80% ethanol containing 0.1 M potassium acetate and store under the same solution at $-20°C$ until required. The quantity of RNA may be determined spectrophotometrically after drying down the pellet and resuspending in distilled water. The optical density should be measured at 260 nm (1 O.D. = 40 μg/ml at 260 nm[b]).

[a]Homogenisation buffer is 0.2 M Tris-HCl pH 8.5, 0.2 M sucrose (RNase free), 30 mM magnesium acetate, 60 mM KCl. This solution should be autoclaved and either used fresh or stored frozen at $-20°C$. Before use add 1% polyvinylpolypyrolidone and 0.31% (v/v) 2-mercaptoethanol.
[b]Scanning spectrophotometric analysis can be used to determine to some extent the purity of the preparation. Clean RNA should have an optical density of 260 – 280 nm ratio of 2.0.

ficiency in an *in vitro* cell-free protein-synthesising system (34). This is especially important if the RNA is to be fractionated further and used for cDNA synthesis. The RNA size may be estimated by running against standards, on a polyacrylamide gel (34).

5. ANALYSIS OF ORGANISATION AND EXPRESSION OF FOREIGN DNA IN PLANT TISSUE

Having transformed plant tissue with foreign DNA by one of the variety of methods described in Section 3, it is now necessary to analyse the organisation and expression of this DNA in the transformed tissue. As methods for such analyses fall into the general domain of molecular biology and have been de-

Table 28. Preparation of Poly(A)$^+$ RNA.

1.	Swell 25 g of oligo(dT) cellulose (Sigma) in sterile binding buffer[a] at room temperature.
2.	Plug the base of a sterile siliconised Pasteur pipette with sterile siliconised glass wool. Load the pipette with a 1 ml column of oligo(dT) cellulose. As a rough guide, 1.0 g of oligo(dT) cellulose will bind 20 − 40 O.D. 260 units of poly(A) mRNA in binding buffer (0.8 − 1.6 mg of RNA).
3.	Wash the column through with five column volumes of sterile distilled water.
4.	Wash the column through with 0.3 M NaOH until the eluent is alkaline (3 − 5 column volumes).
5.	Wash the column with binding buffer (10 column volumes).
6.	Spin down the ethanol precipitated total RNA, pour off the supernatant and dry the pellet under vacuum.
7.	Dissolve the pellet in 0.5 ml of sterile distilled water and heat to 65°C for 4 min. Add an equal volume of 2x binding buffer and allow the sample to cool to room temperature.
8.	Slowly add the RNA solution to the oligo(dT) cellulose column. Wash the column through with 5 − 10 column volumes of binding buffer collecting 1.0 ml fractions in sterile Eppendorf tubes, until the O.D. 260 nm approaches zero.
9.	Wash the column with four volumes of washing buffer[b].
10.	Elute the poly(A) mRNA with five column volumes of low salt eluting buffer[c] and collect 0.5 ml fractions in sterile Eppendorf tubes. Monitor the O.D. 260 nm of each fraction to locate the poly(A) mRNA peak. This can be done conveniently by eluting the column with a peristaltic pump and monitoring continuously the eluent using a spectrophotometer with a microflow cell attachment.
11.	Combine the poly(A) mRNA-containing fraction in a siliconised Corex tube. Add 1/10 volume of 3 M sodium acetate (pH 5.2), two volumes of ethanol and precipitate out the mRNA at − 20°C overnight.
12.	Pellet the precipitated RNA by centrifugation in a swing-out rotor at 8000 *g* for 30 min at 0°C. Carefully decant off the ethanol and then wash the pellet twice with 70% ethanol.
13.	Cover the tube with parafilm, prick several holes in the film and dry under vacuum for 5 min. Resuspend the mRNA in sterile distilled water to a concentration of 0.5 mg/ml and store in aliquots at − 70°C. Alternatively, the mRNA may be stored under 70% ethanol as a precipitate at − 20°C.

[a]Binding buffer is 20 mM Tris-HCl pH 7.5, 0.5 M NaCl, 1.0 mM EDTA, 0.05% SDS. The components (with the exception of SDS) can be mixed from appropriate stock solutions and autoclaved. SDS should be made up as a 10% w/v stock solution and heated to 65°C for 1 h before adding the buffer.
[b]Washing buffer is the same as binding buffer but contains 0.1 M NaCl instead of 0.5 M NaCl.
[c]Low salt eluting buffer is 20 mM Tris-HCl pH 7.5, 1.0 M EDTA, 0.02% SDS.

scribed in detailed elsewhere (15), they will only be outlined here. The organisation of the foreign DNA integrated into the plant genome can best be done by Southern blot analysis using plant DNA material prepared by the rapid mini-prep method discussed in Section 4.1. This will yield information on the copy number of the integrated DNA sequences, whether multiple inserts are tandemly linked or dispersed and on the stability of this DNA in the F1 progeny of transformed regenerated plants. However, it is important to be sure that the plant DNA is prepared from sterile tissue as contamination with *A. tumefaciens* DNA will interfere with the interpretation of the results.

The levels of transcription from the foreign DNA can be estimated by dot blot analysis using poly(A)$^+$ mRNA prepared as described in Section 4.3. This method has recently been used to demonstrate the induction of transcription of a chimaeric gene from the light-inducible promoter of the *ssRubisco* gene by comparing poly(A)$^+$ mRNA from light-grown and dark-grown tissue (13). More detailed studies on levels of transcription can be carried out by Northern blot analysis. The 5' and 3' ends of foreign gene transcripts can be defined by S1 mapping; this is of interest, for example, in defining promoter regions by deletion mutagenesis.

The most direct and immediate approach to studying the expression of foreign genes in transformed plant tissue is by measuring the activity of enzymes encoded by these genes. These enzyme assays include measurements of octopine and nopaline synthase activity and CAT and NPTII activities. In some experiments *npt*II expression will have already been selected for by use of kanamycin. Measurement of enzyme activity provides an estimate of levels of transcription from the appropriate genes but a direct correlation should not be assumed. In situations where the gene product under investigation has no enzymatic activity, expression may be screened by Western blotting, and by standard immunological methods using suitable antibodies.

The repertoire of techniques outlined above has been used on transformed plant tissue, plants regenerated from such tissue and on the F1 progeny of these regenerated plants as has been described recently (30).

5.1 Octopine and nopaline synthase assays

These assays were developed by Otten and Schilperoort (37) and can be used to demonstrate the presence of octopine or nopaline synthase in small amounts of plant tissue.

The reactions are as follows:

arginine + pyruvate + NADH $\overset{ocs}{\rightarrow}$ octopine + NAD

arginine + ketoglutaric acid + NADH $\overset{nos}{\rightarrow}$ nopaline + NAD

Both enzymes actually prefer NADPH to NADH. However, NADPH inhibits these enzymes at high concentrations, in contrast to NADH, and high concentrations have to be used since other enzymes in the tissue extract also use these co-factors. A general protocol is given in *Table 29*, and an assay for octopine synthase is shown in *Figure 11*.

It is possible to assay for both octopine synthase and nopaline synthase together by using a mixture of equal parts of the OCS and NOS assay buffers (see footnotes b and c of *Table 29*). Sometimes it is difficult to see NOS activity owing to endogenous nopaline having already accumulated in the tissue. This nopaline may be removed by passing through Sephadex G100 as follows:

(i)   Take a 0.75 ml microcentrifuge tube and puncture the bottom with a thin needle.

(ii)   Add some siliconised glass beads and fill tube with a slurry of Sephadex G100.

(iii)   Place the small tube into a 1.5 ml microcentrifuge tube and centrifuge at lowest speed for 1 min in a swinging bucket table top centrifuge.

(iv)   Remove run through and add 50 μl of plant extract and repeat centrifugation. The run through contains NOS activity but no nopaline.

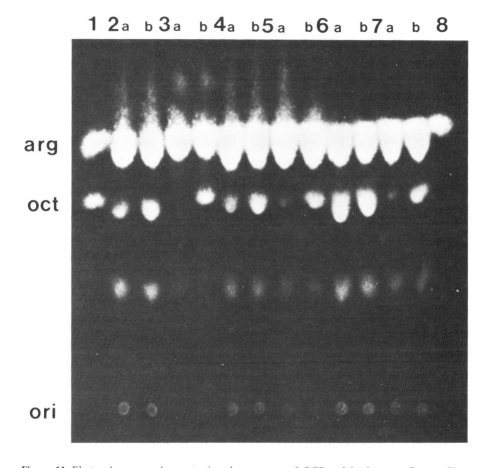

Figure 11. Electropherogram demonstrating the presence of OCS activity in some *P. parodii* + pTiAch5 *in vitro*-transformed colonies. **Lane 1** shows arginine and octopine standards. **Lanes 2−7** are extracts from several different plant tissues. **Lane 2** *P. parodii* + pTiAch5 authentic crown gall. **Lane 3** Non-transformed *P. parodii* tissue. **Lanes 4−7** *P. prodii* hormone independent transformants. **Lane 8** shows an arginine standard. **(a)** Extracts after overnight incubation with 100 mM arginine. **(b)**As **(a)** but with the addition of authentic octopine to the extract. The position of the origin (ori), together with those of the octopine (oct) and arginine (arg) standards, are given at the left-hand side of the figure.

The assay described in *Table 29* has some disadvantages. The concentration of the NADH solution must be checked before use. Some of the components of the octopine synthase reaction mixture condense into a product that comigrates with octopine leading to false positives. A large amount of time is spent preparing and distributing solutions. The following alternative method is based upon the observation that pyruvate or α-ketoglutaric acid rather than arginine are rate limiting for the reaction, and involves measurement of opine accumulation in intact tissue rather than measurement of enzyme activity in crude extract.

(i)   Place a small amount of tissue (at least 10 mg) into a microcentrifuge tube and add 35−50 µl of growth medium (enough to cover about half the

Table 29. Assay for Octopine and Nopaline Synthase.

1.	Put 100 mg or more of plant tissue in a microcentrifuge tube and determine the weight of the tissue. Add an equal weight of extraction buffer[a] and grind plant tissue on ice with a glass rod.
2.	Centrifuge for 2 min in a microcentrifuge in a cold room (4°C) and then put tubes back on ice. Do not keep the extracts on ice for longer than 15 min before testing, as enzyme activities decline rapidly. However, the tissue or extract can be frozen without affecting the enzyme activities.
3.	Add an equal volume of supernatant to OCS[b] or NOS assay[c] buffer (e.g., 5 μl + 5 μl), and incubate at room temperature for 30 – 60 min.
4.	Spot 5 μl of the reaction mix on 3 MM (Whatman) or MN214 (Machereg Nagel, 516 Duren, FRG) paper. Include a lane of methylene green as a marker dye. The samples should be spotted 6 cm from one end of paper cut to a length of 20 cm. Remember to include octopine or nopaline standards (obtainable from Sigma).
5.	Place the paper in an electrophoresis tank with buffer[d] and run at 20 volts/cm (normally 400 volts for 1 h). Stop before the green marker runs off.
6.	Dry the paper thoroughly and stain with phenanthrequinone[e]. Spots are visualized under long wave u.v. illumination. Always hang the paper so that the arginine spot smears out (down) away from the octopine-nopaline spots. It may help to do a second staining which will be indicated by the intensity of the arginine spots. An OCS assay is illustrated in *Figure 11*.

[a]Extraction buffer is made up from a stock solution of 100 mM Tris-HCl pH 8.0, 500 mM sucrose, to which is added 0.1% ascorbic acid and 0.1% cysteine-HCl. The amino acids are added from stock solutions sterilised by filtration. The complete buffer can be stored frozen in 1 ml aliquots.

[b]OCS assay buffer contains 200 mM sodium phosphate pH 6.8, 30 mM arginine-HCl, 75 mM sodium pyruvate, 20 mM NADH. The solution is, however, made up separately as follows. First make RxO buffer which can be stored frozen. RxO buffer is 250 mM sodium phosphate pH 6.8, 40 mM arginine-HCl, 100 mM sodium pyruvate. Mix 75 ml of RxO buffer with 10 ml of 200 mM NADH and 15 ml water to make OCS assay buffer. The NADH solution should be made up fresh from solid stored dry at −20°C. Prepare a small amount of 200 mM NADH in water and determine the true concentration using a spectrophotometer (1 M NADH has an absorbance of 6220 O.D. at 340 nm). Adjust the volume to give the correct NADH concentration.

[c]NOS assay buffer contains 200 mM sodium phosphate pH 6.8, 60 mM arginine-HCl, 60 mM α-ketoglutaric acid, 16 mM NADH. However, first make up RxN buffer containing 250 mM sodium phosphate pH 6.8, 80 mM arginine-HCl, 80 mM α-ketoglutaric acid. Then mix 75 ml of RxN plus 8 ml of 200 mM NADH plus 17 ml of water.

[d]A choice of three electrophoresis buffers exists. 1. 15% acetic acid, 5% HCOOH, 80% water pH 1.8. (Nopaline is near the origin at the positive pole followed by octopine and then arginine.) 2. 50 mM $Na_2B_4O_7$ pH 9.2. (The octopine and nopaline spots are well separated but octopine may comigrate with plant contaminant spots. The order of migration is the same as above.) 3. 50 mM sodium citrate pH 5.0. (Again octopine may comigrate with contaminants. The order of migration from the origin near the positive pole is octopine-nopaline-arginine.)

[e]The staining solution is prepared from 0.02% phenanthrequinone in 95% ethanol and 10% NaOH in 60% ethanol. Stocks must be stored at −20°C. Before use mix equal parts of the stocks and use within 1 – 2 h. Phenanthrequinone is *very poisonous* so work in a fume hood.

tissue and leaving half exposed to air). The growth medium is MSO (*Table 9*) plus 10 mM arginine and either 10 mM pyruvate acid (OCS assays) or 10 mM α-ketoglutaric acid (NOS assays) for analysing onc^+ tissue or MSPI medium for onc^- tissue.

(ii) Incubate overnight in a plant room or at room temperature.

(iii) Remove the tissue from the growth medium, dry on a paper towel and transfer to a fresh tube.

(iv)   Freeze and leave for 10 min at $-20°C$ (to aid crushing) and crush the tissue. Then centrifuge for 5 min in a microcentrifuge.

(v)   Continue from step 4 of *Table 29*.

This method is as reliable and sensitive as the enzymatic assay. Moreover the plant medium can be stored at 4°C for 1 month.

5.2 Chloramphenicol Acetyltransferase Assay

This assay has been used extensively in mammalian transformation systems and is described in detail in Chapter 6 of this volume. Briefly:

(i)   Grind 100 mg of transformed tissue with a glass rod in 100 μl of buffer containing 500 mM sucrose, 250 mM Tris-HCl pH 7.5, 1% ascorbic acid, 0.5 mM leupeptine (Sigma), 10 mM EDTA, 1% cysteine-HCl, 0.5 mM acetyl CoA and 1 μCi[^{14}C]chloramphenicol (50 Ci mmol^{-1} NEN). This can be done in a microcentrifuge tube.

(ii)   Incubate for 30 min at 37°C and centrifuge for 10 min in a microcentrifuge. A suitable positive control is a sonicated extract of *E. coli* containing pBR325 and hence the *cat* gene.

(iii) Extract the reaction mixtures with ethyl acetate and concentrate by evaporation.

(iv)   Subject to ascendant chromatography by spotting the extract on a silica gel thin layer plate with chloroform:methanol (95:5) as eluant to separate acetylated from unacetylated chloramphenicol. These can be visualised by autoradiography.

5.3 Neomycin Phosphotransferase Assay

The best use of *npt*II gene is as a dominant selectable marker to select for plant cell transformants on the basis of kanamycin resistance. Using high resolution gel filtration of a plant cell crude extract Herrera-Estrella *et al.* (7) have shown that transformed plant tissue expressing kanamycin resistance does indeed express an enzyme capable of phosphorylating kanamycin that has the same elution profile as an *E. coli* control extract containing the *npt*II kn^r gene.

Recently a new rapid method has been developed for the detection and quantification of low amounts of NPT in crude cell extracts (41). The assay is based on the electrophoretic separation of the enzyme from the interfering proteins and detection of its enzymatic activity by *in situ* phosphorylation of the antibiotic kanamycin. Kanamycin and [α-^{32}P]ATP, the substrate of the NPT enzyme, are embedded in an agarose gel containing the separated proteins. After the enzymatic reaction, the phosphorylated kanamycin is transferred to P81 phosphocellulose ion exchange paper and the radiolabelled kanamycin visualised by autoradiography. This assay can detect 1 ng of active enzyme from both prokaryotic and eukaryotic sources and can demonstrate changes in size of enzymatically active proteins. The NPT assay is detailed in *Table 30*.

Table 30. Assay for Neomycin Phosphotransferase.

1. Extract 100 mg of plant tissue in 20 μl of buffer A[a] by crushing in microcentrifuge tubes[b].
2. Add 1/10 volume of an extract of 1 g crown gall tissue in 350 μl of buffer B[c] previously incubated at 4°C for 1 h.
3. Centrifuge to remove solids, add bromophenol blue to 0.002% and load into the pockets (40 μl) of a 10% non-denaturing polyacrylamide gel[d]. As the enzyme will run near the dye front do not run the dry front off the bottom.
4. After electrophoresis rinse the gel twice in water and equilibrate in buffer C[e].
5. Transfer the gel to a glass plate and cover it with a 1.5 mm thick layer of 1% agarose containing kanamycin sulphate and [γ-^{32}P]ATP[f]. Incubate 30 min at room temperature.
6. Place one sheet of Whatman P81 on top of the gel followed by two 3MM papers and a stack of filter papers. Transfer for $1-3$ h.
7. Remove the P81 paper from the sandwich taking care to remove the adhering agarose and wash five times in hot water (80 $-$ 90°C) for 5 min each wash.
8. Wrap the paper in 'Cling Film' or 'Saranwrap' and expose to X-ray film overnight or up to one week using intensifying screens at $-$ 70°C.

[a]Buffer A is 40 mM EDTA, 150 mM NaCl, 100 mM NH$_4$Cl, 10 mM Tris-HCl pH 7.5 containing 2.31 mg/ml DTT, 0.12 mg/ml leupeptin and 0.21 mg/ml trypsin inhibitor.
[b]Remember to include controls of uninfected plant and an extract of the NPT enzyme prepared from *E. coli*. The *E. coli* extract is prepared as follows. Grow the cells in 20 ml LB to a density of 4 x 10^8 cells/ml. Centrifuge and resuspend in 50 μl buffer A. Sonicate the cells to obtain a crude extract and centrifuge to remove cellular debris. The supernatant can be added at various dilutions (in buffer A) to the crown gall tissues in buffer B prior to loading on the gel. Coomassie blue staining of a 10% non-denaturing polyacrylamide gel of the *E. coli* sonic extract together with known standards will give an estimation of the amount of NPT enzyme in the extract.
[c]Buffer B is 30 mM NaCl, 15 mM NH$_4$Cl, 3 mM MgCl$_2$, 5 μM EDTA, 3 mM Tris-HCl pH 7.5, 0.031 mg/ml DTT, 36 mg/ml sucrose.
[d]The 10% non-denaturing polyacrylamide gel is made from the following components. A. *Separating gel*. 10 ml acrylamide 29%, bisacrylamide 1%, 12.2 ml H$_2$O, 7.5 ml Tris-HCl pH 8.8 1.5 M, 10 μl TEMED, 200 μl ammonium persulphate 10%. B. *Stacking gel*. 1.5 ml acrylamide 29%, bisacrylamide 1%, 1.25 ml Tris-HCl pH 6.8 1 M, 7.0 ml H$_2$O, 10 μl TEMED, 30 μl ammonium persulphate 10%. A 50x stock of electrophoresis buffer is made by dissolving 288 g of glycine and 60 g Tris base in 400 ml water.
[e]Buffer C is 67 mM Tris-malate pH 7.1, 42 mM MgCl$_2$, 400 mM NH$_4$Cl.
[f]The agarose gel overlay is made by mixing 15 ml of buffer C, 15 ml of molten (45$^{\cup}$C) 2% agarose in water, 100 μCi of [γ-^{32}P]ATP (5$-$7000 Ci/mmol), and 40 μl of kanamycin sulphate (25 mg/ml) at 45°C. This will be enough to cover a 200 cm^2 gel and should be scaled up accordingly for larger gels. There is no improvement of the sensitivity of the assay when kanamycin concentration is increased beyond 50 μl. The use of more ATP does increase the sensitivity.

5.4 **Dihydrofolate Reductase Assay**

Recently it was reported that the cauliflower mosaic virus (CaMV) can be used as a vector to transfer small pieces of DNA to plants (42). Two regions of the CaMV genome, open reading frames (CORF$_s$) II and VII do not seem essential for infection. Thus Brisson *et al.* (42) replaced the CaMV ORF II by the R67 plasmid-encoded dihydrofolate reductase (DHFR) gene (*dhrf*) conferring resistance to trimethoprim in *E. coli* and methotrexate in eukaryotes. The chimaeric viral DNA was shown to be stably propagated in turnip plants where the *dhfr* gene was expressed to produce a functional enzyme. Functional enzyme activity was demonstrated using a rather elegant assay (*Table 31*) based upon incorporation of ^{32}P into DNA in the presence of methotrexate.

Table 31. Assay for Dihydrofolate Reductase.

1. Incuabte 5 g of plant tissue in 5 ml of phosphate-free MSO (for *onc*[+] tissue) or MSPI (for *onc*[−] tissue) medium (*Table 9*) supplemented with methotrexate (Sigma) at 0.1 mg/ml for 22 h at 25°C in the dark shaking gently on a gyrotary shaker[a].

2. Remove the medium and replace with fresh medium containing methotrexate and 35 μCi of [^{32}P]orthophosphate (Amersham). Then continue incubation for a further 22 h.

3. Remove the medium and wash the tissue three times in 15 ml phosphate buffer before freezing in liquid nitrogen.

4. Homogenise the tissue in 0.5 − 1.0 ml of 50 mM NaCl, 10 mM EDTA, 1% sarkosyl, 10 mM Tris-HCl pH 7.5. Extract twice with phenol.

5. Precipitate the nucleic acids with isopropanol and resuspend in 50 μl DNA buffer (10 mM Tris-HCl pH 7.5, 0.1 mM EDTA, 5 mM NaCl) containing 20 μg/ml RNase.

6. After incubation at 37°C for 1 h, add 450 μl of proteinase K (200 μg/ml) in 0.1% SDS and incubate for a further 1 h.

7. After two phenol extractions and isopropanol precipitation, resuspend the pellets in 50 μl DNA buffer and electrophorese on 1% agarose gel.

8. Stain with ethidium bromide and photograph to show plant DNA then place gel with X-ray film to obtain an autoradiograph showing incorporation of ^{32}P into DNA.

[a]Remember to include uninfected tissue incubated with and without methotrexate to show inhibition of DNA synthesis, in addition to the infected plant tissue similarly incubated with and without methotrexate.
[b]The relative intensity of the DNA band will give an indication of the amount of DNA prepared from each plant tissue sample. This can be calculated more accurately using the diphenylamine test (*Table 24*).

6. FUTURE PROSPECTS

Currently the major obstacles to plant genetic engineering for crop improvement are good tissue culture systems and plant regeneration. In addition the problem remains of ensuring the stable integration and expression of transferred genes at the correct stage of development. For the latter a more detailed understanding of plant promoters is required.

The problem also remains of knowing what genes to clone in order to improve plants. We are currently limited to discrete genes that encode storage proteins to improve protein quality of a plant or encode enzymes that degrade herbicides. Complex plant traits like salt tolerance or nodulation for nitrogen fixation are not yet sufficiently understood to enable us to clone the gene or often cluster of genes encoding such traits from one plant species to another. We also need to understand the molecular basis of host pathogen interaction if we wish to clone the gene(s) responsible for pathogen resistance.

Often a crop improvement could be made by removing genes not required that reduce yield by acting as a drain on protein synthesis. As it is not always possible to select for such mutants or to screen variants at the tissue culture level we would like to be able to replace wild-type alleles with cloned mutated ones *via* transformation and homologous recombination.

7. ACKNOWLEDGEMENTS

We would like to thank Ann Depicker for sending us updated opine synthase assays and co-infection protocols developed at the Gent laboratory and for valuable suggestions. We also thank Chris Smith for the protocol on RNA purification and Erik Wiemer who helped to develop the small scale Ti plasmid DNA preparation. Thanks are also due to Luis Herrera-Estrella for providing a restriction map of pLGVneo1103 prior to publication and to Jurek Paszkowski for supplying the protocol for the NPT assay adapted for plant tissue. Finally we thank Glyn Millhouse for photographic services and Ellen Rosier for typing the manuscript.

8. REFERENCES

1. Kosuge,T., Meredith,C.P. and Hollaender,A. eds. (1983) *Genetic Engineering of Plants: An Agricultural Perspective,* published by Plenum Press, New York.
2. Caplan,A., Herrera-Estrella,L., Inze,D., Van Haute,E., Van Montagu,M., Schell,J. and Zambryski,P. (1983) *Science (Wash.), * **222,** 815.
3. Akiyoshi,D.E., Klee,H., Amasino,R.M., Nester,E.W. and Gurdon,M.P. (1984) *Proc. Natl. Acad. Sci. USA,* **81,** 5994.
4. Schröder,G., Waffenschmidt,S., Weiler,E.W. and Schröder,J. (1983) *Eur. J. Biochem.,* **138,** 387.
5. Wang,K., Herrera-Estrella,L., Van Montagu,M. and Zambryski,P. (1984) *Cell,* **38,** 387.
6. Stachel,S. (1984) submitted for publication.
7. Herrera-Estrella,L., De Block,M., Messers,E., Hernalsteens,J.-P., Van Montagu,M. and Schell,J. (1983) *EMBO J.,* **2,** 987.
8. Hooykaas-Van Slogteren,G.M.S., Hooykaas,P.J.J. and Schilperoort,R.A. (1984) *Nature,* **311,** 763.
9. Zambryski,P., Joos,H., Genetello,C., Leemans,J., Van Montagu,M. and Schell,J. (1983) *EMBO J.,* **2,** 2143.
10. Herrera-Estrella,L., Depicker,A., Van Montagu,M. and Schell,J. (1983) *Nature,* **303,** 209.
11. Horsch,R.B., Fraley,R.T., Rogers,S.G., Sanders,P.R., Lloyd,A. and Hoffmann,N. (1983) *Science (Wash.)* **237,** 496.
12. Bevan,M.W., Flavell,R.B. and Chilton,M.-D. (1983) *Nature,* **304,** 184.
13. Herrera-Estrella,L., Van den Broeck,G., Maenhaut,R., Van Montagu,M., Schell,J., Timko,M. and Cashmore,A. (1984) *Nature,* **301,** 115.
14. Hererra-Estrella,L., ... (1984) manuscript in preparation.
15. Maniatis,T., Fritsch,E.F. and Sambrook,J. (eds.), (1982) *Molecular Cloning: A Laboratory Manual,* published by Cold Spring Harbor Laboratory Press, New York.
16. Zambryski,P., Herrera-Estrella,L., De Block,M., Van Montagu,M. and Schell,J. (1984) in *Genetic Engineering, Principles and Methods,* Vol. **6,** Setlow,J. and Hollaender,A. (eds.), Plenum Press, New York, p. 253.
17. Koncz,C., Kreuzaler,F., Kalman,Zs. and Schell,J. (1984), *EMBO J.,* **3,** 1029.
18. Van Haute,E., Joos,H., Maes,M., Warren,G., Van Montagu,M. and Schell,J. (1983) *EMBO J.,* **2,** 411.
19. Klapwijk,P.M. (1979) Ph.D. Thesis University of Leiden, Holland.
20. Garfinkel,D.J., Simpson,R.B., Ream,L.W., White,F.F., Gordon,M.P. and Nester,E.W. (1981) *Cell,* **27,** 143.
21. Currier,T.C. and Nester,E.W. (1976) *J. Bacteriol.,* **126,** 157.
22. Hoekema,A., Hirsch,P.P., Hooykaas,P.J.J. and Schilperoort,R.A. (1983) *Nature,* **303,** 179.
23. Figurski,D.H. and Helinski,D.R. (1979) *Proc. Natl. Acad. Sci. USA,* **76,** 1648.
24. Marton,L., Wullems,G.J., Molendijk,L. and Schilperoort,R.A. (1979) *Nature,* **277,** 129.
25. Vasil,I.K. (ed.) (1980) *International Review of Cytology, Supplements 11A and 11B 'Perspectives in Plant Cell and Tissue Culture',* published by Academic Press.
26. Steinbiss,H.H. and Broughton,W.J. (1983) in *Int. Rev. Cytol.,* Supplement 16 *'Plant Protoplasts',* Giles,K.L. (ed.), Academic Press, p. 191.
27. Draper,J. and Davey,M.R. (1984) in *Plant Cell Culture Technology,* Yeoman,M.M. (ed.), Blackwells, Oxford, in press.
28. Draper,J., Davey,M.R., Freeman,J.P., Cocking,E.C. and Cox,B.J. (1982) *Plant and Cell Physiol.,* **23,** 451.
29. De Greve,H., Leemans,J., Hernalsteens,J.P., Thia-Toong,L., De Beuckeleer,M., Willmitzer, L., Otten,L., Van Montagu,M. and Schell,J. (1982) *Nature,* **300,** 752.

30. De Block,M., Herrera-Estrella,L., Van Montagu,M., Schell,J. and Zambryski,P. (1984) *EMBO J.*, **3**, 1681.
31. Murray,M.G. and Thompson,W.F. (1980) *Nucleic Acids Res.*, **8**, 4321.
32. Hadlaczky,G., Bisztray,G., Praznovszky,T. and Dudits,D. (1983) *Planta*, **157**, 278.
33. Chilton,M.D., Saiki,R.K., Yadav,N., Gordon,M.P. and Quetier,F. (1980) *Proc. Natl. Acad. Sci. USA*, **77**, 4060.
34. Taylor,J.M. (1979) *Annu. Rev. Biochem.*, **48**, 681.
35. Slater,R.J. (1983) in *Techniques in Molecular Biology*, Walker,J.M. and Gaastra,W. (eds.), Croom Helm Ltd., Beckenham, UK, p. 113.
36. Smith,C., personal communication.
37. Otten,L.A.B.M. and Schilperoort,R.A. (1978) *Biochim. Biophys. Acta*, **527**, 497.
38. Vieira,J. and Messing,J. (1982) *Gene*, **19**, 259.
39. Murashige,T. and Skoog,F. (1962) *Physiol. Plant*, **15**, 473.
40. Caboche,M. (1980) *Planta*, **149**, 7.
41. Reiss,B., Sprengel,R., Will,H. and Shaller,H. (1984) *Gene*, in press.
42. Brisson,N., Paszkowski,J., Penswick,J.R., Gronenborn,B., Potrykus,I. and Hohn,T. (1984) *Nature*, **310**, 511.

CHAPTER 5

P Element Mediated Germ Line Transformation of Drosophila

ROGER E. KARESS

1. INTRODUCTION

The *Drosophila* genome is the most thoroughly understood of any metazooan. The great progress in elucidating the mechanisms of development and gene ex- pression in *Drosophila* has been a consequence of the investigator's ability to think about problems in two dimensions, combining both genetic and molecular approaches. The advent of germ line transformation has now added yet a third dimension to our thinking, and thereby vastly broadened our horizons of exploration. Indeed, in the short time since Rubin and Spradling (1,2) published their method of introducing new genes into *Drosophila* with P element vectors, our knowledge of the signals and regulators affecting the spatial and temporal control of expression, dosage compensation, and position effects, has grown immensely.

A partial listing of the genes (or derivatives of genes such as promoter sequences) that have been introduced and expressed in *Drosophila* by P mediated transformation includes *rosy* (1), *white* (3), *actin* (4), larval serum protein genes (5), yolk protein genes (6), *dopadecarboxylase* (7), chorion protein genes (1,8), *alcohol dehydrogenase* (9) and heat shock protein gene *hsp70* (10). In addition, three bacterial genes have been expressed in *Drosophila* under the control of *Drosophila* promoters. These are *chloramphenicol acetyl transferase (CAT)* (5), the *E. coli lacZ* gene (8,10), and *neomycin resistance (neo^R)* (11).

2. BACKGROUND

The transformation procedure exploits the properties of P elements, and their behavior during hybrid dysgenesis, which are discussed below.

2.1 P Elements and Hybrid Dysgenesis

Only the most cursory description of P elements and their role in hybrid dysgenesis is offered here. The reader is directed to several excellent reviews published recently (12,13) for a thorough treatment of this subject.

P-M hybrid dysgenesis is the name given to a syndrome of correlated genetic aberrations that are observed in the F$_1$ offspring (the so-called dysgenic hybrid) of matings between males of certain *Drosophila* strains, called P strains, and females of strains designated M strains. These aberrations include gonadal sterility, high rates of mutation, chromosomal rearrangements and male recombination. Importantly, they are limited almost exclusively to the germ line of the F_1.

P elements are the mediators of P-M hybrid dysgenesis. They are a family of mobile genetic elements, present in all P strains, and absent from most (but not all) M strains. There are two broad classes of P elements, non-defective and defective. Non-defective elements appear to encode at least two activities, a transposase, required for their own transposition and for the transposition of the defective elements; and a factor restricting the transposition of P elements, called the cytotype determining factor. When the elements are stable, they are said to reside in the P cytotype, the cellular environment of P strain flies. The nature of P cytotype is unclear, but one can think of it as the presence of a repressor molecule somewhat like that of bacteriophage lambda. Only when functional P elements carried by the sperm are introduced into an egg of the M cytotype (the cellular environment of M strain flies, which lack functional P elements), are they mobilised to transpose, (just as during zygotic induction a lysogenic lambda phage is suddenly derepressed when introduced into a bacterial cell lacking repressor). Normally this occurs in the cross between P strain males and M strain females, the dysgenic cross. In the reciprocal cross, between an M strain male and a P strain female, or in a P x P cross, the P elements remain in the P cytotype of the egg, and no hybrid dysgenesis is observed. While the inheritance and function of cytotype are of great interest, they do not directly concern us here.

2.2 Structure of P Elements

A number of P elements were cloned and sequenced by O'Hare and Rubin (14). They identified two classes of elements: large 2.9 kb P elements (the non-defective, autonomously transposing elements), and smaller (0.5 – 1.6 kb) defective elements, which appear to have been derived from the larger elements by internal deletion of sequences. The 2.9 kb element has four large open reading frames on a single strand (*Figure 1*), which appear to encode a single polypeptide carrying transposase activity (15). It is flanked by inverted repeats of 31 bp, whose integrity are required in *cis* for transposition of the element (15,16). The smaller, defective P elements are missing some sequences comprising the open reading frames and thus cannot encode the transposase, but by retaining the inverted repeats, they are still competent to transpose if the transposase is supplied from an intact 2.9 kb element.

2.3 P Elements and Germ Line Transformation

The functional integrity of the 2.9 kb P element, and the feasibility of P element mediated germ line transformation, were demonstrated directly by Spradling and Rubin (2). They anticipated that a cloned, intact P element, if introduced into the germ cells of a developing M strain embryo, would mimic the condition of the dysgenic hybrid, in which P elements carried by the sperm are suddenly introduced into a cellular environment of the M cytotype. They found that the 2.9 kb element (carried on a bacterial plasmid) could transpose from the plasmid and integrate into the chromosomal DNA of the fly's germ line. Reasoning that the

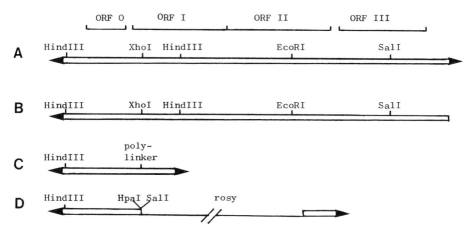

Figure 1. Schematic structure of P elements used in transformation. (**A**) The intact, autonomous 2.9 kb element, such as that present in plasmid pπ25.1, as determined by O'Hare and Rubin (14). The four open reading frames (ORFs) are indicated above. Some landmark restriction endonuclease sites are also shown. (**B**) The 'wings-clipped' P element of plasmid pπ25.7wc. Note that the right hand copy of the inverted repeat has been deleted. (**C**) The basic structure of the Carnegie series of P element vectors, with the cloning region (containing unique restriction nuclease cleavage sites) indicated. (**D**) The Carnegie 20 vector, which contains both a selectable transformation marker (the *rosy* gene) and a few unique restriction cleavage sites for cloning the desired DNA fragment.

transposase supplied by the intact, autonomous element could act as well on a defective element, they repeated the experiment, this time co-injecting the 2.9 kb element and a defective P that had been genetically engineered to carry within it the wild type *rosy* (*ry*) gene (the structural gene for the enzyme xanthine dehydrogenase, a component of the synthetic pathway of eye pigments). The embryos receiving this mixture of defective and non-defective P elements were homozygous *ry* mutants. Among the offspring of some of the injected embryos, a fraction displayed the rosy[+] phenotype, and were subsequently shown to have stably integrated P[*rosy*] transposons in their genomes. That no bacterial plasmid sequences were found associated with the integrated transposons suggested that the transposition was mediated by the inverted repeats of the P elements.

3. STRATEGIC CONSIDERATIONS IN PREPARING FOR TRANSFORMATION

3.1 The Need for a Genetic Marker

The importance of having a phenotypically selectable marker encoded by the transposed DNA cannot be overemphasised. Had Rubin and Spradling examined the offspring of their injected flies by DNA blot hybridisation for evidence of integrated P elements, it is doubtful whether the transformants would have been detected. As it was, they selected transformants on the basis of phenotypic changes induced by the new gene carried by the P element transposon. They were thus able to detect the transformed offspring of an injected fly, even when transformants represented less than 0.5% of the total progeny of that fly (1).

The researcher needs to decide whether the piece of DNA he wishes to transfer will confer an easily scorable or selectable phenotype to the recipient fly. If it does, he might want to directly select for the transformants which have acquired the new phenotype. Often, however, one will be examining a modified gene, or portion of a gene (for example a bit of promoter sequence), where expression is not guaranteed. In these instances, or if the sequence of interest confers no easily recognisable phenotype, the DNA fragment must be linked in *cis* to a gene that will confer such a visible or selectable phenotype. Any gene conferring a visible phenotype in a single dose (i.e., a dominant phenotype) can serve as a marker for transformation. A wild type allele of a gene in an otherwise homozygous mutant stock, can play this role. The two markers that have been used most extensively for this purpose are the wild type structural genes *rosy* (*ry*) and *alcohol dehydrogenase* (*Adh*). More recently, a bacterial drug resistance gene *neo*[R], linked to a *Drosophila* promoter, has been successfully employed as a transformation marker (11). Each of these markers has advantages and disadvantages, which will be discussed below.

3.2 **Markers for Transformation**

3.2.1 *The rosy Gene (the Structural Gene for Xanthine Dehydrogenase)*

The *rosy* gene (*ry*, chromosome 3R, locus 52.0) has been cloned by Bender *et al.* (17). It was the first gene successfully introduced into the *Drosophila* genome by P-mediated transposition, and is still the marker of choice for many applications. The host fly strain can be homozygous for any one of several *ry* mutations, e.g., *ry*[42]. [These, and most other standard fly stocks are described in Lindsley and Grell (18).] Such flies have dark crimson, almost brown eyes, instead of the wild-type brick-red color, a distinction easily identifiable at low magnification. A P[ry] element containing the wild type *rosy* sequence on a 7.3 kb *Hind*III DNA fragment, restores the wild type eye color to the transformant. Even very low levels of *ry* gene activity (1 – 5% of wild type levels) still result in a visible phenotypic change to a nearly wild type eye color in a *ry*[−] stock. Thus, even if the P[ry] element integrates into a transcriptionally unfavourable region of the genome (such as the heterochromatin), the fly bearing that element may still be recognisable as a transformant. Likewise, *ry* is not cell-autonomous, so should it integrate into a chromosomal milieu that restricts its expression only to certain cell types, it would still confer ry[+] phenotype to the eyes. Finally, there exist good third chromosome balancers such as *TM3* (8,15) that carry EMS-induced *ry* alleles, and as such are particularly useful in characterising the transformants and balancing the chromosome bearing the integrated element. (The subject of balancers and their uses will be discussed more fully in Section 7.1.)

One disadvantage to using *rosy* as the transposon marker is that the 7.3 kb *Hind*III fragment can be cleaved by most of the common restriction endonucleases that have hexanucleotide recognition sequences (*Sal*I, *Hpa*I, and *Kpn*I being notable exceptions). This can make manipulations of plasmids containing *ry* a little difficult. Also, the extent of the gene, its transcript and regulatory sequences, have not yet been fully determined.

3.2.2 *Alcohol Dehydrogenase (Adh)*

The *Adh* gene (chromosome 2L, locus 50.1) has been cloned and well characterised (19 − 21). It was also one of the first genes successfully introduced into *Drosophila* (9). Again, numerous stocks of *Adh⁻* mutants are available, which can be used as host strains. The phenotype of the mutant is sensitivity to ethanol. A 6% solution of ethanol is toxic to an *Adh⁻* fly, but a single wild-type allele is sufficient to provide complete protection. One can therefore select for transformants on medium containing ethanol without having to inspect each of the offspring individually. This greatly reduces the time spent tending flies. There are good second chromosome balancers, such as *CyO*, that carry *Adh⁻* alelles (9). The functional gene is found within a 3.5 kb *Xba*I fragment, which has no cleavage sites for *Eco*RI, *Sac*I, *Bgl*II, and *Pvu*I, and is thus relatively convenient to manipulate.

However, since there is no visible phenotype to score, the subsequent mapping and crossing of the *Adh* transformant is slightly more involved than it would otherwise be. Moreover, the selection for ethanol resistance may result in the loss of transformants expressing the *Adh* gene at a low level.

3.2.3 *Neomycin Resistance*

In time, many other markers will of course become available. As each new gene is cloned and successfully reintroduced back into the *Drosophila* genome, it becomes a potential marker for use in subsequent gene transfer studies. Recently, a P element vector has been constructed that contains the bacterial neomycin resistance gene (*neo*ᴿ) under the control of a *Drosophila* heat shock promoter (11). Transformants bearing this element in their genomes are rendered resistant to the drug G418 when present in the food at a concentration of 1 mg/ml, which is otherwise toxic to *Drosophila* larvae. This transposon marker has the distinct advantage of being selectable in virtually any genetic background, including wild type. Thus no special mutations are required in the host strain. However, it suffers from the same limitations as does *Adh*. Transformants expressing the gene at a low level may be missed by the screen, and subsequent manipulations relying on the phenotype require chemical selection rather than visual inspection.

If one is interested only in the transformants that express the transduced genes at a 'normal' level, then the drug resistance selection for transformants, using either *Adh* or *neo*ᴿ, is probably the method of choice. But if the recovery of every transformant is the goal, including any that might be aberrantly expressed (for example, due to position effect), then selection for *rosy* is the way to go.

3.3 **P Element Vectors**

The investigator, having selected a suitable transposon marker, must clone the gene of interest in a P element vector plasmid. These are derivatives of naturally occurring defective P elements that have been cloned in bacterial vectors. They retain all the sequences required in *cis* for transposition, but in addition contain convenient restriction endonuclease cleavage sites for cloning the desired DNA fragment between the inverted repeats of the element. The sequences required in *cis* for a P element to be transposition-competent have not been fully delimited.

They do include the inverted repeats, but probably extend beyond them, at least on the left hand side (see reference 16). Naturally occurring mobile defective P elements bearing deletions within 100 bases of the left end have not yet been found (14).

A number of P element vectors have been described (16) containing synthetic polylinkers inserted between the two inverted repeats. These are the Carnegie series, which have unique cleavage sites for *Pst*I, *Sal*I, *Bam*HI, *Eco*RI, among others (*Figure 1C*). The backbone of these vectors is the pUC8 plasmid. Carnegie 20 (*Figure 1D*) is a P element vector that already carries the 7.3 kb *Hind*III fragment containing the *ry* gene. The cloning region of this vector has a unique *Sal*I and a unique *Hpa*I site, into which can be inserted the DNA fragment under study. More recently *neo*^R containing vectors have been described (11).

The current nomenclature (22) for a recombinant P element uses square brackets to represent the inverted repeats of the element. The construction between the repeats is represented by symbols within the square brackets. Thus, Carnegie 20 may also be represented as P[*rosy*]; and a transposon carrying *Adh* is written P[*Adh*].

3.4 The Need for a Helper Element

As mentioned in Section 2.3, the defective recombinant P transposon will only transpose in the presence of a trans-acting substance supplied by a non-defective P element. The original helper element employed in the transformation procedure was the 2.9 kb element of clone pπ25.1, one of the two intact P elements described and sequenced by O'Hare and Rubin (14). However, because it was an intact, autonomous element, it could itself transpose into the recipient fly genome along with the desired recombinant P element. A double integration of both the defective and non-defective P elements into a single fly genome creates an environment in the fly equivalent to that of a dysgenic hybrid. That is, the recombinant P element is genetically unstable and will continue to excise and transpose to new sites within the genome. Unstable transformants are difficult to characterise properly; repeated crossing and selection is necessary to genetically remove the integrated helper element from the genome. To minimise the risk of a double integration, the transformation procedure has called for the mixture of plasmids containing the desired recombinant P element and the helper element to be injected at a molar ratio of about 5:1 (1).

An improved helper element was made in the course of experiments by Karess and Rubin (15). This element, carried by plasmid pπ25.7wc (and usually called 'wings-clipped'), avoids these problems. While it is able to supply the transposase, it cannot itself transpose, because 23 bases of the 31 base right-hand inverted repeat have been removed. With only a single functional inverted repeat, the wings-clipped P element is rendered immobile. It has proved to be just as effective a helper element as pπ25.1, and to date has never been found integrated in a recipient fly genome.

The optimal defective-to-helper ratio needs to be reassessed, now that wings-clipped is available as the helper element. Unfortunately, no proper study to determine the optimal ratio of recombinant to helper has yet been undertaken. (See Section 4.6 for discussion of the ratios that have been employed.)

3.5 Selecting the Fly Strain for Transformation

There are two basic requirements in choosing the strain to receive the P element by injection (called here the 'host strain'). Firstly the flies must be M strains, and preferably M strains lacking any sequences that hybridise with P elements. Secondly, the flies must have a phenotype that will allow the identification of an individual expressing a gene carried by the recombinant P transposon (see also Section 3.2).

While most standard laboratory stocks are in fact M strains, there are examples of some that do retain a few (or in some cases many) defective P elements in their genomes (23,24). These elements might reduce the efficiency of transformation by competing with the desired recombinant P element for the transposase, and they would also be troublesome in the subsequent molecular analyses of the transformants. The easiest way to test for the presence of P elements in a fly stock that is being contemplated for use as a potential host is by Southern blot analysis (28).

The transformed fly must have a different phenotype from its untransformed siblings. If the gene of interest confers a visible phenotype, and one is confident about the integrity of the gene's structure in the P element vector, then it can be introduced directly into a host strain homozygous for a recessive mutation in the gene under study. The basis of the screen for transformants then becomes the restoration of the wild-type phenotype. In cases where the P element construction confers no visible phenotype, or no assured phenotype (such as a study of promoter activity in a series of deletion mutants), the transformant is selected on the basis of a marker gene such as *Adh* or *rosy*, and the host strain must carry mutations in the corresponding endogenous gene. If selection is based on the drug resistance that the *neo*R gene confers, then the host strain may be of any convenient genetic composition, including wild type.

4. PREPARING MATERIALS FOR MICROINJECTION

4.1 Materials for Microinjection

Dissecting microscope, magnification about 10 – 30 x.

Inverted compound microscope, magnification about 150 – 200 x, equipped with a cold or shielded light source. A cooled stage, or a room maintained at 18°C is desirable, but is not essential.

Micromanipulator and microinstrument holder for holding and positioning the needle. This need not be a fancy model, as the movements required are on the order of tens of micrometers.

Microscope slides.

Coverslips.

Watchmakers' forceps.

Household bleach (sodium hypochlorite).

Double sided sticky cellophane tape (Scotch or Sellotape).

Electric needle puller, such as the vertical pipet puller model 700C from David Kopf Instruments, Tujunga, CA, USA.

Microcapillaries, preferably with inner filament. Available from W.P. Instruments, 60 Fitch St., PO Box 3110, New Haven, CT 06515, USA, Cat. number 1B120F; or from Clark Electromedical Instruments, Box 8 Pangbourne, Reading RG8 7HU, UK, Cat. number GC100F-10. These should be cleaned with filtered absolute ethanol, and dried.

Drierite, calcium chloride, silica gel, or similar desiccant.

Plastic syringe, 25 ml or 50 ml.

Plastic tubing, connecting the syringe with the microinstrument holder and needle.

Disposable Petri dishes.

Filter paper.

Halocarbon oil, series 700 from Halocarbon Products, 82 Burlews Court, Hackensack, NJ, USA; or Voltalef oil grade 10S from Produits Chimique Ugine Kuhlmann, Direction Commerciale: Cedex 21 - 92087 Paris-La Defence, France.

Egg collection chambers, either bottles or small cages.

Egg collection trays, grape juice-agar, or molasses-agar plates (~ 2.5% agar).

4.2 Collecting Staged Embryos

A complete description of the *Drosophila* life cycle and techniques used in *Drosophila* husbandry can be found in reference 25.

Healthy stocks of the host strain will assure a good supply of eggs. Young females (3 − 7 days old) will lay the greatest number of healthy eggs. This is particularly true of *ry* stocks, whose fecundity declines rapidly in flies older than a week or so.

Maintain newly eclosed flies on fresh media, supplemented with a generous supply of yeast paste for 2 − 3 days before beginning egg collections. Flies can be kept in a small population cage, into which collection trays are inserted. Alternatively, a few hundred flies may be kept in a bottle with two openings. An egg collection dish covers one opening, and a cotton plug is inserted in the other to supply adequate ventilation. The bottle top can be covered with a silk mesh, and then inverted upon a dish of food. The females will lay their eggs through the mesh, and the bottle can be easily transferred to a fresh plate for the next collection. It is important to keep these collection bottles clean. As the flies soil the walls with their excreta and with yeast carried by their feet, the walls become more desirable places for oviposition, and will compete with the collection dish.

Egg collection trays can be made of simple grape juice and agar, or molasses and agar, with a little fresh yeast paste dabbed in the center. Since a rough surface can stimulate oviposition, the smooth agar may be lightly scored with a needle, to produce little trenches which become preferential sites for egg-laying, and so facilitate subsequent collecting of the eggs.

The females will lay according to their circadian rhythms, which are often out of phase with those of the investigator. The peak of egg-laying activity is supposed to be in the early evening hours, although this may vary with the particular strain. A reversed light-dark cycling incubator should help entrain the flies to a more accomodating cycle. The flies should, however, be raised in this cycle throughout their life, since rhythms can be established as larvae, and persist into adulthood.

To begin the egg collection, present the flies with a fresh plate containing a small dab of yeast, and, keep them in a dark, quite place at room temperature. Transformation requires that the DNA be injected into early cleavage embryos (0 − 2 h old, at 25°C), before cellularisation of the nuclei has occurred. Generally one-hour egg collections will assure an adequate level of synchrony. But the first collection of the day should be discarded, since the females may store their fertilised eggs for several hours before oviposition, and these will be too far advanced to be of use. One hour is also the approximate time required for one worker to dechorionate, align, desiccate, and inject about 50 − 75 embryos on a coverslip, and still leaves a few minutes for refreshment. Sometimes non-laying flies can be induced to lay by first starving them for yeast (that is, presenting them with only a grape juice-agar plate), and then restoring their yeast supply a few hours later.

4.3 Dechorionating the Embryos

Before the needle can penetrate the egg to dispense the DNA, the tough chorionic membrane (the 'shell') must first be removed, leaving only the much more delicate vitellin membrane to protect the embryo from the outside world. There are two general approaches to dechorionation: mechanical and chemical. It is a good idea, if possible, to perform these and all subsequent steps in a cool (18°C) environment, to slow down the development of the embryos.

4.3.1 *Mechanical Dechorionation*

The newly laid eggs can be easily dechorionated by rolling them gently on a sticky surface, under a dissecting microscope (~ 10 − 15 x magnification). Prepare a work surface by taping two microscope slides together side by side. On one slide place a piece of double-sided sticky cellophane tape. On the other slide, affix a coverslip with tape or a droplet of water, and place a very narrow (1 − 2 mm) strip of double sided tape down the center of the coverslip. It may be easier to obtain the thin strip by scoring a wide one on the coverslip with a scalpel blade and removing the excess.

Harvest the eggs from the food tray, either under the microscope by collecting them with watchmakers' forceps and scooping them up, or with a soft, damp paint brush, and transfer them to the wide strip of tape. Working under magnification, gather a small ball of glue from the surface of the tape with the forceps and hold it tightly. The ball of glue serves to cushion the egg against the hard surface of the forceps, and also to lift the deshelled eggs. Gently stroke the eggs, pushing them about on the sticky surface of the tape until the chorion peels off, exposing the vitellin membrane of the egg. The unshelled egg can be rolled back from the sticky surface of the tape onto the sheet of chorion, from which it can be easily picked up by the ball of glue. Transfer each dechorionated egg to the strip of tape on the coverslip on which the injections will subsequently be carried out, as described in Section 4.3.3 below.

With a little practice, one can easily become skilled at dechorionating and aligning the eggs quickly and with little damage.

4.3.2 *Chemical Dechorionation*

Household bleach, sodium hypochlorite, will quickly remove the chorion from eggs without harm to the developing embryos. However, the viability may vary with the particular brand of bleach. Use a 50% solution of bleach and water that has been cooled slightly (to ~18°C). Gently transfer the eggs from the collection trap to a Millipore type filtration apparatus with a fine damp paint brush or a stream of water from a wash bottle, and wash the eggs down onto a 1 cm Whatman GF-C filter with gentle suction. Try to avoid transferring too much yeast with the eggs, as this may clog the filter. Turn off the suction and add the bleach, letting the eggs sit for 1 – 2 min. Turn on the suction, and gently drain away the bleach. Wash the now dechorionated eggs two or three times with distilled water.

Remove the filter to the dissecting microscope stage. The eggs tend to pile up in the center of the filter. With a damp paint brush, push the eggs off the filter onto a clean glass slide. Carefully transfer the eggs, one by one with a fine ball point sewing needle or a pair of watchmakers' forceps, from the slide onto the sticky tape of the coverslip. A small ball of glue held between the tines of the forceps can be used to carry the eggs.

4.3.3 *Aligning the Embryos*

The dechorionated eggs should be aligned with their posterior ends hanging over a narrow strip of double sided sticky tape, on a coverslip. (The protruding micropyle at the anterior end is a convenient landmark for orienting the eggs.) The coverslip, in turn, may be anchored to a glass slide either with a drop of water, or with tape.

When aligning the embryos, it is advantageous to group them in sets of five. This facilitates counting them, and makes it considerably easier to identify a particular egg later on if, for example, one wishes to remove an uninjected embryo. From 50 to 75 eggs may be set on a single coverslip, over a period of 15 – 25 min. It may be easier, at first, to set up two or three coverslips of 25 eggs each, desiccating each one in turn.

It has been observed that some brands of cellophane tape, and even different batches of the same tape, may have pronounced effects on the viability of the dechorionated embryos. It is a good idea to test a few brands by aligning eggs on several different tapes, covering with halocarbon oil directly, and counting the number that survive to the larval stage.

4.4 **Desiccation**

It is essential that the embryos be desiccated before a significant volume of DNA solution can be introduced. Unfortunately, the amount of desiccation is critical and varies considerably with the prevailing conditions. Over-desiccation severely reduces the viability of the dechorionated embryos, even without injection. Under-desiccation causes the contents of the egg to exude when it is punctuated by the needle. The proper amount must be determined empirically.

Place the coverslip bearing the aligned, dechorionated embryos on top of a Petri dish full of Drierite ($CaCl_2$) or silica gel and cover the dish. In general, one should aim for the minimum amount needed to successfully retain the contents of the egg after the DNA solution is injected. This amount will depend on the humidity of the room, the absorbed water in the Drierite, the time spent aligning the eggs on the tape, etc. However, in most cases this time will range between 5 and 15 min. Following desiccation, immediately cover the embryos with a drop of halocarbon oil. This will be the sole protection for the embryos until the larvae hatch. Several different viscosities of halocarbon oil are available. If it is too runny, the eggs will become exposed to the air and dry out. If too thick, the injection needle becomes difficult to manage. The viscosity of Series 700 (Halocarbon products) is fine. Voltalef Grade 10S is a bit thin, but is adequate. Grade 20S is preferable, but may be difficult to obtain.

4.5 Preparing the Needle

The needle should be as fine as possible to minimise damage to the embryo, but still sufficiently wide to deliver the DNA solution without difficulty.

Pull the microcapillaries with an electric needle puller to a fine point (<1 micron), with a relatively abrupt taper at first, and a more gentle taper towards the end, the length of the point being about half a centimeter or so from the beginning of the taper to the end. A needle with too long a taper will bend in the viscous halocarbon oil with every move of the microscope stage, and the injector will have to wait for its oscillations to damp before proceeding. If the taper is too abrupt, the needle will tear too large a hole in the embryo. Drummond microcaps, or their equivalent, make satisfactory needles, but microcaps that have a fine glass ridge down the inner surface are far superior. These greatly facilitate loading, as a drop of DNA solution placed at the wide end of the needle will find its own way down to the point by capillary action along the ridge.

It is unnecessary to siliconise the microcaps, but they should of course be clean inside, since the smallest speck of dust may clog the opening. The microcaps can be cleaned with filtered absolute ethanol.

Mount the needle in the micromanipulator and observe the taper and the sealed point in the microscope. A satisfactory bevel may be produced by gently jabbing or scraping the point of the needle into the edge of a glass coverslip affixed to a slide. Move the slide and coverslip into focus first, and then slowly bring the needle into the field (using the micromanipulator). Use the stage controls to scrape the glass edge against the needle. The ideal orifice should be small (~1 micron) and slightly beveled. Of course, there is a large element of chance in the production of such a point, but usually a few gentle scrapes of the needle against the glass edge can generate the desired result. Remove the needle from the apparatus; it is now ready for loading. Alternatively, the sealed needle can be loaded with the DNA solution, and the point broken off later.

4.6 **Preparing the DNA for Injection**

The supercoiled plasmids carrying the transposon and the helper P element should be in solutions free of possibly noxious chemicals and buffers, such as CsCl, Tris and EDTA. Precipiate the DNA to be injected, pellet the DNA by centrifugation, and wash the pellets carefully with 70% ethanol in 0.2 M NaCl, and then wash again with 70% ethanol in water. Dissolve the pellets finally in injection buffer (5 mM KCl, 0.1 mM sodium phosphate, pH 6.8) at a final concentration of 1 mg/ml. Clean preparations of DNA are quite stable in this buffer, even when they are subjected to several cycles of freezing and thawing.

Combine the transposon and helper element plasmids in the appropriate ratios, and adjust to the final desired concentration. As mentioned earlier, it is not yet known precisely what the optimal ratio should be. The original 5:1 ratio was suggested to minimise the chances of the helper element transposing. But with the wings-clipped helper, this is clearly not a consideration. The concentration of helper element DNA should not exceed 500 μg/ml, as high concentrations have not been found to improve rates of transformation and may even inhibit (2), possibly by creating a local environment within the embryo of the P (restrictive) cytotype. However, there is no apparent reason to limit the amount of recombinant vector P element DNA. A mixture of 250 μg/ml each of wings-clipped helper and vector has been used successfully to yield a transformation rate of about 30% (5). A 10:1 ratio of vector P to helper P (totalling 500 μg/ml) has also been used with good results, in one case yielding 50% transformants (26). A 20:1 molar ratio of defective vector to helper totalling 1 mg/ml has yielded transformants at about a 30% rate (11). Clearly, the differences in ratio are not so critical to the results (see also Section 8.1).

Briefly centrifuge the solution of DNA in injection buffer in a microcentrifuge to sediment any debris that might clog the needle. Remove about 2 μl from the surface of the solution, and back load it into the needle either by capillarity or by directly loading it with a fine drawn out capillary pipette or a Hamilton-type syringe. The loaded needle should be immediately mounted on the micromanipulator, and the point of the needle carefully lowered into a drop of halocarbon oil on a slide on the microscope stage. This will keep the DNA solution from evaporating, which would result in crystals of salt blocking the opening. If the needle does not break or become clogged, it can be used for several days.

5. INJECTION OF DNA INTO THE EMBRYOS

5.1 **The Proper Developmental Stage for Injection**

The embryonic development of *Drosophila* has been divided into a number of morphologically distinct stages. For a full description of these stages and excellent photomicrographs, see the review by Fullilove and Jacobson (27). The early *Drosophila* embryo is a syncytium of dividing nuclei. About 2 h post fertilisation (at 25°C), cellularisation of the nuclei begins. The first cells to form are the pole cells, at the posterior end of the embryo, the cells destined to become the eggs and sperm of the adult. About 1 h after fertilisation, just before pole cell formation begins, the yolk draws away from the vitellin membrane at the

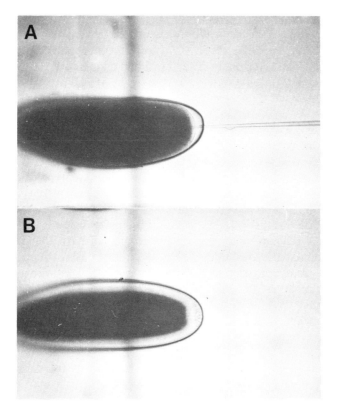

Figure 2. Photomicrographs of developing *Drosophila* embyros. (**A**) A preblastoderm embryo, about 1 h post-fertilisation, being injected. Note the clear space at the posterior end, into which the pole cells will bud. (**B**) An embryo at blastoderm stage, too old for injection. Note the round pole cells occupying the posterior end.

posterior end of the embryo, leaving behind a clear fluid-filled pocket. It is into this pocket that the pole cells will bud. This is an easy stage of development to identify, and such embryos (or younger ones) are ideal for injection (see *Figure 2*).

5.2 **Microinjection**

Successful transformation requires that the P elements be introduced into the developing germ tissue of the embryo. This is accomplished by injecting the plasmid DNA into the posterior end of the precellular embryo, before the pole cells have formed. To slow down the rate of development, injections can be performed at 18°C, in a cooled room, or on a cooled microscope stage, though this is not absolutely necessary.

Mount the embryo-laden coverslip on a slide with a drop of water or a piece of tape, and place it on the stage of the microscope with the posterior ends of the eggs towards the needle. Position the ends of the eggs in the field of view. Now move the needle into the field, using the micromanipulator controls, and crudely align it with the first egg, but keep it slightly above the focal plane of the embryo.

Focus on the midplane of the embryo's posterior. The correct focusing is achieved when the outline of the egg's posterior appears most sharply defined. With the micromanipulator, move the needle slowly into the plane of focus. Once the needle is in focus, it will be properly aligned with the midplane of the embryo. This is the level at which the needle should penetrate. The precise positioning of the needle can be confirmed by gently 'poking' the embryo (using the stage controls), and observing the position of the dent caused by the needle on the egg's surface.

Pierce the embryo's vitellin membrane with the needle, either by moving the stage, or the micromanipulator. Those prone to motion sickness may find that moving the needle to meet the eggs is less nauseating, but better control is achieved by manipulating the stage controls so that the eggs are impaled on the needle. Once the needle has penetrated, pull back until the point is just beyond the boundary of the yolk. Do not inject into the posterior fluid-filled space where the pole cells will form, but rather into the region of the yolk where the pole nuclei reside before they are cellularised. A thin, sharp needle will cause the least damage to the embryo, by requiring less initial force, and thus not penetrating so far into the embryo, and by not leaving such a gaping hole when withdrawn. Try not to disturb the contents of the yolk. Too fat a needle, or too violent a thrust may damage the positional information contained in the egg which is needed for correct embryonic development.

The needle is attached by an air-filled plastic tube to a disposable 25 ml syringe. Squeeze the plunger on the syringe to expel a visible amount of the DNA solution, about 1 − 5% of the egg volume. It should be possible to see the solution flowing into the yolk. If the volume injected is too great, the contents of the embryo may ooze out when the needle is removed. This could also be an indication of insufficient desiccation. Do not despair if a small amount of yolk is extruded when the needle is removed. Such embryos have been known to develop properly into healthy adults.

Invariably, several of the embryos aligned on the coverslip will have developed to the cellular blastoderm stage or beyond. Do not inject these, for the chances of transforming them are small, and their chances for survival are great. The amount of work in tending the flies and their offspring in search of a transformant is much greater than the effort needed to eliminate those embryos. Note the position of any uninjected embryos (either too far advanced or damaged) while injecting down the line of eggs. When the injections are complete, move the coverslip to the stage of a dissecting microscope, and pick off all the viable uninjected embryos with watchmakers' forceps or a fine needle, noting the final tally of 'good' ones.

5.3 Incubation of Injected Embryos

Remove the coverslip bearing the injected embryos to a humid environment at 18 − 20°C until the embryos hatch. One can leave the coverslip covered with its halocarbon oil, and simply place it in a Petri dish with a water-saturated filter paper taped to the lid. Be sure the plate lies flat, or the halocarbon oil will run off the embryos and they will die. When the larvae hatch, about 2 days later, scoop them up with forceps and place them on standard fly medium, about ten to a vial, at 25°C. The larvae are apparently unaffected by several hours of swimming in the halocarbon oil. Alternatively, the coverslip may be trimmed by scoring with a diamond point, and the small shard bearing the embryos placed flat in a shallow trough dug in a Petri dish of fly food. Fill the trough with additional halocarbon oil. This assures that the embryos will not accidentally be exposed to the air. When the larvae hatch, they will crawl right into the food. One should, however, still try to recover these larvae, to determine the number that have hatched, and to place them in a proper vial of fly medium.

6. TESTING FOR TRANSFORMANTS

Each adult derived from an injected embryo (the so-called G_0 generation) should be individually mated with flies of the host strain. It is important to obtain as many G_1 offspring as possible, since the transformants can represent less than one in 200 progeny. Mate each G_0 female to several males, and transfer the parents to new vials every 5 days. G_0 males can be crossed to several different broods of 3−4 females each. These too should be transferred every few days. It is not uncommon to get 150 offspring from a female, and 300 from a male.

The phenotype of the G_1 offspring must be examined. If *Adh* is the selectable marker being employed, the transformed offspring can be selected for their resistance to a 6% ethanol solution (9). If *neo*^{R} is the marker, the larval G_1 offspring should be selected on media containing 1 mg/ml of the drug G418. Such drug selections for transformants are considerable time savers. If *ry* is the marker, the flies should be examined individually for restoration of the ry$^+$ phenotype. While the difference between ry and ry$^+$ eye colour may seem subtle at first, the investigator quickly becomes adept at identifying even small changes in this phenotype.

True transformants with phenotypes intermediate between ry and ry$^+$ have been described, and were shown to be the result of P[ry] integration into transcriptionally inactive or heterochromatic regions of the chromosomes (5,22). Such transformants of course are extremely interesting to study, and would be missed by the selection employed for the P[Adh] or P[neo^R] transformants. Spradling and Rubin (22) found that very few, if any, P[ry] elements integrate at sites that reduce expression of *ry* to the point where the transformants would not be phenotypically detectable.

The fraction of transformed G_1 offspring from a single G_0 parent can vary considerably. Often only one or two transformants will be found, but it is not uncommon to find more. Rarely, transformants will comprise a significant fraction of the total, even 50% or more. When this occurs, the chances are good that a number of different P elements have transposed into the germ line of the injected (G_0) fly, each to a different chromosome site (8). More commonly, though, only a single integration has occurred. Since transposition often occurs premeiotically, several transformed G_1 siblings from one injected parent can be expected to carry the identical P element integration site.

The G_0 flies receiving P[ry] transposons often show an eye phenotype intermediate between ry and ry$^+$. Such 'G$_0$ ry expression' is good assurance that the DNA was successfully introduced into the embryo, and was capable of proper expression. There is, however, no good correlation between G$_0$ ry expression and germline transformation (1,16).

7. CHARACTERISING THE TRANSFORMANTS

The first order of business is to establish a stock from the G$_1$ transformants. The transformed offspring of a given G$_0$ fly can be mated together, or crossed to flies of the host strain, to establish a transformed parent line. As mentioned before, if a large fraction of the offspring have the transformed phenotype, the probability is high that more than one P transposon is present in the genome. If the different integration sites are on different chromosomes, they can be easily isolated later on, once there is no danger of losing the line, by crossing representatives of the parent line to flies carrying dominantly marked balancer chromosomes. Usually, though, only a single insert will be present in the genome. The investigator should be aware that it is considerably less time consuming to generate transformants than it is to characterise and properly examine them for their expected biological properties.

7.1 **Genetic Analysis**

The investigator now needs to establish 'single-insert' lines of his transformant. Standard genetic crosses involving dominantly marked balancer chromosomes will identify and separate the different chromosomes carrying P transposon insertions. Balancer chromosomes contain multiple inversions that tend to suppress recombination with the corresponding homologous chromosome. In addition, they contain one or more dominant genetic markers which can be used to distinguish flies containing the balancer from flies homozygous for the homolog. Examples of useful balancers are *FM7, B* (marked with bar eyes) for the X chromosome, *CyO* (marked with curly wings) for the second, and *TM3, Sb Ser* (marked with stubble bristles and serrated wings) for the third chromosome.

One should have an attached-X stock and a collection of balancer stocks for the three major chromosomes in the genetic background of the host strain. (While this is not absolutely necessary, it will greatly facilitate the subsequent genetic analysis of the transformant). For example, if the P element marker is the wild-type *rosy* gene, one might have stocks of *C(1)DX, yf; ry*^{42}, of *FM7; ry*^{42}, of *CyO; ry*^{42}, and of *TM3, ry/ry*^{42}. Note that since the wild type *ry* gene resides on chromosome 3, one needs a third chromosome balancer that contains a *ry* mutation. Similarly, if the transposition marker employed is *Adh*, one should have a second chromosome balancer such as *CyO, Adh*.

To improve the chances of recovering each of the P transposons in the population, several individuals of the parent line should be crossed separately to each of the balancer stocks. Offspring of these crosses carrying both the transposon marker and the dominant marker of the balancer are then mated to flies of the host strain genotype. The segregation pattern of the transformation marker from the dominant markers of the balancers should indicate the chromosome bearing the integrated element. For example, if a P[ry] transposon integrated on the second chromosome resides in a parental line subjected to the above crosses, one will recover three sets of offspring of the genotype *Fm7, B/+; P[ry]/+; ry/ry* (bar eye and ry$^+$) and $+/+$; *CyO/P[ry]*; *ry/ry* (curly and ry$^+$) and $+/+$; *P[ry]/+; TM3,Sb Ser ry/ry*, (stubble, serrate and ry$^+$). When each class of offspring is crossed to the *ry* host strain, both ry$^+$ and ry$^-$ flies will be found with bar eyes or serrate wings, but all curly winged flies will be ry$^-$, and all non-curly flies will be ry$^+$ (see *Table 1*). This indicates the presence of the P[ry] transposon on the second chromosome. If two P[ry] elements are present on different chromosomes in a single individual, subsequent crosses to the balancer stocks should separate them.

7.2 DNA Blot Analysis

Simultaneously with the beginning of the genetic analysis, one should examine the structure of the integrated P elements by DNA blot hybridisation (28). Genomic DNA is cleaved by a restriction enzyme that cuts once within the integrated P element construction carrying the gene of interest, and the digested DNA is fractionated by gel electrophoresis. The DNA is blotted to nitrocellulose and hybridised with a radioactive probe that only detects one of the two fragments that contain the ends of the P element construct. The number of fragments hybridising will then correspond to the number of P transposons in the population, the size of each fragment being characteristic of the integration site of each element. However, one should be aware that this approach looks at the population of elements within the population of the parent line. An individual transformed fly may have only one of the elements present in the population.

The integrity of the inserted P element should also be confirmed by the appropriate restriction enzyme digests, lest some sequence rearrangement in the element go undetected. Such rearrangements have been known to occur, but are not very common.

7.3 In Situ Hybridisation

Hybridising P element probes to the polytene chromosomes of the transformed line yields the most detailed information about the site of insertion of the element. Briefly, the larval salivary glands of the transformed fly are squashed on a slide to spread the polytene chromosomes. The DNA in the chromosomes is partially denatured and hybridised to a radiolabeled or biotin labeled DNA probe homologous to the integrated P transposon. The site of integration is detected by autoradiography or immunochemical staining (see references 29,30 for a complete description of the procedure).

Obtaining this information is usually not a priority. It is best to wait until the single-insert line has been established and characterised more fully by DNA blotting and by segregation analysis.

Table 1. The Steps Involved in Obtaining and Characterising Transformed Lines.

Step 1	*Example*
Generation G_0	
Host strain embryos are injected with a mixture of helper P element and a P transposon encoding a selectable marker.	*100 G_0 ry^{42} embryos are injected with P[ry] and pπ25.7wc helper.*
Step 2	
Generation G_0	
Surviving adult flies are mated individually to host strain partners to obtain maximum number of offspring.	*Twenty adults survive, and are individually crossed to ry^{42} partners. Ten prove to be fertile, and yield several hundred offspring each.*
Step 3	
Generation G_1	
Offspring of each G_0 are examined (or screened) for the transformed phenotype. The transformed offspring of each individual G_0 are mated together, or mated to host strain partners, to establish the transformed parent lines.	*Two of the 10 G_0 lines produce at least one ry$^+$ offspring. One yields 3 ry$^+$ males and 2 ry$^+$ females, which are mated together to establish parent line A. Another G_0 yields a single ry$^+$ male, which is crossed to ry^{42} to establish parent line B.*
Step 4	
Generation G_2 − and beyond	
The parent lines are examined by DNA blot analysis to determine the number of P transposons that are present in the population of each line.	*DNA blots indicate that each parent line contains a single P transposon in the population.*
Step 5	
Simultaneously with Step 4, representatives of each parent line are mated to various balancer stocks to determine the chromosome linkage of the P element inserts.	*Single ry$^+$ flies from parent line A are crossed to CyO; ry^{42} and TM3, ry/ry^{42} flies. In the next generation, ry$^+$ CyO flies or TM3 flies are individually mated to ry^{42} again. The CyO offspring are noticed to be all ry$^-$, while the TM3 offspring are about equally ry$^+$ and ry$^-$. This suggests that the P[ry] insert segregates from CyO, and therefore resides on the second chromosome.*
	Parent line B is crossed to CyO; ry^{42} and TM3, ry/ry^{42} flies. In the next generation, ry$^+$ CyO flies and ry$^+$ TM3 flies are individually mated to ry^{42}. In the offspring of this cross, ry$^+$ and

Table 1. continued

Step 5 continued	*Example*

ry⁻ eye colours are found with equal frequency among both TM3 *flies and* CyO *flies, suggesting the element is not linked to either autosome. Single ry⁺ males are then mated to attached-X; ry⁴² females. In the next generation, one finds that all the males are ry⁺ and all the females are ry⁻, confirming the X-linkage of the P[ry] element.*

Step 6

From the linkage data, and the DNA blot information, single-insert lines are established. The chromosome bearing the element may be made homozygous, or if the insert causes a lethal mutation, it is maintained over a balancer. The integrity of the P transposon may be checked by DNA blotting, and the position of the transposon by *in situ* hybridisation.

Individuals from parent line A are crossed to CyO; ry⁴² *again, and ry⁺ CyO brothers and sisters (genotype* P[ry]/CyO) *are selected and mated to each other. In the next generation, non-CyO flies are selected, and used to establish the* P[ry]/P[ry] *homozygous stock.*

The P[ry] *insert of parent line B is maintained as a patroclinous attached-X line. Single males from this line are mated to females carrying the X chromosome balancer* FM7. *In the next generation, the* FM7/P[ry] *females are mated to a male from the patroclinous line. The non-FM7 offspring of this cross (brothers and sisters) are used to establish a homozygous stock.*

Step 7

The experimental analysis may begin.

8. SUCCESS RATES

8.1 **Transformation Frequency**

The published frequencies of germ line transformants among surviving injected fertile (G_0) adults vary from $1-2\%$ to 50% (1,3,7,9,10,16,22) with the average being around 10%. A number of parameters affecting the frequency of transformation have been suggested, but most have not been systematically tested. Some of these are as follows.

(i)   The amount of DNA injected. Too high amounts of the helper P element seem to inhibit transformation (2).
(ii) The ratio of defective recombinant to helper element (see Section 4.6).
(iii) The length of the P transposon construction.
(iv) The particular sequences present in the construction. There appears to be a slight tendency for larger P elements to transpose at a lower rate. However, this cannot be the whole story, since transposons of similar size reportedly

transform at different rates. There is no theoretical limit to the size of the P transposon. To date, the largest P element derivative reported to have transformed a fly is 54 kb (16).

The rate of success may well depend on less tangible factors. An individual's technique in injecting embryos plays an important and uncontrollable part in this story. Although early reports (1,16) routinely found 50% transformants among the surviving fertile adults, no subsequent studies have consistently reproduced these high rates.

8.2 Survival Frequency

Only about 15% of injected embryos generally survive to become fertile adults. There are losses at every stage of development. An example is taken from the work of T.Hazelrigg (3,30): of some 3500 embryos injected, 40% hatched, 50% of these became adults, and 65% of these were fertile. Probably the most important factors affecting the viability and fertility of the injected flies are desiccation of the embryo and damage to the embryo caused by the needle. Among beginners, the greatest losses occur at the embryonic stage, but these rapidly decline as experience is gained.

9. REMOBILISING P ELEMENT TRANSPOSON WITHIN THE TRANS-FORMED FLY

If the integrated P transposon is structurally intact, it can be transiently mobilised again, by recreating the environment of a dysgenic hybrid. This is accomplished, as before, by injecting an intact helper P element, such as pπ25.7wc into the developing germ line of the fly, except that the host strain this time is itself a transformed line, containing a defective recombinant P element. The helper element supplies the transposase which remobilises the integrated transposon.

T.Hazelrigg and R.Levis (31) have injected the wings-clipped helper at a concentration of $50-100$ μg/ml into a transformed line containing a single defective P[*white*] transposon. Some $20-40\%$ of the G_0 flies were found to produce at least one offspring showing a new integration site for the P transposon. This frequency of retransposition is higher than the average rate of initial transformation.

The investigator may find it useful to remobilise a P transposon already resident in a fly genome when, for example, it may be necessary to have the gene of interest on a particular chromosome. Or, one may find the occasional transformant that displays a phenotype which is suspected to be a consequence of a position effect. This can be directly tested by moving the P transposon to a new genomic site, and observing a simultaneous change in the expression of the mobilised gene.

10. REFERENCES

1. Rubin,G.M. and Spradling,A.C. (1982) *Science,* **218**, 348.
2. Spradling,A.C. and Rubin,G.M. (1982) *Science,* **218**, 341.
3. Hazelrigg,T., Levis,R.L. and Rubin,G.M. (1983) *Cell,* **36**, 469.

4. Fyrberg,E.A., personal communication.
5. Davies,J., Delaney,S. and Glover,D.M., personal communication.
6. Wensink,P., personal communication.
7. Scholnick,S.B., Morgan,B.A. and Hirsh,J. (1983) *Cell,* **34**, 37.
8. Wakimoto,B., Kalfayan,L. and Spradling,A., personal communication.
9. Goldberg,D., Posakony,J. and Maniatis,T. (1983) *Cell,* **35**, 59.
10. Lis,J.T., Simon,J.A. and Sutton,C.A. (1983) *Cell,* **35**, 403.
11. Steller,H. and Pirrotta,V. (1985) *EMBO J.,* **4**, 163.
12. Engels,W.R. (1983) *Annu. Rev. Genet.,* **17**, 315.
13. Bregliano,J.C. and Kidwell,M.G. (1983) in *Mobile Genetic Elements*, Shapiro,J.A. (ed.), Academic Press, New York, pp. 363.
14. O'Hare,K. and Rubin,G.M. (1983) *Cell,* **34**, 25.
15. Karess and Rubin,G.M. (1984) *Cell,* **38**, 135.
16. Rubin,G.M. and Spradling,A.C. (1983) *Nucleic Acids Res.,* **11**, 6341.
17. Bender,W., Spierer,P. and Hogness,D.S. (1983) *J. Mol. Biol.,* **168**, 17.
18. Lindsley,D. and Grell,R. (1968) *Genetic Variations of Drosophila melanogaster*, published by Carnegie Inst., Washington.
19. Goldberg,D.A. (1980) *Proc. Natl. Acad. Sci. USA,* **77**, 5794.
20. Benyajati,C., Wang,N., Reddy,A., Weinberg,E. and Sofer,W. (1980) *Nucleic Acids Res.,* **8**, 5649.
21. Benyajati,C., Spoerel,N., Hamerle,H. and Ashburner,M. (1983) *Cell,* **33**, 125.
22. Spradling and Rubin (1983) *Cell,* **34**, 47.
23. Bingham,P.M., Kidwell,M.G. and Rubin,G.M. (1982) *Cell,* **29**, 995.
24. Simmons,M., O'Hare,K. and Rubin,G.M., personal communication.
25. Ashburner,M. and Thompson,J.N. (1978) in *The Genetics and Biology of Drosophila*, Vol. **2a**, Academic Press, pp. 2.
26. Zucker,C. and Rubin,G.M., personal communication.
27. Fullilove,S.L. and Jacobson,A.G. (1978) in *The Genetics and Biology of Drosophila*, Vol. **2c**, Academic Press, pp. 106.
28. Southern,E.M. (1975) *J. Mol. Biol.,* **98**, 503.
29. Gall,J.G. and Pardue,M.L. (1971) in *Methods in Enzymology*, Vol. **21D**, Academic Press, pp. 470.
30. Langer-Safer,P.R., Levine,M. and Ward,D.C. (1982) *Proc. Natl. Acad. Sci. USA,* **79**, 4381.
31. Hazelrigg,T., Levis,R. and Rubin,G.M., personal communication.

CHAPTER 6

High Efficiency Gene Transfer into Mammalian Cells

CORNELIA GORMAN

1. INTRODUCTION

As cloning techniques have evolved, a major interest has developed in techniques for gene transfer into eukaryotic cells. This chapter will describe the current techniques used for gene transfer of non-viral vectors, concentrating on the two most widely used methods, the 'calcium phosphate precipitation' and the 'DEAE-dextran' method. Following a discussion of the vectors available for conducting various types of gene transfer experiments, a detailed analysis of these protocols will be presented. Genes can be introduced into cells either transiently or stably. In transient gene transfer, the DNA introduced into cells in culture does not necessarily need to be integrated into the cellular chromatin to be expressed. Expression of incoming DNA can be monitored within 12 h after uptake. This transient level of expression continues for up to 80 h following the introduction of plasmid DNA into mammalian cells. Alternatively, the plasmid DNA can be incorporated into a permanent state to form stably transformed cell-lines. The potential of these two approaches, transient versus stable transformation, will be discussed. The most desirable approach clearly depends upon the type of questions to be answered. Finally, this Chapter will consider how plasmid vectors can be used to give a significant increase in the efficiency of transfer of chromosomal genes, which for many is the ultimate goal.

Many interesting experiments require the use of cellular DNA as the source of genetic material. In these experiments the selection must be based on a phenotypic change which can be scored following the transfer of cellular DNA into a recipient cell. A high stable transformation efficiency is required for this type of experiment to be successful. It is, therefore, often advantageous to use a plasmid marker in a co-transfection with the genomic DNA. Thus by first placing the transfected cells under selection with a dominant vector any background from non-transfected cells can be eliminated. This selection can be removed, if necessary, and the second screening can proceed based on the phenotypic change to be scored. As with many new techniques a 'folklore' has arisen as to the pros and cons of any one approach. The strong and weak points of each approach will be analysed as best as is possible.

2. VECTORS USED IN MAMMALIAN CELL EXPRESSION

All basic plasmid vectors used in gene transfer have four main components. First, the plasmid must contain prokaryotic sequences coding for a bacterial replication origin and an antibiotic resistance marker. These sequences allow the propagation and selection of the plasmid within a bacterial host. Secondly, there must be eukaryotic elements which control initiation of transcription. Specifically these include promoter and perhaps enhancer sequences (1). A third feature of most expression vectors includes sequences involved in the processing of transcripts. While not all vectors contain introns for control of possible splicing, the addition of polyadenylation sequences is required for efficient expression in mammalian cells. Last, is the 'test' gene. As can be seen from the examples shown below the test gene may be a bacterial gene controlled by eukaryotic elements, or complementary DNA's made to a specific RNA population or even a genomic clone.

2.1 The pSV2 Vectors

Many of the most commonly used eukaryotic vectors are based on the prototype vector pSV2 (Bruce Howard and Paul Berg, unpublished results). A diagram of this prototype is shown in *Figure 1*. The prokaryotic sequences contained in the pSV2 vectors include the Amp^R resistance gene and the origin of replication from the bacterial plasmid pBR322. In all the pSV2 vectors these sequences are located on a 2295 base pair fragment encompassing residues 323 to 2618 (see the sequence for pSV2cat provided in the appendix to this chapter). In this vector the SV40 early region promoter is used for the initiation of RNA synthesis in mammalian cells. This region includes the 72 bp repeated enhancer in addition to the start site for transcription. This fragment is a 323 base pair fragment from the early region of SV40 from the *Hind*III site to the *Pvu*II site (on the pSV2cat sequence in the appendix, this region in the pSV2 vectors is from nucleotides 1 to 323). The pSV2

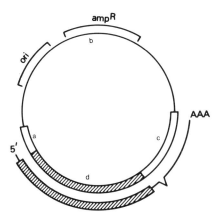

Figure 1. pSV2 Vectors. **a.** The SV40 early region promoter from base pair 1 to 323 on the sequence of pSV2cat. **b.** The region from 324 to 3369 bp which includes the bacterial origin and *amp*^{R} gene from pBR322. **c.** The region from 3370 to 4217 bp which encompasses the polyadenylation site and small t intron from SV40. **d.** The position of the test gene cloned under the transcriptional control of the SV40 early region promoter. In pSV2cat the CAT gene is from region 4218 – 5003 bp.

vectors do include a splice donor-acceptor sequence in the form of the small t-antigen intron from SV40. This 610 base pair fragment is included in the region of nucleotides 3606 to 4217 on the sequence of pSV2cat. Adjacent to these control sequences is the polyadenylation site from the early region of SV40. This 988 based pair fragment comprises the sequence from nucleotides 2618 to 3606 in the pSV2 vectors. The remainder of the plasmid (4217-0) is the region designed for the insertion of any coding sequence.

2.2 Variations

While these general features are representative of the components one needs in a mammalian expression vector the specific details can, of course, vary. The antibiotic resistance gene need not be ampicillin. The genes conveying resistance to tetracycline or chloramphenicol are often used. The choice of promoter used in such a vector may depend on the recipient cell to be used or the experimental design. In the pSV2 vectors, the SV40 early region promoter has been used. This sequence was chosen because this viral control sequence was well studied and thought to be transcriptionally active in many cell types. Interestingly as much more detailed work has been done with viral promoters many have been seen to display a species preference. This preference has been linked with control sequences known as enhancers. The outcome of much work with enhancers now suggests that the promoter sequence used in an expression vector should correlate with the type of recipient cell to be used (1). For example the SV40 early region promoter functions very well in primate cells but is much less efficient in murine cells. While the promoter of choice for high levels of expression in mouse cells is the Moloney murine sarcoma virus (MSV) long terminal repeat (LTR) (2). The most widely applicable promoter appears to be the LTR from the Rous sarcoma virus (RSV) (3). This promoter functions in all mammalian cells tested and even in non-mammalian cells such as those of *Drosophila*. A series of vectors utilising the ubiquitous RSV promoter is shown in *Figure 2*. The sequence for pRSVneo is given in the appendix to this chapter. All the RSV vectors (4) have the same sequence as pRSVneo between nucleotides 1 and 4241.

2.3 Inducible Promoters

It may not always be desirable to have high constitutive expression of an exogenous gene. Very high levels of some gene products may actually be toxic when continuously expressed in a cell. Another way to control the expression of the cloned gene, upon re-introducing it into a mammalian cell, is with the use of an inducible promoter. Such control sequences can be fused to any coding sequence, thereby rendering the gene responsive to the induction stimulus. Examples of inducible promoters which have been successfully used in this way are heat shock protein promoters (5); metallothionien promoters, both mouse and human (6 – 8); growth hormone promoter (9), and the mouse mammary tumour virus (MMTV) LTR (10). These control sequences can be used to regulate expression either transiently or in stable transformants.

It must be remembered that these induction stimuli may effect normal cellular

145

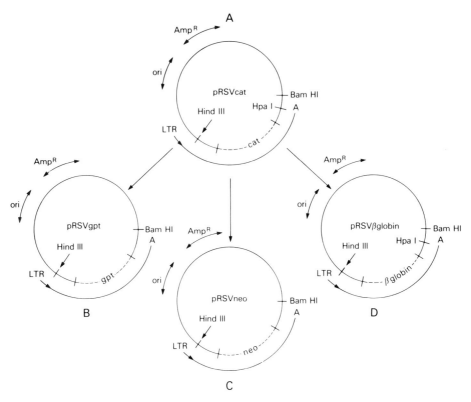

Figure 2. Vectors which use the RSV LTR as a Promoter. **A**, pRSVcat is a transient expression vector. **B**, pRSVgpt and **C**, pRSVneo are vectors used for dominant selection. **D**, pRSVbeta-globin is a vector which can be used to easily insert any coding region under control of the RSV LTR. The sequence for pRSVneo is provided in the appendix and all the above vectors have sequence 1-4241 bp in common.

processes as well. This is especially true of induction with heat shock, since most cellular genes are also regulated either positively or negatively upon heat shock. To some degree this is also true with the induction of the mouse or human metallothionein genes by heavy metal ions. Perhaps a slightly more specific means of increasing transcription is with the use of steroid responsive control sequences such as those contained in the human, but not mouse, metallothionein gene, and in the growth hormone and MMTV promoters. It is important to stress that the vector one should use depends on the goals of the experiment. A control sequence of any one vector can be changed or modified according to these specific goals.

2.4 Transient Versus Stable Expression Vectors

Basically there are two types of experiments using mammalian cell gene transfer techniques. Transient expression allows the investigator to look at gene products, either RNA or protein, within hours of DNA uptake. Up to 80 h following uptake, the plasmid vectors are nuclear but have not become integrated into the host cell chromatin. It has become popular to use vectors coding for proteins which are either unique to mammalian cells or easily distinguished from any host cell function. This approach allows rapid screening for the production of plasmid coded gene products. For this purpose a variety of bacterial genes have been used in eukaryotic expression vectors. The two most widely used transient assay systems utilise two bacterial genes, the gene encoding chloramphenicol acetyl-transferase (CAT) or the *E. coli lacZ* gene that encodes β-galactosidase. The pSV2 derivatives carrying each of these genes are commonly used for transient expression. (All the pSV2 vectors share the sequence from nucleotide 1 to 4127 in the sequence of pSV2cat given in the appendix. In pSV2cat the CAT gene is encoded by nucleotides 4218 to 5003.) A detailed discussion of the assays for each of these enzymes is given in Sections 7.1 and 7.2.

Similarly, two of the most popular dominant selectable vectors make use of the *de novo* synthesis of bacterial gene products which are the basis of selective survival in mammalian cell culture. These selections are based on the expression of the *E. coli* gene *gpt* or *neo*R. A detailed description of the rational behind each selection is provided in Sections 8.2 and 8.3

3. PREPARATION OF DNA

In addition to the well-being of the cells (see Section 4) an important parameter in these experiments, is the state of the plasmid DNA. A check by gel electrophoresis will assure that DNA preparations are free of RNA and chromosomal DNA. Periodic checks will ensure that plasmid has remained supercoiled. It is useful to band the DNA in cesium chloride twice to assure the purity of the plasmid preparations. There are many protocols for making plasmid DNA, but the method giving reproducibly pure DNA for transfections uses Triton lysis. The procedure is given in *Table 1*.

4. CELL CULTURE

Regardless of the technique chosen, the first step one should take is to make a commitment to learn how best to grow the cells which are to be the recipients of the foreign genes. While this may seem too obvious to mention, it is of utmost importance. The culture conditions of cells grown for transfection can be the biggest source of variation in gene transfer experiments and some knowledge of the growth properties of the cells helps immensely when it comes to trouble-shooting one's experiments.

Certainly cells that will grow in monolayers, as a single layer of cells on an adherent surface, are the easiest to transfect. Such cells are somewhat easier to use for transfection experiments than cells which grow in suspension culture. Established cell lines are also preferable to using freshly explanted cells known as primary cultures. This is due to the fact that primary cells have not become adapted to growth *in vitro* and therefore their growth cycle slows greatly with extended time in culture. There is a high correlation between growth rate in culture and the ability to take up DNA. For a first attempt at gene transfer, the suggested cell types include lines established from human cervical carcinomas such as HeLa

Table 1. Protocol to Prepare Plasmid DNA.

1.	Culture a 5 ml innoculum from a single bacterial colony containing plasmid DNA at 37°C overnight. Inoculate an 800 ml culture of superbroth[a] with the overnight culture and incubate for 36 h at 37°C with vigorous shaking.
2.	Spin down the bacteria at 4°C, at 5000 r.p.m. in a Sorval G30 rotor (or its equivalent) for 10 min.
3.	Resuspend the bacteria in 100 ml of TE[b]. Centrifuge again to pellet the bacteria. (At this point the bacterial pellet may be stored at −20°C.)
4.	Resuspend the pellet in 9 ml of TES[c]. Mix thoroughly. This and subsequent steps of the lysis procedure are carried out on ice.
5.	Add 0.9 ml of 10 mg/ml lysozyme dissolved in TES[c] and incubate on ice for 5 min.
6.	Add 3.7 ml of 0.25 M EDTA pH 8.0. Keep on ice for 5 min.
7.	Add 14.5 ml of cold Triton solution[d]. Mix and leave in the ice-bucket for 10 min.
8.	Centrifuge at 25 000 r.p.m. in a SW27 rotor (or its equivalent) for 30 min at 4°C.
9.	Decant the supernatant into a 50 ml polypropylene tube and adjust the weight of the supernatant to 30.17 g with TE[b]. Add 28.14 g CsCl and mix throughly. Add 4.5 ml of 10 mg/ml ethidium bromide dissolved in 10 mM Tris-HCl pH 7.5.
10.	Pipette the mixture into a polyallomer quick seal tube and seal following the manufacturer's instructions. One preparation will fill one quick seal tube for either the Beckman VTi 50 or Ti 60 rotor. The vertical rotor should be centrifuged for at least 18 h at 45 000 r.p.m. at 20°C. The Ti 60 rotor should be centrifuged for at least 60 h at 35 000 r.p.m.
11.	Following centrifugation, illuminate the tube with long wave u.v. light. Pierce the tube from the side using a 20 ml syringe with a 19 guage needle and collect 4 − 5 ml containing the lower plasmid band.
12.	Either reband (see Step 13) or extract the ethidium with isopropanol saturated with CsCl solution at the concentration used for banding. Dialyse against TE to remove the CsCl. Add 1/10 volume of 5 M sodium chloride and two volumes of ethanol. Leave overnight at −20°C. Collect the precipitate by centrifugation. Drain away the supernatant and dry to remove all traces of ethanol.
13.	To reband the plasmid, prepare a CsCl solution using 1.08 g of CsCl per ml of TE and add 0.17 ml of ethidium bromide (10 mg/ml) for each ml of TE used. This solution is then used to top off the tubes for the second banding in either a VTi 50 or VTi 65 rotor.

[a]Superbroth is made from stock solutions A and B. Solution A is made by mixing 120 g of tryptone, 240 g of yeast extract, 50 ml of glycerol, and 9000 ml of water. Solution B contains 125 g of K_2HPO_4, and 38 g of KH_2PO_4 made up to 1 l with water. Each solution should be autoclaved separately. Mix together 900 ml of A and 100 ml of B for 1 liter of Superbroth, pH of 7.2.
[b]TE is 10 mM Tris-HCl pH 7.9, 1 mM EDTA.
[c]TES is 50 mM Tris-HCl pH 7.5, 40 mM EDTA, 25% (w/v) sucrose.
[d]Triton solution is made up by mixing together 1 ml of 10% Triton, 31.5 ml of 25 mM EDTA, 5 ml of 1 M Tris-HCl pH 7.9, and 62.5 ml of water, this solution should be filter sterilised.

cells (epithelial type); mouse embryo fibroblasts such as 3T3 or 3T6 cells and mouse liver cells commonly known as L cells.

4.1 Solutions for Cell Culture

4.1.1 *Media*

Various media can be used for transfection; the most common medium is Dulbecco's modified Eagle's medium (DMEM). The recipe is provided in *Table 2*. The only precaution that should be noted is that any media containing $CaCl_2$, such as the RPMI series of media, cannot be used for $CaPO_4$ transfection. The excess calcium will cause a dense precipitate to form.

Table 2. Dulbecco's Modified Eagle's Medium[a].

Components	Mg/ml
Inorganic Salts	
$CaCl_2$ (anhydrous)	200.00
$Fe(NO_3)_3:9H_2O$	0.10
KCl	400.00
$MgSO_4.7H_2O$	200.00
NaCl	6400.00
$NaHCO_3$	3700.00
$NaH_2PO_4.H_2$	125.00
Other components	
Glucose	4500.00
Phenol red	15.00
Amino Acids	
L-arginine:HCl	84.00
L-cysteine	48.00
L-glutamine	580.00
Glycine	30.00
L-histidine:HCl:H_2	42.00
L-isoleucine	105.00
L-leucine	105.00
L-lysine:HCl	146.00
L-methionine	30.00
L-phenylalanine	66.00
L-serine	42.00
L-threonine	95.00
L-tryptophan	16.00
L-tyrosine	72.00
L-valine	94.00
Vitamins	
D-calcium panotothenate	4.00
Choline chloride	4.00
Folic acid	4.00
i-inositol	7.20
Nicotinamide	4.00
Pyridoral HCl	4.00
Riboflavin	0.40
Thiamine HCl	4.00

[a]See the text for additional comments upon the purity of the water, buffering of the medium, the addition of antibotics and serum.

The water used to make the tissue culture medium must be double distilled. It is also preferable that the water has first been deionised before distilling. Be sure the water has cooled after distillation before making the medium. The pH of the distilled water should measure approximately 5.5. Each chemical should be very pure and dissolved separately before being added to the mixing vat. The pH should be checked at this time. For DMEM the pH should be between 7.5 and 7.7; this will adjust to $7.3-7.4$ in an atmosphere of 10% CO_2. The pH of the media is very important for success in transfections. The optimum for the DMEM is as just stated between 7.3 and 7.4. An alternative approach to achieving this value is to add Hepes buffer pH 7.2 to a final concentration of 50 mM. For sterilisation the medium is passed through a sterile filter (0.2 micron) either in a special sterile room or under a laminar air flow hood. The medium should be stored air tight as made. Just prior to use, penicillin (100 μ/ml) and streptomycin (100 mg/ml) are added along with serum. The most commonly used sera for tissue culture are new born calf or fetal bovine.

4.1.2 *Trypsin*

Trypsin is used to remove cells from monlayers as described in Section 4.2.1. Trypsin is used as 0.025% solution made in normal sterile saline. It is often convenient to purchase a 10 x solution and dilute for use. Be sure to use only tissue culture grade trypsin. Some cells are easier to subculture using a trypsin-EDTA soution. The final solution contains 0.5 g of trypsin and 0.2 g EDTA per liter of normal saline.

4.1.3 *Phosphate-buffered Saline (PBS)*

PBS is used as an isotonic solution for washing cells. A 10 x solution may be made up by dissolving 80 g of NaCl, 20 g of KCl, 15 g of Na_2HPO_4, and 20 g of KH_2PO_4 in one litre of pure water.

4.2 Subculturing Monolayer Cells

When cells have grown to almost cover the surface of the dish or flask they are said to be approaching confluency. Cells which are contact inhibited will stop growing once the density of the cells is sufficient for the cells to touch or contact other cells. Cells which continue to divide at this point will form dense areas of three dimensional growth or foci. These cells are not contact inhibited. For gene transfer experiments it is advisable to subculture the cells at a point before they become confluent and therefore before growth is inhibited. Subculture involves removing the cells from the plate by trypsinisation, counting the cells and re-plating them at a lower cell density.

4.2.1 *Trypsinisation of cells*

Warm all solutions to 37°C using a water bath. Aspirate off the media from the cells. Add trypsin (1 x) solution. For a 9 cm plate add 1 ml; for a 25 cm^2 flask use 0.5 ml of trypsin. Replace the cells into the incubator to maintain the temperature at 37°C. After $1-2$ min, remove the culture from the incubator and examine the cells under a microscope. The cells should be rounded and singular. They can be easily dislodged from the plate by gentle shaking. If the cells still adhere to the plate, replace the culture into the incubator and wait another minute. When the cells no longer adhere, stop the trypsinisation by adding 0.5 ml to 1 ml of DMEM containing 10% serum. It is usually best to transfer the cells to a 50 ml sterile tube at this point. Add sufficient medium to give a volume equal to that in which the cells were grown. If the cells were grown on a 9 cm dish, increase the volume to 10 ml. The cells can now be counted and replated.

4.2.2 *Counting Cells*

The cells are now counted so that an accurate, reproducible number of cells can be used in subculturing. This is very important for transfection experiments since the cells should be in log phase growth at the time they come in contact with the DNA. For many cells in monolayer culture, a cell density of $3 - 5 \times 10^5$ per 9 cm dish or 1×10^4 cells per cm^2 is suitable for transfection experiments. If a Coulter counter is available it is easy to learn to use. However, if this equipment is not available, a simple method for counting cells involves the use of a haemocytometer. This method consists of microscopically counting the number of cells in a very small volume. The haemocytometer is a microscope slide with grids on it to form squares of various sizes. There is a uniform 0.1 nm space between the slide and the coverslip. This space allows the addition of a suspension of cells by capillary action. Then cells within certain squares are counted microscopically with the aid of a hand counter. If one knows the size of a square and therefore the volume, the concentration of cells in suspension can be calculated. A haemocytometer will have nine large squares.

A	B	A
B	C	B
A	B	A

When the slide has been filled correctly each large square lies under 10^{-4} ml of liquid. To count the cells, take a small drop of a uniform cell suspension. This can be easily done using a sterile Pasteur pipette. The liquid should fill the space between the slide and coverslip but not overflow it. Remember not to reuse the pipette as it is no longer sterile. Routinely the cells in the C square and B squares are counted and averaged. This number is then multiplied by 10^4 to give the number of cells in 1 ml of media. Once this concentration is known, the dilution factor to yield the correct number of cells per dish can be calculated. For routine subculturing, most cells can be plated at between 5×10^4 and 1×10^5 cells/9 cm dish or 1×10^3 per cm^2. For transfection experiments cells should be plated a little more densely so that they are in the log phase of growth. This is usually a density of 5×10^5 cells per 9 cm plate or 1×10^4 cm^2.

Most cells will reach confluency by $4 - 7$ days when plated as described above. It is good practice to record the number of times a cell line has been subcultured. This is easily done by including this changing number as part of the identification of the cells. As cells are continually subcultured their ability to take up DNA, either in plasmid or viral form, changes. For most transfection experiments cells should not be used for more than 15 consecutive subcultures.

5. THE 'CALCIUM PHOSPHATE METHOD'

This is the most widely used method (11). It can be applied very successfully to most cells in monolayer and some cells in suspension. The method can be used very efficiently for either transient expression or stable transformation. The versatility of this technique has made it the most popular method. However, it can be a little more difficult to define the parameters important to achieve success with this procedure than with the 'DEAE-Dextran method'.

5.1 Stock Solutions for the 'Calcium Phosphate Method'

(i)   10 x Hepes-buffered saline (HBS). The 10 x stock solution contains 8.18% NaCl (w/v), 5.94% Hepes (w/v), 0.2% Na_2HPO_4 (w/v). The solution is stored at 4°C, in 50 ml aliquots. The solid Hepes used for making this solution is best stored desiccated at 4°C. For transfection use the 10 x solution to prepare a 2 x HBS solution and adjust the pH to 7.12 with 1 N NaOH. Sterilise the solution by filtration through a nitrocellulose membrane. Great care is needed in making up this buffer since the pH is very critical for these experiments.

(ii)   2 M $CaCl_2$. This is sterilised by filtration through a nitrocellulose membrane and stored at 4°C.

(iii)   15% glycerol/HBS is made by mixing 30 ml to 50% glycerol (w/v), 50 ml of 2 x HBS (pH 7.12) and 20 ml of water. The solution is sterilised by filtration and stored at 4°C.

5.2 The Basic Protocols

There are many variations to the technique that uses calcium phosphate. Two of the most widely used variations of the protocol are described below. It is best to test both for any given cell-line. This preliminary testing can be done using one of the transient assay systems described below (Section 7).

(i)   On the day before transfection (day 1), replate the cells to be used at a density of $10^4/cm^2$.

(ii)   On day 2, replace the culture medium with fresh medium containing 10% fetal calf serum. DNA is added to the cells 3 h later.

(iii)   Prepare the calcium phosphate-DNA precipitate using the stock solutions warmed-up to room temperature. For a 25 mm² flask containing 5 ml of medium, set up the following solutions. In tube A, place a solution containing 5 μg of DNA together with 31 μl of 2 M $CaCl_2$ and bring the final volume to 0.25 ml with water. Add 0.25 ml of 2 x HBS to tube B. To make the precipitate, the contents of tube A must be added to the HBS in tube B. This order of addition is crucial. Add the DNA solution dropwise to the HBS. The precipitate will form immediately. If the precipitate looks dense and opaque, rather than translucent, the HBS has not been prepared at the correct pH.

(iv) Pipette the precipitate onto the cells by slightly tilting the dish and adding the precipitate to the medium. Put the cells back into incubator immediately to assure that the pH does not change.

(v) Incubate the cells for 3.5 − 4 h.

(vi) Examine the precipitate which has formed. The best precipitate resembles small grains which cover the cells. If the pH of the medium is too acidic, no precipitate will form, whereas if it is too alkaline, large spheres of precipitate will float in the medium. The state of the DNA also has a great influence on the appearance of the precipitate (Section 3).

(vii) Wash the cells in serum-free medium. A glycerol shock may also be carried out at this time to improve the transfection efficiency. The shock is performed by adding 0.5 ml of 15% glycerol/HBS per flask and incubating the cells at 37°C. The optimum time varies between 30 sec and 3 min for different cell types. NIH/3T3 and CV1 cells require 2 min, for example, whereas HeLa cells require 30 sec. The 'shock solution' should be removed before the cells start to shrink. The cells should then be washed and fed with complete medium.

(vii) Harvest the cells on day 4. A test of transient expression can be carried out at this stage (see Section 7).

5.3 Variations on the Basic Protocol

5.3.1 *Extended Contact Between the Cells and the Calcium Phosphate Precipitate*

Another commonly used technique allows the calcium phosphate precipitate to remain on the cells for 18 h. The precipitate is then washed off thoroughly and fresh medium is added to the cells. When the precipitates have been allowed to remain on the cells for this extended time, the addition of glycerol seems to have less effect at increasing expression of transfected DNA as measured in the transient expression system (Section 7).

5.3.2 *Butyrate Treatment*

Treating the transfected cells with butyrate has been shown to increase the number of cells which can express the incoming DNA at least three-fold. This can result in 40% of the cells in the population giving transient expression (12). Furthermore, this treatment can specifically increase the transcriptional activity of particular plasmids containing SV40, polyoma or papilloma virus control regions. The procedure follows the protocol in Section 5.2 through to step (vii). After the glycerol shock, fresh medium is added to the cells followed by the sodium butyrate solution which is allowed to remain in contact with the cells overnight. The next morning the butyrate is removed and the cells are washed and re-fed. The amount of butyrate added varies with the cell type; for example, 10 mM butyrate is added to CVI cells, 5 mM butyrate to NIH/3T3 and HeLa cells and 2 mM butyrate to Chinese hamster ovary (CHO) cells. It is most convenient to use a stock solution of 0.5 M sodium butyrate prepared in one of the following ways:

(i)   Take 1.5 ml of butyric acid (MW 88) and add NaOH solution until the solution has a pH of 7.0. Adjust the total volume to 25 ml and sterilise the solution by filtration.

(ii)   Dissolve 1.375 g of sodium butyrate (MW 109) in water, adjust the pH to 7.0 and bring to a final volume of 25 ml. Sterilise the solution by filtration.

For a 10 mM solution, add 200 μl of the stock solution to a 10 cm plate containing 10 ml of medium.

5.3.3 *Chloroquine Treatment*

Chloroquine has also been used to enhance transfection efficiency (13). However, since this method can be used with either calcium phosphate or DEAE-dextran it will be discussed below.

6. THE 'DEAE-DEXTRAN METHOD'

This technique has been widely used to increase the efficiency of virus infection. More recently, modified procedures have also been used to facilitate the uptake of plasmid DNA (14). However, it should be remembered that no one has been able to make permanent transformed cell lines using this method. For transient expression it can be quite useful and has been successfully used for lymphocytes (15) in addition to cells in monolayer.

6.1 **Stock Solutions**

(i)   The DEAE-dextran stock solution contains 100 mg DEAE-dextran per ml in Tris-buffered saline (TBS). The solution is sterilised by filtration and stored in aliquots at $-20°C$.

(ii)   TBS is 25 mM Tris-HCl pH 7.4, 137 mM NaCl, 5 mM KCl, 0.7 mM $CaCl_2$, 0.5 mM $MgCl$, 0.6 mM Na_2HPO_4.

6.2 **Protocol for the Transfection of a Monolayer Cell Culture on a 10 cm Plate**

(i)   Wash the cells with DMEM. Be sure to use serum free medium and wash four times. The presence of serum during this time can inhibit the transfection. Wash a further time with TBS.

(ii)   Prepare a solution containing 20 μg of DNA in 250 μl of TBS and 80 μl of the DEAE-dextran stock.

(iii) Add the solution to 4 ml of DMEM (serum free) and pipette on to the washed cells.

(iv) Incubate the cells for 3 h at 37°C. This is the optimal length of time for most monolayer cell cultures.

(v)   Wash the cells twice in TBS and add fresh medium.

6.3 **Variations on the Basic Protocol**

6.3.1 *Transfection of Lymphocytes*

(i)   From the 100 mg/ml stock of DEAE-dextran (Section 6.1), make up a new stock solution of 1 mg/DEAE-dextran per ml in TBS and store at 4°C.

(ii)   Proceed as in the protocol of Section 6.2 except that in step (ii) use 250 μl of TBS containing 5 mg of DNA and add this to 250 μl of DEAE-dextran (1 mg/ml).

(iii)   Suspend 10^7 cells in the 0.5 ml DNA-DEAE dextran solution. Incubate the cell suspension at room temperature for 30 min.

(iv)   Add 4.5 ml of TBS and pellet the cells by centrifugation.

(v)   Wash the cells twice with normal medium and seed into multiwell plates.

The time course of transient expression may vary depending on the DNA. Following the transfection with DEAE-dextran the optimum harvest time is 24−48 h.

6.3.2 *Chloroquine Treatment*

Chloroquine has been used to increase the efficiency of expression of plasmid DNA following transfection using either the calcium phosphate or DEAE-dextran methods (13). The amount of chloroquine may need to be titrated for specific types of cell, but the following protocol is commonly used. Chloroquine (Sigma) is kept as a 2 mg/ml stock solution, that is stored at 4°C in the dark for up to one week. Use a 1:100 dilution to give a final concentration of 200 μg/ml in DMEM. Good results have been achieved when the chloroquine is added with the DNA solution, provided that exposure to the cells is not prolonged. Increased toxicity often occurs about 4 h of exposure.

6.3.3 *Glycerol Shock Following DEAE-dextran*

Recently (16) it has been shown that indeed the use of a glycerol shock following transfection with DEAE-dextran does increase the transient expression levels. The shock is as described in Section 5.2 with the difference that the optimum time for harvest is 72 h following transfection. It is also necessary to use more test plasmid DNA (between 15 and 25 μg per 60 mm plate of cells).

7. TRANSIENT EXPRESSION

As has been mentioned above there can often be more than one experimental protocol which should be tested for transfection efficiency of various cell types. The ideal method for testing protocols is transient expression because it is relatively easy and can be used to define various parameters quickly. By assaying for expression of a particular enzyme encoded specifically by the plasmid DNA, one can test the various protocols and determine the best one for a particular system. With this approach it has proven useful to adopt methodology long used in prokaryotic research. Control sequences can be fused to a readily assayable protein coding region which allows quantitation of a functional gene by enzyme assay. This method can be more sensitive and quantitative than measuring actual

Table 3. Chloramphenicol Acetyl Transferase Assay.

1.	Following transfection with one of the transient methods described above, the cells are washed wtih PBS, harvested and transferred to a 1.5 ml microcentrifuge tube.
2.	To the cell pellet add 100 μl of 0.25 mM Tris-HCl pH 7.8.
3.	Disrupt the cells by freezing and thawing. To freeze-thaw, immerse the tube in an ethanol-dry ice bath for 5 min and then transfer to a 37°C bath. Repeat the cycle three times.
4.	Spin down the debris and save the supernatant to test for enzyme activity. Samples may be saved at this point by storage at -20°C.
5.	Depending on the cell type and promoter to be assayed the amount of extract assayed may vary. The reaction mixture contains:

35   70 μl 0.25 M Tris-HCl pH 7.8

             35 μl water

20   20 μl cell extract

146 0.5   1 μ[^{14}C]Chloramphenicol (40 – 50 Ci/mmol) (NEN, or Amersham)

    10   20 μl of 4 mM acetyl Co-A[a].

6.	Incubate the reaction mixture for 10 – 30 min at 37°C. The incubation time can be increased up to 60 min provided enough active acetyl Co-A[b] is added to keep the assay linear.
7.	Extract chloramphenicol with 1 ml ethyl acetate by vortexing for 30 sec.
8.	Spin in a microcentrifuge and save the top organic phase which will contain all forms of chloramphenicol (the two forms of monoacetate, the di-acetate and unconverted chloramphenicol, *Figure 3*).
9.	Dry down the ethyl acetate under vacumm. This step requires about 2 h.
10.	Resuspend the chloramphenicol samples in 20 μl ethyl acetate[c] and spot into silica gel thin layer chromatography plates. (These can be obtained from either Baker or Merck.)
11.	The plates are subjected to ascending chromatography[d] with a 95:5 mixture of chloroform:methanol.
12.	After air drying, expose the chromatography plate to X-ray film overnight. The percentage of the total chloramphenicol which has been converted to the monoacetate form gives an estimate of transcriptional activity and therefore efficiency of transfection.

[a]4 mM acetyl CoA is made by dissolving 1.5 mg in 0.5 ml water. = 3 ~g/~l
[b]Acetyl CoA is very unstable it should be made up fresh or kept at -20°C for not more than 10 days.
[c]Ethyl acetate should only be pipetted with glass pipettes.
[d]The t.l.c. tank should be lined with filter paper around the inside to assist equilibration. The solvent should be made fresh each day, since chloroform is very volatile.

transcripts. Another advantage to this approach is that it allows one to investigate variations in a basic protocol with relative ease. The most widely used assay systems utilise bacterial genes. The strength of this approach is that the bacterial product can be easily distinguished from mammalian isozymes. In cases where the gene for CAT is used, then there is no mammalian counterpart. The two most prevalent assay systems are detailed below. Other genes which have been used as assay systems are the *E. coli* galactokinase (17), *gpt* (18), and *neo*^{R} (19) genes. These are less widely used due to difficulties with the assays. Another slightly different means of determining transfection efficiencies is by antibody staining for expression of the introduced gene. The bacterial gene-products CAT and β-galactosidase can readily be detected by antibody staining. The expression of viral antigens is also widely used as a measure of transfection.

7.1 Chloramphenicol Acetyl Transferase Assay

This gene confers chloramphenicol resistance to bacteria. It has been studied ex-

CAT assay of extract from 5 x 10⁶ cells

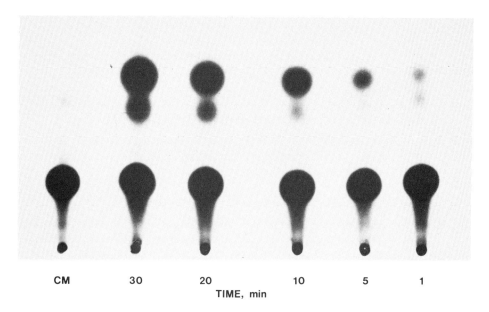

Figure 3. CAT Assay. A time course of a CAT assay is shown. The extract tested was made from CV-1 cells transfected with pRSVcat. The amount of actylated chloramphenicol (CM) can easily be seen to increase with an increase in assay time. This assay shown uses only 1/20th of the extract made from a 9 cm dish. The CAT assay is very sensitive, detecting less than 1 picogram of CAT enzyme.

tensively and sensitive assays of its enzymatic activity are available. Since the coding region is rather small (700 bp), vectors carrying this gene are easily modified. The expression of the bacterial gene CAT in mammalian cells permits an easy assay for promoters in eukaryotic vectors (20). The prototype vectors pSV2cat (20) (*Figure 1*) and pRSVcat (3) (*Figure 2*) utilise either the SV40 promoter or the Rous sarcoma virus long terminal repeat to direct transcription of this bacterial gene.

The assay, which follows the conversion of [¹⁴C]chloramphenicol to its acetylation derivatives by thin layer chromatography, is given in *Table 3*. Control CAT can be purchased from PL Biochemicals. A time course of a CAT assay is shown in *Figure 3*.

7.2 β-Galactosidase Assay

The bacterial gene for β-galactosidase is another coding region used as a transient marker protein (21). Construction of vectors containing this gene are a little more cumbersome to manipulate due to the large coding region and this assay, while quite simple, is rather less sensitive than the CAT assay. It serves as a good control to monitor the competence of cells. The two enzymes, CAT and β-galactosidase, can be assayed in the same cellular extract.

(i)   To assay for β-galactosidase, lyse the cells in 0.25 M Tris-HCl pH 7.8, 5 mM dithiothreitol.

157

(ii)   Set up the following reaction mixture:
   $10 - 150 \ \mu l$ of cell lysate
    $1 \ \mu l$ of 60 mM Na_2HPO_4, 40 mM NaH_2PO_4, 10 mM KCl, 1 mM
    $MgCl_2$, 50 mM β-mercaptoethanol
    0.2 ml of 2 mg/ml ONPG (o-nitrophenyl-β-D-galactopyrano-
    side) dissolved in 60 mM Na_2HPO_4, 40 mM NaH_2PO_4.

(iii) Incubate the reaction at 37°C until a visible yellow colour is achieved.

(iv)   Stop the reaction by adding 0.5 ml of 1 M $NaCO_3$. Measure the color-
  metric change in a spectrophotometer at 420 nm.

It is important to remember that mammalian cells contain a eukaryotic isozyme for β-galactosidase. Therefore a blank containing a non-transfected cellular lysate should be included in the experiment.

7.3 Assay of Transfection Efficiency by Antibody Staining

The efficiency of transfection can be estimated from an assay of transcriptional activity such as that described above. In addition the actual number of cells which are expressing the transfected DNA can be determined by staining the cells with an antibody that recognises an antigen encoded by the plamsid DNA. This technique, although not quantitative, is a little more flexible, since any protein that is not normally expressed in the recipient cell can be used as a marker protein provided that antibodies are available. In addition to the two bacterial genes described above other widely used marker genes are β-globin, and various virus antigens. A staining method using peroxidase-coupled antibodies is much easier to use and more sensitive than methods using fluorescent antibodies. This technique, which is described below, can be varied to use fluorescent reagents if desired.

The indirect antibody staining of transfected cells is carried out as follows:

(i)   The cells may be grown on plates of any size. However, 35 mm dishes of six
  well dishes are convenient.

(ii) Fix the cells for 2 min with a freshly made solution containing equal
  volumes of methanol and acetone and wash with PBS.

(iii) The proper dilution of the antibody should be determined by first conducting an antibody dilution curve. In many cases it is convenient to use the
  supernatant from a hybridoma culture either undiluted or diluted 10-fold.
  Ascites fluid dilutions should be tested ranging from 100 to 5000. Dilutions
  of all antibodies at this and at subsequent steps should be carried out in
  10% foetal calf serum, 1% bovine serum albumin sterile PBS.

(iv) For a 35 mm dish add 1 ml of the first antibody and incubate at room
  temperature for $2-3$ h.

(v)   Remove the antibody solution. Wash for 10 min with four changes of PBS.

(vi) Add the second antibody-peroxidase conjugate diluted in the solution
  described above. Again it may be necessary to carry out a dilution curve to
  determine the proper dilution for the second antibody. The suggested range
  for the titration is 1:50 to 1:1000. The incubation period for the second
  antibody can be as short as one hour at room temperature or as long as
  over night at 4°C.

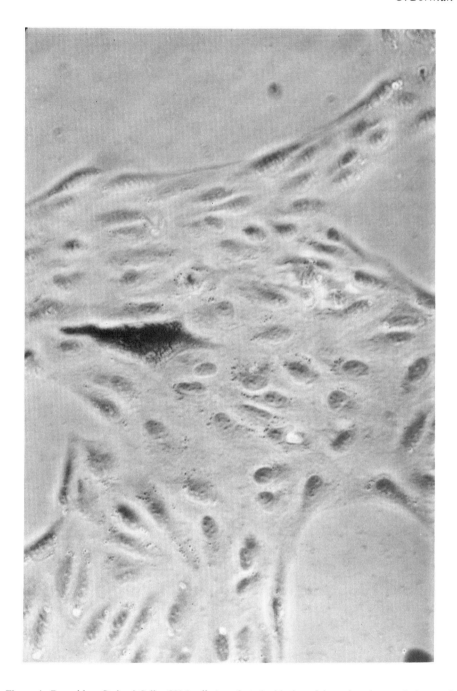

Figure 4. Peroxidase Stained Cells. CV-1 cells transfected with the calcium phosphate technique and pSV2cat plasmid, and stained with a monoclonal antibody for CAT followed by a secondary antibody conjugated with peroxidase.

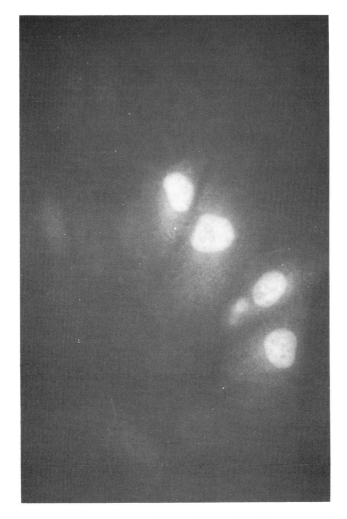

Figure 5. Fluorescent Stained Cells. CV-1 cells transfected with pRSVcat DNA using the calcium phosphate technique followed by butyrate treatment. The first antibody was a polyclonal antibody for CAT followed by a rhodamine conjugated second antibody.

(vii) Remove the antibody solution and wash as in step (v).

(viii) The substrate for the peroxidase is now added. O-Diansidine (Sigma) has been found to give the best cyto-chemical staining with a very low background. To prepare the substrate solution, first make a saturated solution of O-diansidine in absolute ethanol. After allowing the undissolved substrate to settle, add 1 ml of this solution to 99 ml of PBS, followed by 10 μl of hydrogen peroxide (a dilution of 1:10 000). A small amount of the substrate (1 ml) may be tested by adding 1 μl of the peroxidase conjugated antibody, following which a dark brown-purple colour should appear. Add the solution of substrate to the plates and allow 1 hour at room temperature for the colour reaction to develop. Rinse the plates well with distilled water and mount under coverslips. The color change should be permanent (*Figure 4*).

For fluorescence follow the above steps (i) − (v). Then use fluorescent second antibody rather than the peroxidase antibody. Remember the use of this kind of reagent requires that a fluorescent microscope is available. It is also somewhat limited in that the plates cannot be kept permanently as the fluorescence fades. However, very satisfactory results can be achieved in this way (*Figure 5*).

It is possible to determine the best transfection protocol for a particular cell-line by either assaying for the enzyme activity present after transfection or by determining the number of transfected cells in a population. The task is then 2-fold in choosing which gene transfer system to use. It is best to try different vectors as well as variations of the basic transfection protocols. Time carefully spent in determining the best protocol at this stage will save both time and frustration later in long-term experiments.

8. STABLE EXPRESSION

This section will describe the use of possible selection protocols: the thymidine kinase selection, based on the expression of exogenously added thymidine kinase (*tk*) gene in mutant cells lacking the endogenous activity, and the two most widely uesd dominant selectable markers which use the expression of the *E. coli* genes *gpt* or *neo*R. The use of a dominant marker means that the recipient cells need not be genetically marked and so these latter two systems are very versatile. Transient expression experiments carried out as described above can be used to determine which controlling elements should be included in a vector for expression within a particular cell type. The general protocol to be followed to achieve stable transformation involves subculturing the cells into the selective medium two days following transfection. The cells are seeded into the appropriate selective medium at a density of 5 x 10⁴ cells/10 cm dish. The medium is changed once every 4 − 5 days until cells on control plates, that have not received DNA, die under selection. Discrete colonies can then be seen in the transfected plates.

8.1 Thymidine Kinase Selection

This selection can be used only with cells which are deficient in the expression of TK (22). By placing these mutant cells into the appropriate media cell death will occur. The cells lack TK and therefore the pathway using thymidine to make inosine monophosphate (IMP) is blocked. By growing the cells in medium with aminopterin to inhibit the conversion of other precursors to IMP death will occur. To rescue these cells by the thymidine pathway, the *tk* gene is supplied by the plasmid DNA. This, with the addition of thymidine and hypoxanthine, allows only the mutant cells which are expressing the exogenously added *tk* gene to survive. The selective medium is colloquially referred to as HAT medium. It contains 10^{-4} M hypoxanthine, 2×10^{-5} M amenopterin and 10^{-4} M thymidine. Stock solutions which can then be easily diluted upon addition to DMEM are given below:

(i)   Dilute a 10^{-2} M hypoxanthine stock solution 400-fold.

(ii) Dilute a 10^{-3} M amenopterin stock solution 1000-fold.
(iii) Dilute a 0.04 M thymidine stock solution 100-fold.

Since this selection can only be used with mutant cells, one is limited to the recipient cell types that are available. Mouse L tk − cells are the most commonly used. The two dominant selections described below are a little more versatile.

8.2 GPT Selection

In a way, dominant selection using the *gpt* gene, which utilises (23) the expression of the bacterial enzyme hypoxanthine phosphoribosyltransferase, is a variation of the tk selection. Again, the pathway converting precursors to IMP is blocked by amenopterin as with the HAT selection. Mycophenolic acid, an inhibitor of IMP dehydrogenase, prevents the formation of guanosine monophosphate. With both these inhibitors present, supplementing the medium with hypoxanthine is not sufficient. However, inhibition can be overcome if both guanine and hypoxanthine are provided. Instead of directly supplying guanine the cells can be selected by their ability to convert xanthine to guanine. Mammalian enzymes cannot make this conversion. Cells expressing the bacterial gene can, however, use exogenously added xanthine, converted to guanine, for survival. The prototype vectors for this selection are pSV2gpt (18) and pRSVgpt (4). The selection works very well for NIH 3T3, rat cells, monkey kidney CV1 cells and CHO cells. A description of the selective medium is given in *Table 4*.

8.3 Selection by Neomycin Resistance

This dominant selection is based on the idea of survival due to drug resistance. Neomycin is a bacterial antibiotic which interfers with prokaryotic ribosomes. While mammalian cells are not affected by neomycin or kanamycin, an analogue to these drugs will effect eukaryotic ribosomes. This analogue G418 is now available through Gibco as Geneticin. The *neo*^R gene codes for a phosphotransferase which inactivates the G418. Therefore cells which express this bacterial resistance marker, under eukaryotic control, can survive in selective media. This selection can be used on almost any cell type. Prototype vectors pSV2neo (24), pRSVneo (4).

Geneticin (Gibco) is added to the culture medium at a concentration that is determined empirically. A titration curve is necessary to determine the amount of geneticin necessary to kill untransfected cells. The geneticin can be kept as a stock solution at 100 mg/ml. Try a range between 200 μg/ml and 1 mg/ml. Cells can escape selection either if too low a dose of antibiotic is used or if the plating density is too high initially. Ideally control cells should die in one week allowing colonies to form by 10−14 days.

9. GENE TRANSFER OF CELLULAR DNA

This chapter has concentrated on gene transfer experiments using plasmid vectors. These vectors should be used to define the best technique required for gene transfer into a particular cell system. Plasmid vectors can also be used to study controls of gene expression. However, an aspect of gene transfer not mentioned

Table 4. Medium for gpt Selection.

A. *Stock Solutions*

1. Xanthine is dissolved in 0.1 M NaOH at a concentration of 5 mg/ml. Adjust the pH to 10.5 with HCl. Sterilise by filtration and store at room temperature.
2. Hypoxanthine solution contains 136 mg of hypoxanthine dissolved in 80 ml of water. Add 5.0 ml of 0.1 M NaOH and then adjust the pH to 9.5. Make-up to 100 ml with water. Sterilise by filtration and store at room temperature. This solution may be difficult to dissolve at first and so should be stirred for $10-15$ min.
3. Thymidine is dissolved at a concentration of 59 μg/100 ml of water. Sterilise by filtration. Store at 4°C.
4. Dissolve glycine at 500 mg/100 ml of water. Sterilise by filtration and store at 4°C.
5. Aminopterin. Dissolve 8.8 mg of aminopterin in 10 ml of PBS. Sterilise by filtration and freeze 1 ml aliquots. This solution should be protected from the light.
6. Mycophenolic acid. This is obtained from Eli Lilly by special request. It is dissolved in absolute alcohol at 25 mg/ml glass and tube.
7. Dialysed serum. Dialyse 100 ml of fetal calf serum against 2 l of 0.9% NaCl at 4°C for 2 days. Change the saline and dialyse for a further 2 days. Repeat with two further changes. Sterilise by filtration under positive pressure.

B. *GPT Selective Media*

This is made up fresh immediately before use with the stock solutions described above. A 100 ml solution is made up as follows:

DMEM (with antiboitcs as described in section 4)	80 ml
Dialysed FCS	10 ml
Xanthine stock (see 8.2.1)	5 ml
Glycine	0.2 ml
Hypoxanthine	1 ml
Thymidine	2 ml
Aminopterin	0.225 ml
Mycophenolic acid	0.1 ml

Table 5. Preparation of Cellular DNA Suitable for Transfection.

1. Prepare enough dishes of cells to yield about 10^8 cells. This is about four 75 cm^2 flasks or eight 130 mm dishes.
2. Wash the cells with PBS. Repeat this wash twice.
3. Add 5 ml of proteinase K solution[a] to the cells. Pool the cells of two dishes or flasks into a 50 ml tube, and digest with the proteinase K for 6 h at 47°C.
4. Extract the DNA solution twice with 10 ml of phenol. Use only redistilled phenol buffered with 50 mM Tris-HCl pH 7.5.
5. Dialyse the aqueous phase from the phenol extraction against 4 liters of TEN[b] for 24 h. Dialysis should be carried out in such a manner that the volume of the DNA solution is allowed to increase by three fold. Repeat this step once.
6. Prepare the DNA solution for CsCl gradient banding by adding 1.058 g CsCl for each ml of solution. Add 0.84 ml of 10 mg/ml ethidium bromide.
7. The DNA is banded at 34 000 r.p.m. for 72 h using a Ti 60 rotor or 24 h using a VTi 50.
8. To remove the band of cellular DNA follow the instruction given in Section 3.2, but be sure to use a large gauge needle such as a 16 gauge to prevent shearing the DNA. Prepare the DNA for precipitation as described in Section 3.2.

[a]The proteinase K solution is made by dissolving 5 mg of proteinase K in 23.8 ml 0.5 M EDTA pH 8.0 and 1.25 ml of 10% sarkosyl.
[b]TEN is 5 mM Tris-HCl pH 7.5, 2.5 mM EDTA, 20 mM NaCl.

yet is the transfer of cellular genes from one cell into another. The number of recipient cells which integrate donor cellular DNA is few. Therefore it is necessary to have an experimental protocol which allows the selection of these cells from within a population. Two basic approaches have been used in isolating clones which have received cellular DNA transferred from a donor cell. The first is based on the selection of a particular phenotype conveyed by a cellular gene which has been transferred. For this type of experiment the recipient cells must lack the phenotypic function provided by the donor DNA. An example of such an experiment is described by Lowy *et al.* (25). Here, mutant cells lacking adenine phosphoribosyl transferase (APRT) were used as the recipient cells. Cellular DNA taken from normal cells was used as donor DNA. Clones were selected for their ability to survive in HAT medium. Only cells which have the *aprt* gene can grow in this medium. Therefore cells which survive contain a functional *aprt* gene from the donor cellular DNA. The transfection efficiency in these experiment was 10^{-4}. That means for every 10 000 cells which were exposed to the donor DNA, one cell was able to survive in selection and contained the *aprt* gene. The limitations of this approach are that the gene one hopes to clone must convey a selectable phenotype and there must be an appropriate recipient cell lacking the expression of the gene is question.

If however, the gene of interest does not convey a selectable phenotype the technique of co-transfection can be used. Co-transfection (26) means using a selectable plasmid marker in addition to the genomic DNA. This allows the selection of cells which have integrated donor DNA based on the expression of the selectable plasmid. These clones can then be screened for the expression of the desired gene. Transfection efficiencies using plasmid vector can be quite high (10^{-1}) increasing the probability that one of the clones will also carry a specific cellular gene. The two approaches can be combined by selecting first for plasmid gene function to isolate any cell expressing foreign DNA. A secondary selection can then be used for specific phenotypic expression. Using the co-transfection techniques, 25% of the clones have been seen to contain both the plasmid vector and genomic DNA (24).

The protocols for a genomic transfer do not differ from those described for plasmid vectors. However, again, the state of the DNA is very important in achieving efficient transfection. With this in mind, a suggested protocol for preparing high molecular weight genomic DNA is provided in *Table 5*.

10. REFERENCES

1. Khoury,G. and Gruss,P. (1983) *Cell,* **33**, 313.
2. Laimons,L. *et al.* (1982) *Proc. Natl. Acad. Sci. USA,* **79**, 6453.
3. Gorman,C. *et al.* (1982) *Proc. Natl. Acad. Sci. USA,* **79**, 6777.
4. Gorman,C., Padmanabhan,R. and Howard,B. (1983) *Science,* **221**, 551.
5. Pelham,H. and Bienz,M. (1982) *EMBO J.,* **1**, 1473.
6. Mayo,K., Warren,R. and Palmiter,R. (1982) *Cell,* **29**, 99.
7. Hamer,D. and Walling,M.J. (1982) *J. Mol. Appl. Genet.,* **1**, 273.
8. Karin,M. and Richards,R. (1982) *Nature,* **299**, 797.
9. Robins,D. *et al.* (1982) *Cell,* **29**, 623.
10. Lee,F. *et al.* (1981) *Nature,* **294**, 228.
11. Graham,F. and van der Eb,A. (1973) *Virology,* **52**, 456.

12. Gorman,C., Howard,B. and Reeves,R. (1983) *Nucleic Acids Res.,* **11**, 7631.
13. Luthman,H. and Magnusson,G. (1983) *Nucleic Acids Res.,* **11**, 1295.
14. Sompayrac,L. and Danna,K. (1981) *Proc. Natl. Acad. Sci. USA,* **12**, 7575.
15. Banerji,J., Olson,L. and Schaffner,W. (1983) *Cell,* **33**, 729.
16. Lopata,M., Cleveland,D. and Sollinar-Webb,B. (1984) *Nucleic Acids Res.,* **12**, 5707.
17. Schumperli,D., Howard,B. and Rosenberg,M. (1982) *Proc. Natl. Acad. Sci. USA,* **79**, 257.
18. Mulligan,R. and Berg,P. (1980) *Science,* **209**, 1422.
19. Scholer,H. and Gruss,P. (1984) *Cell,* **36**, 403.
20. Gorman,C., Moffat,L. and Howard,B. (1982) *Mol. Cell. Biol.,* **2**, 1044.
21. Hall,C. *et al.* (1983) *J. Mol. Appl. Genet.,* **2**, 101.
22. Wigler,M. *et al.* (1977) *Cell,* **11**, 223.
23. Mulligan,R. and Berg,P. (1981) *Proc. Natl. Acad. Sci. USA,* **78**, 2072.
24. Southern,P. and Berg,P. (1982) *J. Mol. Appl. Genet.,* **1**, 327.
25. Wigler,M. *et al.* (1979) *Cell,* **16**, 777.
26. Lowy,I. *et al.* (1980) *Cell,* **22**, 817.
27. Queen,C. and Korn,L. (1979) *Methods in Enzymology,* **65**, pp.

APPENDIX
The Sequence of pSV2cat

```
          10         20         30         40         50         60
CTTTTTGCAA AAGCCTAGGC CTCCAAAAAA GCCTCCTCAC TACTTCTGGA ATAGCTCAGA

          70         80         90        100        110        120
GGCCGAGGCG GCCTCGGCCT CTGCATAAAT AAAAAAAATT AGTCAGCCAT GGGGCGGAGA

         130        140        150        160        170        180
ATGGGCGGAA CTGGGCGGAG TTAGGGGCGG GATGGGCGGA GTTAGGGGCG GGACTATGGT

         190        200        210        220        230        240
TGCTGACTAA TTGAGATGCA TGCTTTGCAT ACTTCTGCCT GCTGGGGAGC CTGGTTGCTG

         250        260        270        280        290        300
ACTAATTGAG ATGCATGCTT TGCATACTTC TGCCTGCTGG GGAGCCTGGG GACTTTCCAC

         310        320        330        340        350        360
ACCCTAACTG ACACACATTC CACAGCTGCC TCGCGCGTTT CGGTGATGAC GGTGAAAACC

         370        380        390        400        410        420
TCTGACACAT GCAGCTCCCG GAGACGGTCA CAGCTTGTCT GTAAGCGGAT GCCGGGAGCA

         430        440        450        460        470        480
GACAAGCCCG TCAGGGCGCG TCAGCGGGTG TTGGCGGGTG TCGGGGCGCA GCCATGACCC

         490        500        510        520        530        540
AGTCACGTAG CGATAGCGGA GTGTATACTG GCTTAACTAT GCGGCATCAG AGCAGATTGT

         550        560        570        580        590        600
ACTGAGAGTG CACCATATGC GGTGTGAAAT ACCGCACAGA TGCGTAAGGA GAAAATACCG

         610        620        630        640        650        660
CATCAGGCGC TCTTCCGCTT CCTCGCTCAC TGACTCGCTG CGCTCGGTCG TTCGGCTGCG

         670        680        690        700        710        720
GCGAGCGGTA TCAGCTCACT CAAAGGCGGT AATACGGTTA TCCACAGAAT CAGGGGATAA

         730        740        750        760        770        780
CGCAGGAAAG AACATGTGAG CAAAAGGCCA GCAAAAGGCC AGGAACCGTA AAAAGGCCGC

         790        800        810        820        830        840
GTTGCTGGCG TTTTTCCATA GGCTCCGCCC CCCTGACGAG CATCACAAAA ATCGACGCTC

         850        860        870        880        890        900
AAGTCAGAGG TGGCGAAACC CGACAGGACT ATAAAGATAC CAGGCGTTTC CCCCTGGAAG

         910        920        930        940        950        960
CTCCCTCGTG CGCTCTCCTG TTCCGACCCT GCCGCTTACC GGATACCTGT CCGCCTTTCT
```

165

```
      970        980  _       990       1000       1010  _      1020
CCCTTCGGGA AGCGTGGCGC TTTCTCAATG CTCACGCTGT AGGTATCTCA GTTCGGTGTA

     1030        1040       1050  _      1060       1070       1080
GGTCGTTCGC TCCAAGCTGG GCTGTGTGCA CGAACCCCCC GTTCAGCCCG ACCGCTGCGC

     1090       1100       1110       1120       1130       1140
CTTATCCGGT AACTATCGTC TTGAGTCCAA CCCGGTAAGA CACGACTTAT CGCCACTGGC

     1150       1160       1170       1180       1190       1200
AGCAGCCACT GGTAACAGGA TTAGCAGAGC GAGGTATGTA GGCGGTGCTA CAGAGTTCTT

     1210       1220       1230       1240       1250       1260
GAAGTGGTGG CCTAACTACG GCTACACTAG AAGGACAGTA TTTGGTATCT GCGCTCTGCT

     1270       1280       1290       1300       1310       1320
GAAGCCAGTT ACCTTCGGAA AAAGAGTTGG TAGCTCTTGA TCCGGCAAAC AAACCACCGC

     1330       1340       1350       1360       1370       1380
TGGTAGCGGT GGTTTTTTTG TTTGCAAGCA GCAGATTACG CGCAGAAAAA AAGGATCTCA

     1390       1400       1410       1420       1430       1440
AGAAGATCCT TTGATCTTTT CTACGGGGTC TGACGCTCAG TGGAACGAAA ACTCACGTTA

     1450       1460       1470       1480       1490       1500
AGGGATTTTG GTCATGAGAT TATCAAAAAG GATCTTCACC TAGATCCTTT TAAATTAAAA

     1510       1520       1530       1540       1550       1560
ATGAAGTTTT AAATCAATCT AAAGTATATA TGAGTAAACT TGGTCTGACA GTTACCAATG

     1570       1580       1590       1600       1610       1620
CTTAATCAGT GAGGCACCTA TCTCAGCGAT CTGTCTATTT CGTTCATCCA TAGTTGCCTG

     1630       1640       1650       1660       1670       1680
ACTCCCCGTC GTGTAGATAA CTACGATACG GGAGGGCTTA CCATCTGGCC CCAGTGCTGC

     1690       1700       1710       1720       1730       1740
AATGATACCG CGAGACCCAC GCTCACCGGC TCCAGATTTA TCAGCAATAA ACCAGCCAGC

     1750       1760       1770       1780       1790       1800
CGGAAGGGCC GAGCGCAGAA GTGGTCCTGC AACTTTATCC GCCTCCATCC AGTCTATTAA

     1810       1820       1830       1840       1850       1860
TTGTTGCCGG GAAGCTAGAG TAAGTAGTTC GCCAGTTAAT AGTTTGCGCA ACGTTGTTGC

     1870       1880       1890       1900       1910       1920
CATTGCTGCA GGCATCGTGG TGTCACGCTC GTCGTTTGGT ATGGCTTCAT TCAGCTCCGG

     1930       1940       1950       1960       1970       1980
TTCCCAACGA TCAAGGCGAG TTACATGATC CCCCATGTTG TGCAAAAAAG CGGTTAGCTC

     1990       2000       2010       2020       2030       2040
CTTCGGTCCT CCGATCGTTG TCAGAAGTAA GTTGGCCGCA GTGTTATCAC TCATGGTTAT

     2050       2060       2070       2080       2090       2100
GGCAGCACTG CATAATTCTC TTACTGTCAT GCCATCCGTA AGATGCTTTT CTGTGACTGG

     2110       2120       2130       2140       2150       2160
TGAGTACTCA ACCAAGTCAT TCTGAGAATA GTGTATGCGG CGACCGAGTT GCTCTTGCCC

     2170       2180       2190       2200       2210       2220
GGCGTCAACA CGGGATAATA CCGCGCCACA TAGCAGAACT TTAAAAGTGC TCATCATTGG

     2230       2240       2250       2260       2270       2280
AAAACGTTCT TCGGGGCGAA AACTCTCAAG GATCTTACCG CTGTTGAGAT CCAGTTCGAT

     2290       2300       2310       2320       2330       2340
GTAACCCACT CGTGCACCCA ACTGATCTTC AGCATCTTTT ACTTTCACCA GCGTTTCTGG

     2350       2360       2370       2380       2390       2400
GTGAGCAAAA ACAGGAAGGC AAAATGCCGC AAAAAAGGGA ATAAGGGCGA CACGGAAATG

     2410       2420       2430       2440       2450       2460
TTGAATACTC ATACTCTTCC TTTTTCAATA TTATTGAAGC ATTTATCAGG GTTATTGTCT
```

```
      2470       2480       2490       2500       2510       2520
CATGAGCGGA TACATATTTG AATGTATTTA GAAAAATAAA CAAATAGGGG TTCCGCGCAC

      2530       2540       2550       2560       2570       2580
ATTTCCCCGA AAAGTGCCAC CTGACGTCTA AGAAACCATT ATTATCATGA CATTAACCTA

      2590       2600       2610       2620       2630       2640
TAAAAATAGG CGTATCACGA GGCCCTTTCG TCTTCAAGAA TTCCTTTGCC TAATTTAAAT

      2650       2660       2670       2680       2690       2700
GAGGACTTAA CCTGTGGAAA TATTTTGATG TGGGAAGCTG TTACTGTTAA AACTGAGGTT

      2710       2720       2730       2740       2750       2760
ATTGGGGTAA CTGCTATGTT AAACTTGCAT TCAGGGACAC AAAAAACTCA TGAAAATGGT

      2770       2780       2790       2800       2810       2820
GCTGGAAAAC CCATTCAAGG GTCAAATTTT CATTTTTTTG CTGTTGGTGG GGAACCTTTG

      2830       2840       2850       2860       2870       2880
GAGCTGCAGG GTGTGTTAGC AAACTACAGG ACCAAATATC CTGCTCAAAC TGTAACCCCA

      2890       2900       2910       2920       2930       2940
AAAAATGCTA CAGTTGACAG TCAGCAGATG AACACTGACC ACAAGGCTGT TTTGGATAAG

      2950       2960       2970       2980       2990       3000
GATAATGCTT ATCCAGTGGA GTGCTGGGTT CCTGATCCAA GTAAAAATGA AAACACTAGA

      3010       3020       3030       3040       3050       3060
TATTTTGGAA CCTACACAGG TGGGGAAAAT GTGCCTCCTG TTTTGCACAT TACTAACACA

      3070       3080       3090       3100       3110       3120
GCAACCACAG TGCTTCTTGA TGAGCAGGGT GTTGGGCCCT TGTGCAAAGC TGACAGCTTG

      3130       3140       3150       3160       3170       3180
TATGTTTCTG CTGTTGACAT TTGTGGGCTG TTTACCAACA CTTCTGGAAC ACAGCAGTGG

      3190       3200       3210       3220       3230       3240
AAGGGACTTC CCAGATATTT TAAAATTACC CTTAGAAAGC GGTCTGTGAA AAACCCCTAC

      3250       3260       3270       3280       3290       3300
CCAATTTCCT TTTTGTTAAG TGACCTAATT AACAGGAGGA CACAGAGGGT GGATGGGCAG

      3310       3320       3330       3340       3350       3360
CCTATGATTG GAATGTCCTC TCAAGTAGAG GAGGTTAGGG TTTATGAGGA CACAGAGGAG

      3370       3380       3390       3400       3410       3420
CTTCCTGGGG ATCCAGACAT GATAAGATAC ATTGATGAGT TTGGACAAAC CACAACTAGA

      3430       3440       3450       3460       3470       3480
ATGCAGTGAA AAAAATGCTT TATTTGTGAA ATTTGTGATG CTATTGCTTT ATTTGTAACC

      3490       3500       3510       3520       3530       3540
ATTATAAGCT GCAATAAACA AGTTAACAAC AACAATTGCA TTCATTTTAT GTTTCAGGTT

      3550       3560       3570       3580       3590       3600
CAGGGGGAGG TGTGGGAGGT TTTTTAAAGC AAGTAAAACC TCTACAAATG TGGTATGGCT

      3610       3620       3630       3640       3650       3660
GATTATGATC TCTAGTCAAG GCACTATACA TCAAATATTC CTTATTAACC CCTTTACAAA

      3670       3680       3690       3700       3710       3720
TTAAAAAGCT AAAGGTACAC AATTTTTGAG CATAGTTATT AATAGCAGAC ACTCTATGCC

      3730       3740       3750       3760       3770       3780
TGTGTGGAGT AAGAAAAAAC AGTATGTTAT GATTATAACT GTTATGCCTA CTTATAAAGG

      3790       3800       3810       3820       3830       3840
TTACAGAATA TTTTTCCATA ATTTTCTTGT ATAGCAGTGC AGCTTTTTCC TTTGTGGTGT

      3850       3860       3870       3880       3890       3900
AAATAGCAAA GCAAGCAAGA GTTCTATTAC TAAACACAGC ATGACTCAAA AAACTTAGCA

      3910       3920       3930       3940       3950       3960
ATTCTGAAGG AAAGTCCTTG GGGTCTTCTA CCTTTCTCTT CTTTTTTGGA GGAGTAGAAT
```

167

```
        3970         3980  __    3990         4000        _____4010        4020
     GTTGAGAGTC AGCAGTAGCC TCATCATCAC TAGATGGCAT TTCTTCTGAG CAAAACAGGT

       __4030         4040         4050         4060         4070        _4080
     TTTCCTCATT AAAGGCATTC CACCACTGCT CCCATTCATC AGTTCCATAG GTTGGAATCT

         4090        _____4100  __    4110         4120         4130 _____ 4140
     AAAATACACA AACAATTAGA ATCAGTAGTT TAACACATTA TACACTTAAA AATTTTATAT

       __4150        _____4160         4170         4180 ___    4190         4200
     TTACCTTAGA GCTTTAAATC TCTGTAGGTA GTTTGTCCAA TTATGTCACA CCACAGAAGT

         4210        _4220  __  __4230         4240 ___    4250         4260
     AAGGTTCCTT CACAAAGATC CGGCGAATTT CTGCCATTCA TCCGCTTATT ATCACTTATT

         4270        _____4280        _4290         4300         4310 __    4320
     CAGGCGTAGC ACCAGGCGTT TAAGGGCACC AATAACTGCC TTAAAAAAAT TACGCCCCGC

         4330        _4340        _4350 _    4360         4370         4380
     CCTGCCACTC ATCGCAGTAC TGTTGTAATT CATTAAGCAT TCTGCCGACA TGGAAGCCAT

         4390         4400        _____4410        _____4420         4430         4440
     CACAGACGGC ATGATGAACC TGAATCGCCA GCGGCATCAG CACCTTGTCG CCTTGCGTAT

        4450        _____4460        4470 ___    4480        _4490 _        _4500
     AATATTTGCC CATGGTGAAA ACGGGGGCGA AGAAGTTGTC CATATTGGCC ACGTTTAAAT

        4510 __        _4520        _____ 4530        _____ 4540         4550         4560
     CAAAACTGGT GAAACTCACC CAGGGATTGG CTGAGACGAA AAACATATTC TCAATAAACC

         4570        _4580        _4590         4600         4610         4620
     CTTTAGGGAA ATAGGCCAGG TTTTCACCGT AACACGCCAC ATCTTGCGAA TATATGTGTA

        4630 _        4640         4650         4660         4670         4680
     GAAACTGCCG GAAATCGTCG TGGTATTCAC TCCAGAGCGA TGAAAACGTT TCAGTTTGCT

         4690         4700        _____4710         4720_        _4730 __    4740
     CATGGAAAAC GGTGTAACAA GGGTGAACAC TATCCCATAT CACCAGCTCA CCGTCTTTCA

         4750        _____4760 __    4770         4780         4790        _4800
     TTGCCATACG GAATTCCGGA TGAGCATTCA TCAGGCGGGC AAGAATGTGA ATAAAGGCCG

     _        4810         4820         4830        _____4840        _____ 4850        _____4860
     GATAAAACTT GTGCTTATTT TTCTTTACGG TCTTTAAAAA GGCCGTAATA TCCAGCTGAA

         4870        _4880         4890         4900 ___    4910         4920
     CGGTCTGGTT ATAGGTACAT TGAGCAACTG ACTGAAATGC CTCAAAATGT TCTTTACGAT

       ___ 4930         4940         4950         4960         4970        _____4980
     GCCATTGGGA TATATCAACG GTGGTATATC CAGTGATTTT TTTCTCCATT TTAGCTTCCT

     _____ 4990         5000 ___
     TAGCTCCTGA AAATCTCGCC AAG
```

```
     MISSRES=          0        REFLECT=          1;        FORM='C';

     P= A OR G,   Q= C OR T,   R= A OR T,   S= C OR G,   V= A OR C,   W=G OR T
```

	# OF SITES	SITES	FRAGMENTS	FRAGMENT ENDS	
AAT 2 (GACGTC)	1				
		2543	5003 (99.9)	2543	2543
ACC 1 (GTVWAC)	1				
		503	5003 (99.9)	503	503

168

	# OF SITES	SITES	FRAGMENTS		FRAGMENT ENDS	
ACY 1 (GPCGQC)	2					
		2161	4621	(92.4)	2543	2161
		2543	382	(7.6)	2161	2543
AHA 3 (TTTAAA)	9					
		1489	1659	(33.2)	4833	1489
		1508	692	(13.8)	1508	2200
		2200	590	(11.8)	3563	4153
		2634	565	(11.3)	2634	3199
		3199	434	(8.7)	2200	2634
		3563	364	(7.3)	3199	3563
		4153	341	(6.8)	4153	4494
		4494	339	(6.8)	4494	4833
		4833	19	(0.4)	1489	1508
ALU 1 (AGCT)	25					
		53	700	(14.0)	1976	2676
		324	575	(11.5)	4150	4725
		373	521	(10.4)	1292	1813
		392	329	(6.6)	3821	4150
		673	286	(5.7)	2822	3108
		899	281	(5.6)	392	673
		1035	271	(5.4)	53	324
		1292	257	(5.1)	1035	1292
		1813	244	(4.9)	3115	3359
		1913	226	(4.5)	673	899
		1976	180	(3.6)	3487	3667
		2676	154	(3.1)	3667	3821
		2822	146	(2.9)	2676	2822
		3108	136	(2.7)	899	1035
		3115	129	(2.6)	4725	4854
		3359	128	(2.6)	3359	3487
		3487	119	(2.4)	4854	4973
		3667	100	(2.0)	1813	1913
		3821	63	(1.3)	1913	1976
		4150	54	(1.1)	5002	53
		4725	49	(1.0)	324	373
		4854	20	(0.4)	4982	5002
		4973	19	(0.4)	373	392
		4982	9	(0.2)	4973	4982
		5002	7	(0.1)	3108	3115
APA 1 (GGGCCC)	1					
		3094	5003	(99.9)	3094	3094
AVA 2 (GGACC)	3					
		1763	3917	(78.3)	2849	1763
		1985	864	(17.3)	1985	2849
		2849	222	(4.4)	1763	1985
AVA 3 (ATGCAT)	2					
		196	4948	(98.9)	251	196
		251	55	(1.1)	196	251
AVR 2 (CCTAGG)	1					
		14	5003	(99.9)	14	14
BAL 1 (TGGCCA)	1					
		4486	5003	(99.9)	4486	4486
BAM H1 (GGATCC)	1					
		3369	5003	(99.9)	3369	3369

	# OF SITES	SITES	FRAGMENTS		FRAGMENT ENDS	
BBV 1 (GCTGC)	16					
		325	1509	(30.2)	3819	325
		371	781	(15.6)	2042	2823
		468	474	(9.5)	2823	3297
		637	419	(8.4)	655	1074
		655	331	(6.6)	3488	3819
		1074	328	(6.6)	1348	1676
		1139	206	(4.1)	1142	1348
		1142	191	(3.8)	3297	3488
		1348	189	(3.8)	1676	1865
		1676	177	(3.5)	1865	2042
		1865	169	(3.4)	468	637
		2042	97	(1.9)	371	468
		2823	65	(1.3)	1074	1139
		3297	46	(0.9)	325	371
		3488	18	(0.4)	637	655
		3819	3	(0.1)	1139	1142
BGL 1 (GCCNNNNNGGC)	2					
		62	3326	(66.5)	1739	62
		1739	1677	(33.5)	62	1739
BVU 1 (GPGCQC)	1					
		3094	5003	(99.9)	3094	3094
DDE 1 (CTNAG)	14					
		55	683	(13.7)	3211	3894
		542	540	(10.8)	1582	2122
		1007	518	(10.4)	2693	3211
		1416	487	(9.7)	55	542
		1582	465	(9.3)	542	1007
		2122	448	(9.0)	4531	4979
		2548	426	(8.5)	2122	2548
		2693	409	(8.2)	1007	1416
		3211	386	(7.7)	4145	4531
		3894	166	(3.3)	1416	1582
		4006	145	(2.9)	2548	2693
		4145	139	(2.8)	4006	4145
		4531	112	(2.2)	3894	4006
		4979	79	(1.6)	4979	55
ECO R1 (GAATTC)	2					
		2618	2870	(57.4)	4751	2618
		4751	2133	(42.6)	2618	4751
ECO R1* (NAATTN)	25					
		96	1249	(25.0)	243	1492
		188	565	(11.3)	2053	2618
		243	419	(8.4)	2784	3203
		1492	405	(8.1)	4346	4751
		1798	348	(7.0)	4751	96
		2053	306	(6.1)	1492	1798
		2618	255	(5.1)	1798	2053
		2631	194	(3.9)	3899	4093
		2784	183	(3.7)	3266	3449
		3203	153	(3.1)	2631	2784
		3242	145	(2.9)	3513	3658
		3266	119	(2.4)	3680	3799
		3449	100	(2.0)	3799	3899
		3513	92	(1.8)	96	188
		3658	82	(1.6)	4225	4307
		3680	64	(1.3)	3449	3513
		3799	55	(1.1)	188	243
		3899	48	(1.0)	4130	4178
		4093	47	(0.9)	4178	4225
		4130	39	(0.8)	3203	3242
		4178	39	(0.8)	4307	4346

	# OF SITES	SITES	FRAGMENTS		FRAGMENT ENDS	
		4225	37	(0.7)	4093	4130
		4307	24	(0.5)	3242	3266
		4346	22	(0.4)	3658	3680
		4751	13	(0.3)	2618	2631
ECO R2 (CCTGG)	9					
		230	2471	(49.4)	893	3364
		285	908	(18.1)	3364	4272
		759	657	(13.1)	4576	230
		880	474	(9.5)	285	759
		893	248	(5.0)	4272	4520
		3364	121	(2.4)	759	880
		4272	56	(1.1)	4520	4576
		4520	55	(1.1)	230	285
		4576	13	(0.3)	880	893
FNU4H 1 (GCNGC)	25					
		68	660	(13.2)	4411	68
		325	592	(11.8)	3819	4411
		371	474	(9.5)	2823	3297
		468	457	(9.1)	2366	2823
		521	331	(6.6)	3488	3819
		637	328	(6.6)	1348	1676
		655	257	(5.1)	68	325
		658	229	(4.6)	2137	2366
		776	206	(4.1)	1142	1348
		931	191	(3.8)	3297	3488
		1074	189	(3.8)	1676	1865
		1139	155	(3.1)	776	931
		1142	150	(3.0)	1865	2015
		1348	143	(2.9)	931	1074
		1676	118	(2.4)	658	776
		1865	116	(2.3)	521	637
		2015	97	(1.9)	371	468
		2042	95	(1.9)	2042	2137
		2137	65	(1.3)	1074	1139
		2366	53	(1.1)	468	521
		2823	46	(0.9)	325	371
		3297	27	(0.5)	2015	2042
		3488	18	(0.4)	637	655
		3819	3	(0.1)	655	658
		4411	3	(0.1)	1139	1142
FOK 1 (GGATG)	8					
		150	1218	(24.3)	2073	3291
		407	1198	(23.9)	407	1605
		1605	948	(18.9)	3291	4239
		1786	519	(10.4)	4239	4758
		2073	395	(7.9)	4758	150
		3291	287	(5.7)	1786	2073
		4239	257	(5.1)	150	407
		4758	181	(3.6)	1605	1786
GDI 1 (RGGCCR)	6					
		17	3278	(65.5)	1208	4486
		745	728	(14.6)	17	745
		756	452	(9.0)	756	1208
		1208	447	(8.9)	4573	17
		4486	87	(1.7)	4486	4573
		4573	11	(0.2)	745	756
HAE 1 (RGGCCR)	6					
		17	3278	(65.5)	1208	4486
		745	728	(14.6)	17	745
		756	452	(9.0)	756	1208
		1208	447	(8.9)	4573	17
		4486	87	(1.7)	4486	4573
		4573	11	(0.2)	745	756

171

	# OF SITES	SITES	FRAGMENTS		FRAGMENT ENDS	
HAE 2 (PGCGCQ)	2					
		606	4633	(92.6)	976	606
		976	370	(7.4)	606	976
HAE 3 (GGCC)	17					
		18	1392	(27.8)	3095	4487
		61	670	(13.4)	76	746
		70	587	(11.7)	2014	2601
		76	494	(9.9)	2601	3095
		746	458	(9.2)	1209	1667
		757	434	(8.7)	775	1209
		775	267	(5.3)	1747	2014
		1209	222	(4.4)	4574	4796
		1667	180	(3.6)	4841	18
		1747	87	(1.7)	4487	4574
		2014	80	(1.6)	1667	1747
		2601	45	(0.9)	4796	4841
		3095	43	(0.9)	18	61
		4487	18	(0.4)	757	775
		4574	11	(0.2)	746	757
		4796	9	(0.2)	61	70
		4841	6	(0.1)	70	76
HGA 1 (GACGC)	4					
		438	3279	(65.5)	2162	438
		834	750	(15.0)	1412	2162
		1412	578	(11.6)	834	1412
		2162	396	(7.9)	438	834
		1483	258	(5.2)	1993	2251
		1588	237	(4.7)	3370	3607
		1929	105	(2.1)	1483	1588
		1947	78	(1.6)	1393	1471
		1993	75	(1.5)	1299	1374
		2251	46	(0.9)	1947	1993
		2268	36	(0.7)	2268	2304
		2304	18	(0.4)	1929	1947
		2974	17	(0.3)	2251	2268
		3370	12	(0.2)	1471	1483
		3607	11	(0.2)	1374	1385
		4217	8	(0.2)	1385	1393
MBO 2 (GAAGA)	11					
		611	1313	(26.2)	2611	3924
		1382	1145	(22.9)	4469	611
		1473	771	(15.4)	611	1382
		2228	755	(15.1)	1473	2228
		2306	467	(9.3)	4002	4469
		2415	196	(3.9)	2415	2611
		2611	109	(2.2)	2306	2415
		3924	91	(1.8)	1382	1473
		3937	78	(1.6)	2228	2306
		4002	65	(1.3)	3937	4002
		4469	13	(0.3)	3924	3937
MNL 1 (CCTC)	35					
		20	876	(17.5)	4024	4900
		32	611	(12.2)	1988	2599
		35	400	(8.0)	1171	1571
		59	370	(7.4)	3579	3949
		65	339	(6.8)	2695	3034
		72	267	(5.3)	904	1171
		78	262	(5.2)	359	621
		329	251	(5.0)	78	329
		359	242	(4.8)	3034	3276
		621	226	(4.5)	621	847
		847	206	(4.1)	1782	1988
		904	192	(3.8)	3355	3547
		1171	130	(2.6)	1652	1782

	# OF SITES	SITES	FRAGMENTS		FRAGMENT ENDS	
		1571	123	(2.5)	4900	20
		1652	81	(1.6)	1571	1652
		1782	57	(1.1)	847	904
		1988	54	(1.1)	2641	2695
		2599	45	(0.9)	3979	4024
		2641	42	(0.8)	2599	2641
		2695	32	(0.6)	3285	3317
		3034	30	(0.6)	329	359
		3276	30	(0.6)	3949	3979
HGI A1 (GRGCRC)	4					
		548	3259	(65.1)	2292	548
		1046	1161	(23.2)	1046	2207
		2207	498	(10.0)	548	1046
		2292	85	(1.7)	2207	2292
HGI C1 (GGQPCC)	2					
		1573	2712	(54.2)	1573	4285
		4285	2291	(45.8)	4285	1573
HGI D1 (GPCGQC)	2					
		2161	4621	(92.4)	2543	2161
		2543	382	(7.6)	2161	2543
HHA 1 (GCGC)	14					
		333	2821	(56.4)	2515	333
		436	393	(7.9)	1360	1753
		466	337	(6.7)	1846	2183
		607	332	(6.6)	2183	2515
		640	270	(5.4)	640	910
		910	174	(3.5)	1077	1251
		977	141	(2.8)	466	607
		1077	109	(2.2)	1251	1360
		1251	103	(2.1)	333	436
		1360	100	(2.0)	977	1077
		1753	93	(1.9)	1753	1846
		1846	67	(1.3)	910	977
		2183	33	(0.7)	607	640
		2515	30	(0.6)	436	466
HIND 2 (GTQPAC)	4					
		2164	3665	(73.3)	3502	2164
		2893	729	(14.6)	2164	2893
		3133	369	(7.4)	3133	3502
		3502	240	(4.8)	2893	3133
HIND 3 (AAGCTT)	1					
		5001	5003	(99.9)	5001	5001
HINF 1 (GANTC)	9					
		632	2263	(45.2)	1620	3883
		707	1233	(24.6)	4402	632
		1103	517	(10.3)	1103	1620
		1620	396	(7.9)	707	1103
		3883	303	(6.1)	4099	4402
		3966	109	(2.2)	3966	4075
		4075	83	(1.7)	3883	3966
		4099	75	(1.5)	632	707
		4402	24	(0.5)	4075	4099
HPH 1 (GGTGA)	14					
		342	2114	(42.3)	2340	4454
		351	1125	(22.5)	351	1476
		1476	617	(12.3)	4728	342

	# OF SITES	SITES	FRAGMENTS		FRAGMENT ENDS	
		1703	396	(7.9)	1703	2099
		2099	227	(4.5)	1476	1703
		2325	226	(4.5)	2099	2325
		2340	118	(2.4)	4584	4702
		4454	68	(1.4)	4516	4584
		4508	54	(1.1)	4454	4508
		4516	18	(0.4)	4702	4720
		4584	15	(0.3)	2325	2340
		4702	9	(0.2)	342	351
		4720	8	(0.2)	4508	4516
		4728	8	(0.2)	4720	4728
HPA 1 (GTTAAC)	1					
		3502	5003	(99.9)	3502	3502
HPA 2 (CCGG)	15					
		378	2061	(41.2)	2159	4220
		412	583	(11.7)	4798	378
		939	527	(10.5)	412	939
		1086	408	(8.2)	4220	4628
		1112	404	(8.1)	1302	1706
		1302	242	(4.8)	1917	2159
		1706	190	(3.8)	1112	1302
		1740	147	(2.9)	939	1086
		1807	128	(2.6)	4628	4756
		1917	110	(2.2)	1807	1917
		2159	67	(1.3)	1740	1807
		4220	42	(0.8)	4756	4798
		4628	34	(0.7)	378	412
		4756	34	(0.7)	1706	1740
		4798	26	(0.5)	1086	1112
MBO 1 (GATC)	17					
		1299	2085	(41.7)	4217	1299
		1374	670	(13.4)	2304	2974
		1385	610	(12.2)	3607	4217
		1393	396	(7.9)	2974	3370
		1471	341	(6.8)	1588	1929
		3285	24	(0.5)	35	59
		3317	23	(0.5)	3556	3579
		3328	15	(0.3)	3331	3346
		3331	12	(0.2)	20	32
		3346	11	(0.2)	3317	3328
		3355	9	(0.2)	3276	3285
		3547	9	(0.2)	3346	3355
		3556	9	(0.2)	3547	3556
		3579	7	(0.1)	65	72
		3949	6	(0.1)	59	65
		3979	6	(0.1)	72	78
		4024	3	(0.1)	32	35
		4900	3	(0.1)	3328	3331
MST 1 (TGCGCA)	1					
		1845	5003	(99.9)	1845	1845
NCI 1 (CCSGG)	5					
		377	3222	(64.4)	2158	377
		412	699	(14.0)	412	1111
		1111	696	(13.9)	1111	1807
		1807	351	(7.0)	1807	2158
		2158	35	(0.7)	377	412
NCO 1 (CCATGG)	2					
		107	4343	(86.8)	107	4450
		4450	660	(13.2)	4450	107

	# OF SITES	SITES	FRAGMENTS		FRAGMENT ENDS	
NDE 1 (CATATG)	1					
		554	5003	(99.9)	554	554
PST 1 (CTGCAG)	2					
		1866	4045	(80.9)	2824	1866
		2824	958	(19.1)	1866	2824
PVU 1 (CGATCG)	1					
		1992	5003	(99.9)	1992	1992
PVU 2 (CAGCTG)	2					
		323	4530	(90.5)	323	4853
		4853	473	(9.5)	4853	323
RSA 1 (GTAC)	5					
		539	1571	(31.4)	2104	3675
		2104	1565	(31.3)	539	2104
		3675	667	(13.3)	4875	539
		4337	662	(13.2)	3675	4337
		4875	538	(10.8)	4337	4875
SAU96 1 (GGNCC)	8					
		1667	3575	(71.5)	3095	1667
		1746	616	(12.3)	1985	2601
		1763	248	(5.0)	2601	2849
		1985	245	(4.9)	2849	3094
		2601	222	(4.4)	1763	1985
		2849	79	(1.6)	1667	1746
		3094	17	(0.3)	1746	1763
		3095	1	(0.0)	3094	3095
SFA N1 (GATGC)	13					
		195	1145	(22.9)	2312	3457
		250	1052	(21.0)	820	1872
		408	957	(19.1)	3457	4414
		524	504	(10.1)	4414	4918
		579	280	(5.6)	4918	195
		600	230	(4.6)	2082	2312
		820	220	(4.4)	600	820
		1872	210	(4.2)	1872	2082
		2082	158	(3.2)	250	408
		2312	116	(2.3)	408	524
		3457	55	(1.1)	195	250
		4414	55	(1.1)	524	579
		4918	21	(0.4)	579	600
SPH 1 (GCATGC)	2					
		198	4948	(98.9)	253	198
		253	55	(1.1)	198	253
STU 1 (AGGCCT)	1					
		17	5003	(99.9)	17	17
TAQ 1 (TCGA)	2					
		832	3559	(71.1)	2276	832
		2276	1444	(28.9)	832	2276
THA 1 (CGCG)	8					
		332	2821	(56.4)	2514	332
		334	581	(11.6)	778	1359

	# OF SITES	SITES	FRAGMENTS		FRAGMENT ENDS	
		437	493	(9.9)	1689	2182
		778	341	(6.8)	437	778
		1359	332	(6.6)	2182	2514
		1689	330	(6.6)	1359	1689
		2182	103	(2.1)	334	437
		2514	2	(0.0)	332	334
TTH111 I (GACNNNGTC)	1					
		476	5003	(99.9)	476	476
XHO 2 (PGATCQ)	8					
		1373	2160	(43.2)	4216	1373
		1384	1102	(22.0)	2267	3369
		1470	847	(16.9)	3369	4216
		1482	768	(15.4)	1482	2250
		2250	86	(1.7)	1384	1470
		2267	17	(0.3)	2250	2267
		3369	12	(0.2)	1470	1482
		4216	11	(0.2)	1373	1384
XMN 1 (GAANNNNTTC)	1					
		2220	5003	(99.9)	2220	2220

THE FOLLOWING SITES DO NOT APPEAR:

ASU 2 (TTCGAA)	AVA 1 (CQCGPG)	BCL 1 (TGATCA)
BGL 2 (AGATCT)	BSSH 2 (GCGCGC)	BST E2 (GGTNACC)
BST X1 (CCANNNNNNTGG)	CLA 1 (ATCGAT)	ECA 1 (GGTNACC)
ECO RV (GATATC)	KPN 1 (GGTACC)	MLU 1 (ACGCGT)
MST 2 (CCTNAGG)	NAE 1 (GCCGGC)	NAR 1 (GGCGCC)
NRU 1 (TCGCGA)	PAER7 1 (CTCGAG)	SAL 1 (GTCGAC)
SMA 1 (CCCGGG)	SST 1 (GAGCTC)	SST 2 (CCGCGG)
XBA 1 (TCTAGA)	XHO 1 (CTCGAG)	XMA 1 (CCCGGG)
XMA 3 (CGGCCG)		

APPENDIX
The Sequence of pRSVneo

BST E2 (GGTNACC)	BST X1 (CCANNNNNNTGG)	CLA 1 (ATCGAT)
ECA 1 (GGTNACC)	ECO RV (GATATC)	KPN 1 (GGTACC)
PAER7 1 (CTCGAG)	SAL 1 (GTCGAC)	SST 1 (GAGCTC)
SST 2 (CCGCGG)	STU 1 (AGGCCT)	XBA 1 (TCTAGA)
XHO 1 (CTCGAG)		

```
        10        20        30        40        50        60
CTTGGAGGTG CACACCAATG TGGTGAATGG TCAAATGGCG TTTATTGTAT CGAGCTAGGC

        70        80        90       100       110       120
ACTTAAATAC AATTATCTCT GCAATGCGGA ATTCAGTGGT TCGTCCAATC CATGTCAGAC

       130       140       150       160       170       180
CTGTCTGTTG CCTTCCTAAT AAGGCACGAT CGTACCACCT TACTTCCACC AATCGGCATG

       190       200       210       220       230       240
CACGGTGCTT TTTCTCTCCT TGTAAGGCAT GTTGCTAACT CATCGTTACC ATGTTGCAAG
```

176

```
        250        260        270       _280        290        300
ACTACAAGTG TATTGCATAA GACTACATTT CCCCCTCCCT ATGCAAAAGC GAAACTACTA

       _310       _320       _330        340        350        360
TATCCTGAGG GGACTCCTAA CCGCGTACAA CCGAAGCCCC GCTTTTCGCC TAAACACACC

       _370  ___ _380  _       _390   ____ _400  ___ _410  _       420
CTAGTCCCCT CAGATACGCG TATATCTGGC CCGTACATCG CGAAGCAGCG CAAAACGCCT

       _430       _440       _450        460        470       _480
AACCCTAAGC AGATTCTTCA TGCAATTGTC GGTCAAGCCT TGCCTTGTTG TAGCTTAAAT

    __ _490  _       _500       _510        520        530        540
TTTGCTCGCG CACTACTCAG CGACCTCCAA CACACAAGCA GGGAGCAGAT ACTGGCTTAA

       _550   __      560       _570       _580  ___     590        600
CTATGCGGCA TCAGAGCAGA TTGTACTGAG AGTGCACCAT ATGCGGTGTG AAATACCGCA

       _610        620       _630       _640       _650       _660
CAGATGCGTA AGGAGAAAAT ACCGCATCAG GCGCTCTTCC GCTTCCTCGC TCACTGACTC

       _670       680 _       690       _700        710        720
GCTGCGCTCG GTCGTTCGGC TGCGGCGAGC GGTATCAGCT CACTCAAAGG CGGTAATACG

       730       _740        750        760        770  ____  780
GTTATCCACA GAATCAGGGG ATAACGCAGG AAAGAACATG TGAGCAAAAG GCCAGCAAAA

       _790       800 _      810        820        830        840
GGCCAGGAAC CGTAAAAAGG CCGCGTTGCT GGCGTTTTTC CATAGGCTCC GCCCCCCTGA

       _850       _860  __     870  ____    880        890        900
CGAGCATCAC AAAAATCGAC GCTCAAGTCA GAGGTGGCGA AACCCGACAG GACTATAAAG

       _910       _920  _  ____ _930  _  ____ 940        950       _960
ATACCAGGCG TTTCCCCCTG GAAGCTCCCT CGTGCGCTCT CCTGTTCCGA CCCTGCCGCT

    ____ 970        980        990        1000 ____ 1010        1020
TACCGGATAC CTGTCCGCCT TTCTCCCTTC GGGAAGCGTG GCGCTTTCTC AATGCTCACG

      1030  ____ 1040        1050       1060  __      1070  _      1080
CTGTAGGTAT CTCAGTTCGG TGTAGGTCGT TCGCTCCAAG CTGGGCTGTG TGCACGAACC

      1090       1100       1110  __     1120       1130  _      1140
CCCCGTTCAG CCCGACCGCT GCGCCTTATC CGGTAACTAT CGTCTTGAGT CCAACCCGGT

      1150       1160       1170        1180       1190       1200
AAGACACGAC TTATCGCCAC TGGCAGCAGC CACTGGTAAC AGGATTAGCA GAGCGAGGTA

      1210       1220       1230       1240       1250       1260
TGTAGGCGGT GCTACAGAGT TCTTGAAGTG GTGGCCTAAC TACGGCTACA CTAGAAGGAC

      1270       1280       1290       1300       1310       1320
AGTATTTGGT ATCTGCGCTC TGCTGAAGCC AGTTACCTTC GGAAAAAGAG TTGGTAGCTC

      1330       1340       1350       1360       1370       1380
TTGATCCGGC AAACAAACCA CCGCTGGTAG CGGTGGTTTT TTTGTTTGCA AGCAGCAGAT

      1390       1400  __     1410       1420       1430       1440
TACGCGCAGA AAAAAAGGAT CTCAAGAAGA TCCTTTGATC TTTTCTACGG GGTCTGACGC

    ____ 1450       1460       1470       1480       1490       1500
TCAGTGGAAC GAAAACTCAC GTTAAGGGAT TTTGGTCATG AGATTATCAA AAAGGATCTT

    __ _1510  _  ____ _1520  _      1530       _1540        1550       1560
CACCTAGATC CTTTTAAATT AAAAATGAAG TTTTAAATCA ATCTAAAGTA TATATGAGTA

      1570       1580       1590       1600       1610       1620
AACTTGGTCT GACAGTTACC AATGCTTAAT CAGTGAGGCA CCTATCTCAG CGATCTGTCT

      1630  __      1640       1650       1660       1670       1680
ATTTCGTTCA TCCATAGTTG CCTGACTCCC CGTCGTGTAG ATAACTACGA TACGGGAGGG

      1690       1700       1710       1720       1730       1740
CTTACCATCT GGCCCCAGTG CTGCAATGAT ACCGCGAGAC CCACGCTCAC CGGCTCCAGA

      1750       1760       1770       1780       1790  _      1800
TTTATCAGCA ATAAACCAGC CAGCCGGAAG GGCCGAGCGC AGAAGTGGTC CTGCAACTTT
```

```
        1810 ____    1820    _____1830 _____   1840       1850        1860
  ATCCGCCTCC ATCCAGTCTA TTAATTGTTG CCGGGAAGCT AGAGTAAGTA GTTCGCCAGT

        1870 ____    1880       1890 _____  1900       1910        1920
  TAATAGTTTG CGCAACGTTG TTGCCATTGC TGCAGGCATC GTGGTGTCAC GCTCGTCGTT

        1930       1940 ____   1950     ____1960       1970 ____    1980
  TGGTATGGCT TCATTCAGCT CCGGTTCCCA ACGATCAAGG CGAGTTACAT GATCCCCCAT

        1990       2000 ___    2010     _____2020 _      2030        2040
  GTTGTGCAAA AAAGCGGTTA GCTCCTTCGG TCCTCCGATC GTTGTCAGAA GTAAGTTGGC

  ____    2050       2060       2070       2080 ___    2090        2100
  CGCAGTGTTA TCACTCATGG TTATGGCAGC ACTGCATAAT TCTCTTACTG TCATGCCATC

  _       2110       2120       2130 _      2140       2150        2160
  CGTAAGATGC TTTTCTGTGA CTGGTGAGTA CTCAACCAAG TCATTCTGAG AATAGTGTAT

  _____   2170       2180       2190 ___    2200       2210        2220
  GCGGCGACCG AGTTGCTCTT GCCCGGCGTC AACACGGGAT AATACCGCGC CACATAGCAG

        2230     ____2240       2250     ____2260       2270        2280
  AACTTTAAAA GTGCTCATCA TTGGAAAACG TTCTTCGGGG CGAAAACTCT CAAGGATCTT

        2290 _____    2300 ___    2310       2320 _      2330 _____   2340
  ACCGCTGTTG AGATCCAGTT CGATGTAACC CACTCGTGCA CCCAACTGAT CTTCAGCATC

        2350 ___    2360       2370       2380       2390 _____   2400
  TTTTACTTTC ACCAGCGTTT CTGGGTGAGC AAAAACAGGA AGGCAAAATG CCGCAAAAAA

        2410       2420       2430       2440 ___    2450        2460
  GGGAATAAGG GCGACACGGA AATGTTGAAT ACTCATACTC TTCCTTTTTC AATATTATTG

        2470       2480       2490       2500       2510        2520
  AAGCATTTAT CAGGGTTATT GTCTCATGAG CGGATACATA TTTGAATGTA TTTAGAAAAA

        2530       2540 ___    2550       2560       2570 _____2580
  TAAACAAATA GGGGTTCCGC GCACATTTCC CCGAAAAGTG CCACCTGACG TCTAAGAAAC

        2590       2600       2610       2620       2630        2640
  CATTATTATC ATGACATTAA CCTATAAAAA TAGGCGTATC ACGAGGCCCT TTCGTCTTCA

  _____   2650       2660 ___    2670       2680       2690        2700
  AGAATTCCTT TGCCTAATTT AAATGAGGAC TTAACCTGTG GAAATATTTT GATGTGGGAA

  ___    2710       2720 ___    2730       2740       2750        2760
  GCTGTTACTG TTAAAACTGA GGTTATTGGG GTAACTGCTA TGTTAAACTT GCATTCAGGG

        2770       2780       2790       2800       2810 ___    2820
  ACACAAAAAA CTCATGAAAA TGGTGCTGGA AAACCCATTC AAGGGTCAAA TTTTCATTTT

        2830       2840       2850 ___    2860       2870        2880
  TTTGCTGTTG GTGGGGAACC TTTGGAGCTG CAGGGTGTGT TAGCAAACTA CAGGACCAAA

        2890       2900       2910       2920 ___    2930        2940
  TATCCTGCTC AAACTGTAAC CCCAAAAAAT GCTACAGTTG ACAGTCAGCA GATGAACACT

        2950       2960       2970       2980       2990        3000
  GACCACAAGG CTGTTTTGGA TAAGGATAAT GCTTATCCAG TGGAGTGCTG GGTTCCTGAT

  _       3010       3020       3030       3040       3050        3060
  CCAAGTAAAA ATGAAAACAC TAGATATTTT GGAACCTACA CAGGTGGGGA AAATGTGCCT

  _       3070       3080       3090       3100       3110        3120
  CCTGTTTTGC ACATTACTAA CACAGCAACC ACAGTGCTTC TTGATGAGCA GGGTGTTGGG

  ____    3130 _____   3140 ___    3150       3160 ___    3170        3180
  CCCTTGTGCA AAGCTGACAG CTTGTATGTT TCTGCTGTTG ACATTTGTGG GCTGTTTACC

        3190       3200       3210       3220       3230 ___    3240
  AACACTTCTG GAACACAGCA GTGGAAGGGA CTTCCCAGAT ATTTTAAAAT TACCCTTAGA

        3250       3260       3270 _      3280       3290 _____   3300
  AAGCGGTCTG TGAAAAACCC CTACCCAATT TCCTTTTTGT TAAGTGACCT AATTAACAGG
```

```
___  3310 __  ____3320 _____ 3330     3340 ____ 3350    _____3360
AGGACACAGA GGGTGGATGG GCAGCCTATG ATTGGAATGT CCTCTCAAGT AGAGGAGGTT

      3370 ____  3380    _____3390     ____3400     3410       3420
AGGGTTTATG AGGACACAGA GGAGCTTCCT GGGGATCCAG ACATGATAAG ATACATTGAT

      3430      3440      3450      3460      3470    _____3480
GAGTTTGGAC AAACCACAAC TAGAATGCAG TGAAAAAAAT GCTTTATTTG TGAAATTTGT

_____ 3490      3500      3510  _____3520     __3530 _    3540
GATGCTATTG CTTTATTTGT AACCATTATA AGCTGCAATA AACAAGTTAA CAACAACAAT

 __   3550      3560      3570 _____ 3580 ____   3590 __    3600
TGCATTCATT TTATGTTTCA GGTTCAGGGG GAGGTGTGGG AGGTTTTTTA AAGCAAGTAA

_____3610      3620      3630 _____ 3640      3650      3660
AACCTCTACA AATGTGGTAT GGCTGATTAT GATCTCTAGT CAAGGCACTA TACATCAAAT

      3670      3680 _____3690 _____ 3700 __   __3710      3720
ATTCCTTATT AACCCCTTTA CAAATTAAAA AGCTAAAGGT ACACAATTTT TGAGCATAGT

      3730      3740      3750      3760      3770      3780
TATTAATAGC AGACACTCTA TGCCTGTGTG GAGTAAGAAA AAACAGTATG TTATGATTAT

      3790      3800      3810      3820 _____3830      3840
AACTGTTATG CCTACTTATA AAGGTTACAG AATATTTTTC CATAATTTTC TTGTATAGCA

_____3850      3860      3870      3880      3890      3900
GTGCAGCTTT TTCCTTTGTG GTGTAAATAG CAAAGCAAGC AAGAGTTCTA TTACTAAACA

_____3910 _    3920 _____3930      3940      3950 __   3960
CAGCATGACT CAAAAAACTT AGCAATTCTG AAGGAAAGTC CTTGGGGTCT TCTACCTTTC

_____ 3970 _____3980      3990 _____ 4000 _____4010      4020
TCTTCTTTTT TGGAGGAGTA GAATGTTGAG AGTCAGCAGT AGCCTCATCA TCACTAGATG

_____4030 _____ 4040      4050 _    4060      4070      4080
GCATTTCTTC TGAGCAAAAC AGGTTTTCCT CATTAAAGGC ATTCCACCAC TGCTCCCATT

      4090      4100 ___  4110      4120 _____4130      4140
CATCAGTTCC ATAGGTTGGA ATCTAAAATA CACAAACAAT TAGAATCAGT AGTTTAACAC

      4150 _____4160      4170 _____4180 __   4190      4200
ATTATACACT TAAAAATTTT ATATTTACCT TAGAGCTTTA AATCTCTGTA GGTAGTTTGT

_____4210      4220      4230      4240 _____4250 _    4260
CCAATTATGT CACACCACAG AAGTAAGGTT CCTTCACAAA GATCCGGGAC CAAAGCGGCC

      4270      4280 ___  4290      4300 ___   4310      4320
ATCGTGCCTC CCCACTCCTG CAGTTCGGGG GCATGGATGC GCGGATAGCC GCTGCTGGTT

_____4330      4340      4350      4360 ___   4370      4380
TCCTGGATGC CGACGGATTT GCACTGCCGG TAGAACTCCG CGAGGTCGTC CAGCCTCAGG

_____4390      4400 _____4410      4420      4430      4440
CAGCAGCTGA ACCAACTCGC GAGGGGATCG AGCCCGGGGT GGGCGAAGAA CTCCAGCATG

_____4450 __   4460 _____4470      4480 _____4490      4500
AGATCCCCGC GCTGGAGGAT CATCCAGCCG GCGTCCCGGA AAACGATTCC GAAGCCCAAC

      4510 _____4520 _____4530      4540      4550 _____ 4560
CTTTCATAGA AGGCGGCGGT GGAATCGAAA TCTCGTGATG GCAGGTTGGG CGTCGCTTGG

      4570 _____ 4580 _____4590      4600      4610      4620
TCGGTCATTT CGAACCCCAG AGTCCCGCTC AGAAGAACTC GTCAAGAAGG CGATAGAAGG

      4630 _____4640      4650      4660      4670      4680
CGATGCGCTG CGAATCGGGA GCGGCGATAC CGTAAAGCAC GAGGAAGCGG TCAGCCCATT

_____4690 _____4700      4710      4720      4730 _____4740
CGCCGCCAAG CTCTTCAGCA ATATCACGGG TAGCCAACGC TATGTCCTGA TAGCGGTCCG

      4750 _____4760 _____4770      4780 _____4790      4800
CCACACCCAG CCGGCCACAG TCGATGAATC CAGAAAAGCG GCCATTTTCC ACCATGATAT
```

```
        4810 _____4820 ____   4830 _____4840     4850 _____4860
    TCGGCAAGCA GGCATCGCCA TGGGTCACGA CGAGATCCTC GCCGTCGGGC ATGCGCGCCT

     ___4870     4880 _____4890      4900 _____4910 _____4920
    TGAGCCTGGC GAACAGTTCG GCTGGCGCGA GCCCCTGATG CTCTTCGTCC AGATCATCCT

    _____4930 ____ 4940 ____ 4950 _____4960     4970      4980
    GATCGACAAG ACCGGCTTCC ATCCGAGTAC GTGCTCGCTC GATGCGATGT TTCGCTTGGT

    _____4990     5000__ ___  5010 ____ 5020     5030__      5040
    GGTCGAATGG GCAGGTAGCC GGATCAAGCG TATGCAGCCG CCGCATTGCA TCAGCCATGA

        5050      5060 _____5070      5080 ____   5090 __    5100
    TGGATACTTT CTCGGCAGGA GCAAGGTGAG ATGACAGGAG ATCCTGCCCC GGCACTTCGC

     ___5110 _    5120      5130 _____5140 _____5150      5160
    CCAATAGCAG CCAGTCCCTT CCCGCTTCAG TGACAACGTC GAGCACAGCT GCGCAAGGAA

        5170 _____5180 _____5190 _____5200 _____5210 _____5220
    CGCCCGTCGT GGCCAGCCAC GATAGCCGCG CTGCCTCGTC CTGCAGTTCA TTCAGGGCAC

    ___   5230      5240 _____5250 _____5260 __   5270 ___  5280
    CGGACAGGTC GGTCTTGACA AAAAGAACCG GGCGCCCCTG CGCTGACAGC CGGAACACGG

    _____5290 _____ 5300     5310      5320      5330 ___  5340
    CGGCATCAGA GCAGCCGATT GTCTGTTGTG CCCAGTCATA GCCGAATAGC CTCTCCACCC

    _____5350     5360      5370      5380      5390 _____5400
    AAGCGGCCGG AGAACCTGCG TGCAATCCAT CTTGTTCAAT CATGCGAAAC GATCCTCATC

    _   5410 _____5420 _____ 5430 __   5440 __   5450      5460
    CTGTCTCTTG ATCAGATCTT GATCCCCTGC GCCATCAGAT CCTTGGCGGC AAGAAAGCCA

    ___  5470     5480      5490 _____5500 _____ 5510 _    5520
    TCCAGTTTAC TTTGCAGGGC TTCCCAACCT TACCAGAGGG CGCCCCAGCT GGCAATTCCG

    _   5530      5540      5550 ____5560      5570      5580
    GTTCGCTTGC TGTCCATAAA ACCGCCCAGT CTAGCTATCG CCATGTAAGC CCACTGCAAG

    __   5590     5600 _    5610      5620      5630 _____ 5640
    CTACCTGCTT TCTCTTTGCG CTTGCGTTTT CCCTTGTCCA GATAGCCCAG TAGCTGACAT

    _____5650     5660      5670      5680      5690      5700
    TCATCCGGGG TCAGCACCGT TTCTGCGGAC TGGCTTTCTA CGTGTTCCGC TTCCTTTAGC

    ___   5710 _ _____5720 _____5730 ____
    AGCCCTTGCG CCCTGAGTGC TTGCGGCAGC GTGAAG
```

```
    MISSRES=         0        REFLECT=         1;       FORM='C';

    P= A OR G,   Q= C OR T,   R= A OR T,   S= C OR G,   V= A OR C,   W=G OR T
```

	# OF SITES	SITES	FRAGMENTS		FRAGMENT ENDS	
AAT 2 (GACGTC)	1					
		2567	5736	(99.9)	2567	2567
ACY 1 (GPCGQC)	6					
		2185	2422	(42.2)	5499	2185
		2567	1903	(33.2)	2567	4470
		4470	702	(12.2)	4549	5251
		4549	382	(6.7)	2185	2567
		5251	248	(4.3)	5251	5499
		5499	79	(1.4)	4470	4549

	# OF SITES	SITES	FRAGMENTS		FRAGMENT ENDS	
AHA 3 (TTTAAA)	7					
		1513	3072	(53.6)	4177	1513
		1532	692	(12.1)	1532	2224
		2224	590	(10.3)	3587	4177
		2658	565	(9.9)	2658	3223
		3223	434	(7.6)	2224	2658
		3587	364	(6.3)	3223	3587
		4177	19	(0.3)	1513	1532
ALU 1 (AGCT)	26					
		53	700	(12.2)	2000	2700
		472	521	(9.1)	1316	1837
		697	458	(8.0)	4689	5147
		923	419	(7.3)	53	472
		1059	360	(6.3)	5147	5507
		1316	329	(5.7)	3845	4174
		1837	304	(5.3)	4385	4689
		1937	286	(5.0)	2846	3132
		2000	257	(4.5)	1059	1316
		2700	244	(4.3)	3139	3383
		2846	226	(3.9)	697	923
		3132	225	(3.9)	472	697
		3139	211	(3.7)	4174	4385
		3383	180	(3.1)	3511	3691
		3511	154	(2.7)	3691	3845
		3691	146	(2.5)	2700	2846
		3845	136	(2.4)	923	1059
		4174	128	(2.2)	3383	3511
		4385	103	(1.8)	5632	5735
		4689	100	(1.7)	1837	1937
		5147	63	(1.1)	1937	2000
		5507	54	(0.9)	5735	53
		5553	53	(0.9)	5579	5632
		5579	46	(0.8)	5507	5553
		5632	26	(0.5)	5553	5579
		5733	7	(0.1)	3132	3139
APA 1 (GGGCCC)	1					
		3118	5736	(99.9)	3118	3118
ASU 2 (TTCGAA)	1					
		4569	5736	(99.9)	4569	4569
AVA 1 (CQCGPG)	1					
		4413	5736	(99.9)	4413	4413
AVA 2 (GGACC)	5					
		1787	2788	(48.6)	4735	1787
		2009	1374	(24.0)	2873	4247
		2873	864	(15.1)	2009	2873
		4247	488	(8.5)	4247	4735
		4735	222	(3.9)	1787	2009
BAL 1 (TGGCCA)	1					
		5170	5736	(99.9)	5170	5170
BAM H1 (GGATCC)	1					
		3393	5736	(99.9)	3393	3393
BBV 1 (GCTGC)	25					
		405	781	(13.6)	2066	2847
		661	474	(8.3)	2847	3321
		679	468	(8.2)	3843	4311

	# OF SITES	SITES	FRAGMENTS		FRAGMENT ENDS	
		1098	419	(7.3)	679	1098
		1163	415	(7.2)	5726	405
		1166	408	(7.1)	5291	5699
		1372	387	(6.7)	4627	5014
		1700	331	(5.8)	3512	3843
		1889	328	(5.7)	1372	1700
		2066	256	(4.5)	405	661
		2847	244	(4.3)	4383	4627
		3321	206	(3.6)	1166	1372
		3512	191	(3.3)	3321	3512
		3843	189	(3.3)	1700	1889
		4311	177	(3.1)	1889	2066
		4380	101	(1.8)	5190	5291
		4383	93	(1.6)	5014	5107
		4627	69	(1.2)	4311	4380
		5014	65	(1.1)	1098	1163
		5107	42	(0.7)	5148	5190
		5148	41	(0.7)	5107	5148
		5190	27	(0.5)	5699	5726
		5291	18	(0.3)	661	679
		5699	3	(0.1)	1163	1166
		5726	3	(0.1)	4380	4383
BCL 1 (TGATCA)	1					
		5409	5736	(99.9)	5409	5409
BGL 1 (GCCNNNNNGGC)	1					
		1763	5736	(99.9)	1763	1763
BGL 2 (AGATCT)	1					
		5414	5736	(99.9)	5414	5414
BSSH 2 (GCGCGC)	1					
		4853	5736	(99.9)	4853	4853
BVU 1 (GPGCQC)	3					
		3118	3965	(69.1)	4889	3118
		4410	1292	(22.5)	3118	4410
		4889	479	(8.4)	4410	4889
DDE 1 (CTNAG)	18					
		305	1125	(19.6)	4588	5713
		369	683	(11.9)	3235	3918
		425	540	(9.4)	1606	2146
		496	518	(9.0)	2717	3235
		566	465	(8.1)	566	1031
		1031	426	(7.4)	2146	2572
		1440	409	(7.1)	1031	1440
		1606	328	(5.7)	5713	305
		2146	213	(3.7)	4375	4588
		2572	206	(3.6)	4169	4375
		2717	166	(2.9)	1440	1606
		3235	145	(2.5)	2572	2717
		3918	139	(2.4)	4030	4169
		4030	112	(2.0)	3918	4030
		4169	71	(1.2)	425	496
		4375	70	(1.2)	496	566
		4588	64	(1.1)	305	369
		5713	56	(1.0)	369	425
ECO R1 (GAATTC)	2					
		89	3183	(55.5)	2642	89
		2642	2553	(44.5)	89	2642

182

	# OF SITES	SITES	FRAGMENTS		FRAGMENT ENDS	

ECO R1* (NAATTN)   23

SITES	FRAGMENTS		FRAGMENT ENDS	
70	1311	(22.9)	4202	5513
89	1039	(18.1)	477	1516
443	565	(9.9)	2077	2642
477	419	(7.3)	2808	3227
1516	354	(6.2)	89	443
1822	306	(5.3)	1516	1822
2077	293	(5.1)	5513	70
2642	255	(4.4)	1822	2077
2655	194	(3.4)	3923	4117
2808	183	(3.2)	3290	3473
3227	153	(2.7)	2655	2808
3266	145	(2.5)	3537	3682
3290	119	(2.1)	3704	3823
3473	100	(1.7)	3823	3923
3537	64	(1.1)	3473	3537
3682	48	(0.8)	4154	4202
3704	39	(0.7)	3227	3266
3823	37	(0.6)	4117	4154
3923	34	(0.6)	443	477
4117	24	(0.4)	3266	3290
4154	22	(0.4)	3682	3704
4202	19	(0.3)	70	89
5513	13	(0.2)	2642	2655

ECO R2 (CCTGG)   6

SITES	FRAGMENTS		FRAGMENT ENDS	
783	2471	(43.1)	917	3388
904	1654	(28.8)	4865	783
917	934	(16.3)	3388	4322
3388	543	(9.5)	4322	4865
4322	121	(2.1)	783	904
4865	13	(0.2)	904	917

FNU4H 1 (GCNGC)   45

SITES	FRAGMENTS		FRAGMENT ENDS	
405	474	(8.3)	2847	3321
545	457	(8.0)	2390	2847
661	415	(7.2)	5726	405
679	412	(7.2)	3843	4255
682	331	(5.8)	3512	3843
800	328	(5.7)	1372	1700
955	253	(4.4)	5446	5699
1098	236	(4.1)	4778	5014
1163	229	(4.0)	2161	2390
1166	206	(3.6)	1166	1372
1372	191	(3.3)	3321	3512
1700	189	(3.3)	1700	1889
1889	155	(2.7)	800	955
2039	150	(2.6)	1889	2039
2066	143	(2.5)	955	1098
2161	140	(2.4)	405	545
2390	130	(2.3)	4383	4513
2847	118	(2.1)	682	800
3321	116	(2.0)	545	661
3512	114	(2.0)	4513	4627
3843	103	(1.8)	5343	5446
4255	96	(1.7)	4682	4778
4308	95	(1.7)	2066	2161
4311	90	(1.6)	5190	5280
4380	87	(1.5)	5020	5107
4383	69	(1.2)	4311	4380
4513	65	(1.1)	1098	1163
4627	53	(0.9)	4255	4308
4641	52	(0.9)	5291	5343
4682	41	(0.7)	4641	4682
4778	41	(0.7)	5107	5148
5014	37	(0.6)	5148	5185
5017	27	(0.5)	2039	2066
5020	24	(0.4)	5699	5723
5107	18	(0.3)	661	679

183

	# OF SITES	SITES	FRAGMENTS		FRAGMENT ENDS	
		5148	14	(0.2)	4627	4641
		5185	11	(0.2)	5280	5291
		5190	5	(0.1)	5185	5190
		5280	3	(0.1)	679	682
		5291	3	(0.1)	1163	1166
		5343	3	(0.1)	4308	4311
		5446	3	(0.1)	4380	4383
		5699	3	(0.1)	5014	5017
		5723	3	(0.1)	5017	5020
		5726	3	(0.1)	5723	5726
FOK 1 (GGATG)	12					
		1629	1723	(30.0)	5642	1629
		1810	1218	(21.2)	2097	3315
		2097	980	(17.1)	3315	4295
		3315	457	(8.0)	4940	5397
		4295	454	(7.9)	4461	4915
		4325	287	(5.0)	1810	2097
		4461	183	(3.2)	5459	5642
		4915	181	(3.2)	1629	1810
		4940	136	(2.4)	4325	4461
		5397	62	(1.1)	5397	5459
		5459	30	(0.5)	4295	4325
		5642	25	(0.4)	4915	4940
GDI 1 (RGGCCR)	4					
		769	3938	(68.7)	1232	5170
		780	1335	(23.3)	5170	769
		1232	452	(7.9)	780	1232
		5170	11	(0.2)	769	780
HAE 1 (RGGCCR)	4					
		769	3938	(68.7)	1232	5170
		780	1335	(23.3)	5170	769
		1232	452	(7.9)	780	1232
		5170	11	(0.2)	769	780
HAE 2 (PGCGCQ)	4					
		630	4251	(74.1)	1000	5251
		1000	867	(15.1)	5499	630
		5251	370	(6.5)	630	1000
		5499	248	(4.3)	5251	5499
HAE 3 (GGCC)	15					
		388	1138	(19.8)	3119	4257
		770	779	(13.6)	5345	388
		781	587	(10.2)	2038	2625
		799	496	(8.6)	4257	4753
		1233	494	(8.6)	2625	3119
		1691	458	(8.0)	1233	1691
		1771	434	(7.6)	799	1233
		2038	391	(6.8)	4780	5171
		2625	382	(6.7)	388	770
		3119	267	(4.7)	1771	2038
		4257	174	(3.0)	5171	5345
		4753	80	(1.4)	1691	1771
		4780	27	(0.5)	4753	4780
		5171	18	(0.3)	781	799
		5345	11	(0.2)	770	781
HGA 1 (GACGC)	5					
		858	2285	(39.8)	2186	4471
		1436	2044	(35.6)	4550	858
		2186	750	(13.1)	1436	2186
		4471	578	(10.1)	858	1436
		4550	79	(1.4)	4471	4550

	# OF SITES	SITES	FRAGMENTS		FRAGMENT ENDS	
HGI A1 (GRGCRC)	7					
		8	2635	(45.9)	2316	4951
		572	1161	(20.2)	1070	2231
		1070	603	(10.5)	5141	8
		2231	564	(9.8)	8	572
		2316	498	(8.7)	572	1070
		4951	190	(3.3)	4951	5141
		5141	85	(1.5)	2231	2316
HGI C1 (GGQPCC)	4					
		1597	3619	(63.1)	1597	5216
		5216	1834	(32.0)	5499	1597
		5251	248	(4.3)	5251	5499
		5499	35	(0.6)	5216	5251
HGI D1 (GPCGQC)	6					
		2185	2422	(42.2)	5499	2185
		2567	1903	(33.2)	2567	4470
		4470	702	(12.2)	4549	5251
		4549	382	(6.7)	2185	2567
		5251	248	(4.3)	5251	5499
		5499	79	(1.4)	4470	4549
HHA 1 (GCGC)	27					
		408	1760	(30.7)	2539	4299
		488	436	(7.6)	5708	408
		631	393	(6.9)	1384	1777
		664	337	(5.9)	1870	2207
		934	332	(5.8)	2207	2539
		1001	270	(4.7)	664	934
		1101	266	(4.6)	4885	5151
		1275	228	(4.0)	4625	4853
		1384	176	(3.1)	4449	4625
		1777	174	(3.0)	1101	1275
		1870	169	(2.9)	5260	5429
		2207	150	(2.6)	4299	4449
		2539	143	(2.5)	488	631
		4299	110	(1.9)	5598	5708
		4449	109	(1.9)	1275	1384
		4625	100	(1.7)	1001	1101
		4853	98	(1.7)	5500	5598
		4855	93	(1.6)	1777	1870
		4885	80	(1.4)	408	488
		5151	71	(1.2)	5429	5500
		5188	67	(1.2)	934	1001
		5252	64	(1.1)	5188	5252
		5260	37	(0.6)	5151	5188
		5429	33	(0.6)	631	664
		5500	30	(0.5)	4855	4885
		5598	8	(0.1)	5252	5260
		5708	2	(0.0)	4853	4855
HIND 2 (GTQPAC)	4					
		2188	4398	(76.7)	3526	2188
		2917	729	(12.7)	2188	2917
		3157	369	(6.4)	3157	3526
		3526	240	(4.2)	2917	3157
HIND 3 (AAGCTT)	1					
		5734	5736	(99.9)	5734	5734
HINF 1 (GANTC)	15					
		312	2263	(39.5)	1644	3907
		432	1282	(22.4)	4766	312
		656	517	(9.0)	1127	1644

185

	# OF SITES	SITES	FRAGMENTS		FRAGMENT	ENDS
		731	396	(6.9)	731	1127
		1127	362	(6.3)	4123	4485
		1644	224	(3.9)	432	656
		3907	134	(2.3)	4632	4766
		3990	120	(2.1)	312	432
		4099	109	(1.9)	3990	4099
		4123	83	(1.4)	3907	3990
		4485	75	(1.3)	656	731
		4522	58	(1.0)	4522	4580
		4580	52	(0.9)	4580	4632
		4632	37	(0.6)	4485	4522
		4766	24	(0.4)	4099	4123
HPH 1 (GGTGA)	7					
		22	2701	(47.1)	2364	5065
		1500	1478	(25.8)	22	1500
		1727	693	(12.1)	5065	22
		2123	396	(6.9)	1727	2123
		2349	227	(4.0)	1500	1727
		2364	226	(3.9)	2123	2349
		5065	15	(0.3)	2349	2364
HPA 1 (GTTAAC)	1					
		3526	5736	(99.9)	3526	3526
HPA 2 (CCGG)	24					
		963	2061	(35.9)	2183	4244
		1110	1054	(18.4)	5645	963
		1136	404	(7.0)	1326	1730
		1326	275	(4.8)	4476	4751
		1730	242	(4.2)	1941	2183
		1764	190	(3.3)	1136	1326
		1831	181	(3.2)	4751	4932
		1941	171	(3.0)	5347	5518
		2183	147	(2.6)	963	1110
		4244	131	(2.3)	5089	5220
		4347	127	(2.2)	5518	5645
		4414	110	(1.9)	1831	1941
		4468	103	(1.8)	4244	4347
		4476	90	(1.6)	4999	5089
		4751	77	(1.3)	5270	5347
		4932	67	(1.2)	1764	1831
		4999	67	(1.2)	4347	4414
		5089	67	(1.2)	4932	4999
		5220	54	(0.9)	4414	4468
		5248	34	(0.6)	1730	1764
		5270	28	(0.5)	5220	5248
		5347	26	(0.5)	1110	1136
		5518	22	(0.4)	5248	5270
		5645	8	(0.1)	4468	4476
MBO 1 (GATC)	31					
		148	1175	(20.5)	148	1323
		1323	670	(11.7)	2328	2998
		1398	610	(10.6)	3631	4241
		1409	446	(7.8)	5438	148
		1417	396	(6.9)	2998	3394
		1495	376	(6.6)	4458	4834
		1507	341	(5.9)	1612	1953
		1612	311	(5.4)	5080	5391
		1953	258	(4.5)	2017	2275
		1971	237	(4.1)	3394	3631
		2017	165	(2.9)	4241	4406
		2275	105	(1.8)	1507	1612
		2292	81	(1.4)	4921	5002
		2328	78	(1.4)	1417	1495
		2998	78	(1.4)	4834	4912
		3394	78	(1.4)	5002	5080

# OF SITES	SITES	FRAGMENTS		FRAGMENT ENDS	
	3631	75	(1.3)	1323	1398
	4241	46	(0.8)	1971	2017
	4406	36	(0.6)	2292	2328
	4442	36	(0.6)	4406	4442
	4458	19	(0.3)	5391	5410
	4834	18	(0.3)	1953	1971
	4912	17	(0.3)	2275	2292
	4921	17	(0.3)	5421	5438
	5002	16	(0.3)	4442	4458
	5080	12	(0.2)	1495	1507
	5391	11	(0.2)	1398	1409
	5410	9	(0.2)	4912	4921
	5415	8	(0.1)	1409	1417
	5421	6	(0.1)	5415	5421
	5438	5	(0.1)	5410	5415
MBO 2 (GAAGA) 15					
	435	1313	(22.9)	2635	3948
	635	1269	(22.1)	4902	435
	1406	771	(13.4)	635	1406
	1497	755	(13.2)	1497	2252
	2252	399	(7.0)	4026	4425
	2330	210	(3.7)	4692	4902
	2439	200	(3.5)	435	635
	2635	196	(3.4)	2439	2635
	3948	167	(2.9)	4425	4592
	3961	109	(1.9)	2330	2439
	4026	100	(1.7)	4592	4692
	4425	91	(1.6)	1406	1497
	4592	78	(1.4)	2252	2330
	4692	65	(1.1)	3961	4026
	4902	13	(0.2)	3948	3961
MLU 1 (ACGCGT) 1					
	376	5736	(99.9)	376	376
MNL 1 (CCTC) 41					
	5	611	(10.7)	2012	2623
	274	400	(7.0)	1195	1595
	307	370	(6.5)	3603	3973
	368	357	(6.2)	4837	5194
	504	339	(5.9)	2719	3058
	645	269	(4.7)	5	274
	871	267	(4.7)	928	1195
	928	245	(4.3)	5496	5
	1195	242	(4.2)	3058	3300
	1595	226	(3.9)	645	871
	1676	219	(3.8)	4048	4267
	1806	206	(3.6)	1806	2012
	2012	206	(3.6)	4455	4661
	2623	192	(3.3)	3379	3571
	2665	176	(3.1)	4661	4837
	2719	141	(2.5)	504	645
	3058	136	(2.4)	368	504
	3300	136	(2.4)	5194	5330
	3309	130	(2.3)	1676	1806
	3341	102	(1.8)	5394	5496
	3352	95	(1.7)	4267	4362
	3355	81	(1.4)	1595	1676
	3370	64	(1.1)	5330	5394
	3379	61	(1.1)	307	368
	3571	57	(1.0)	871	928
	3580	54	(0.9)	2665	2719
	3603	54	(0.9)	4401	4455
	3973	45	(0.8)	4003	4048
	4003	42	(0.7)	2623	2665
	4048	33	(0.6)	274	307
	4267	32	(0.6)	3309	3341
	4362	30	(0.5)	3973	4003
	4374	27	(0.5)	4374	4401

	# OF SITES	SITES	FRAGMENTS		FRAGMENT ENDS	
		4401	23	(0.4)	3580	3603
		4455	15	(0.3)	3355	3370
		4661	12	(0.2)	4362	4374
		4837	11	(0.2)	3341	3352
		5194	9	(0.2)	3300	3309
		5330	9	(0.2)	3370	3379
		5394	9	(0.2)	3571	3580
		5496	3	(0.1)	3352	3355
MST 1 (TGCGCA)	2					
		1869	3281	(57.2)	1869	5150
		5150	2455	(42.8)	5150	1869
MST 2 (CCTNAGG)	2					
		304	4070	(71.0)	304	4374
		4374	1666	(29.0)	4374	304
NAE 1 (GCCGGC)	2					
		4467	5453	(95.1)	4750	4467
		4750	283	(4.9)	4467	4750
NAR 1 (GGCGCC)	2					
		5251	5488	(95.7)	5499	5251
		5499	248	(4.3)	5251	5499
NCI 1 (CCSGG)	10					
		1135	2062	(35.9)	2182	4244
		1831	1226	(21.4)	5645	1135
		2182	696	(12.1)	1135	1831
		4244	613	(10.7)	4475	5088
		4413	397	(6.9)	5248	5645
		4414	351	(6.1)	1831	2182
		4475	169	(2.9)	4244	4413
		5088	160	(2.8)	5088	5248
		5248	61	(1.1)	4414	4475
		5645	1	(0.0)	4413	4414
NCO 1 (CCATGG)	1					
		4818	5736	(99.9)	4818	4818
NDE 1 (CATATG)	1					
		578	5736	(99.9)	578	578
NRU 1 (TCGCGA)	2					
		398	3999	(69.7)	398	4397
		4397	1737	(30.3)	4397	398
PST 1 (CTGCAG)	4					
		1890	2425	(42.3)	5201	1890
		2848	1430	(24.9)	2848	4278
		4278	958	(16.7)	1890	2848
		5201	923	(16.1)	4278	5201
PVU 1 (CGATCG)	2					
		147	3867	(67.4)	2016	147
		2016	1869	(32.6)	147	2016
PVU 2 (CAGCTG)	3					
		4384	4614	(80.4)	5506	4384
		5146	762	(13.3)	4384	5146
		5506	360	(6.3)	5146	5506

	# OF SITES	SITES	FRAGMENTS		FRAGMENT ENDS	
RSA 1 (GTAC)	7					
		152	1571	(27.4)	2128	3699
		325	1565	(27.3)	563	2128
		393	1248	(21.8)	3699	4947
		563	941	(16.4)	4947	152
		2128	173	(3.0)	152	325
		3699	170	(3.0)	393	563
		4947	68	(1.2)	325	393
SAU96 1 (GGNCC)	11					
		388	1389	(24.2)	4735	388
		1691	1303	(22.7)	388	1691
		1770	1128	(19.7)	3119	4247
		1787	616	(10.7)	2009	2625
		2009	488	(8.5)	4247	4735
		2625	248	(4.3)	2625	2873
		2873	245	(4.3)	2873	3118
		3118	222	(3.9)	1787	2009
		3119	79	(1.4)	1691	1770
		4247	17	(0.3)	1770	1787
		4735	1	(0.0)	3118	3119
SFA N1 (GATGC)	16					
		548	1145	(20.0)	2336	3481
		603	1052	(18.3)	844	1896
		624	1001	(17.5)	5283	548
		844	815	(14.2)	3481	4296
		1896	296	(5.2)	4326	4622
		2106	255	(4.4)	5028	5283
		2336	230	(4.0)	2106	2336
		3481	220	(3.8)	624	844
		4296	210	(3.7)	1896	2106
		4326	190	(3.3)	4622	4812
		4622	85	(1.5)	4812	4897
		4812	67	(1.2)	4961	5028
		4897	64	(1.1)	4897	4961
		4961	55	(1.0)	548	603
		5028	30	(0.5)	4296	4326
		5283	21	(0.4)	603	624
SMA 1 (CCCGGG)	1					
		4413	5736	(99.9)	4413	4413
SPH 1 (GCATGC)	2					
		176	4673	(81.5)	176	4849
		4849	1063	(18.5)	4849	176
TAQ 1 (TCGA)	11					
		50	2108	(36.8)	2300	4408
		856	1444	(25.2)	856	2300
		2300	806	(14.1)	50	856
		4408	647	(11.3)	5139	50
		4525	191	(3.3)	4570	4761
		4570	162	(2.8)	4761	4923
		4761	156	(2.7)	4983	5139
		4923	117	(2.0)	4408	4525
		4959	45	(0.8)	4525	4570
		4983	36	(0.6)	4923	4959
		5139	24	(0.4)	4959	4983
THA 1 (CGCG)	16					
		322	1762	(30.7)	2538	4300
		377	871	(15.2)	5187	322
		399	581	(10.1)	802	1383
		487	493	(8.6)	1713	2206

	# OF SITES	SITES	FRAGMENTS		FRAGMENT ENDS	
		802	406	(7.1)	4448	4854
		1383	332	(5.8)	2206	2538
		1713	330	(5.8)	1383	1713
		2206	315	(5.5)	487	802
		2538	301	(5.2)	4886	5187
		4300	88	(1.5)	399	487
		4359	59	(1.0)	4300	4359
		4398	55	(1.0)	322	377
		4448	50	(0.9)	4398	4448
		4854	39	(0.7)	4359	4398
		4886	32	(0.6)	4854	4886
		5187	22	(0.4)	377	399
TTH111 I (GACNNNGTC)	1					
		5132	5736	(99.9)	5132	5132
XHO 2 (PGATCQ)	13					
		1397	1696	(29.6)	5437	1397
		1408	1102	(19.2)	2291	3393
		1494	847	(14.8)	3393	4240
		1506	768	(13.4)	1506	2274
		2274	392	(6.8)	4441	4833
		2291	335	(5.8)	5079	5414
		3393	246	(4.3)	4833	5079
		4240	201	(3.5)	4240	4441
		4441	86	(1.5)	1408	1494
		4833	23	(0.4)	5414	5437
		5079	17	(0.3)	2274	2291
		5414	12	(0.2)	1494	1506
		5437	11	(0.2)	1397	1408
XMA 1 (CCCGGG)	1					
		4413	5736	(99.9)	4413	4413
XMA 3 (CGGCCG)	1					
		5344	5736	(99.9)	5344	5344
XMN 1 (GAANNNNTTC)	1					
		2244	5736	(99.9)	2244	2244

THE FOLLOWING SITES DO NOT APPEAR:

ACC 1 (GTVWAC)   AVA 3 (ATGCAT)   AVR 2 (CCTAGG)

CHAPTER 7

The Construction and Characterisation of Vaccinia Virus Recombinants Expressing Foreign Genes

MICHAEL MACKETT, GEOFFREY L. SMITH and BERNARD MOSS

1. INTRODUCTION

DNA sequence studies coupled with *in vitro* expression systems have yielded a wealth of information regarding eukaryotic gene organisation and expression. However, many questions about regulation of gene expression, post-transcriptional and post-translational modifications cannot be answered by these studies. The use of eukaryotic viruses to re-introduce genes into cultured eukaryotic cells should overcome some of the limitations of the *in vitro* system. Initially the papovavirus SV40 was used as a vector, at least in part because of the extensive information regarding its genome organisation and the fact that the small size of its genome facilitated the *in vitro* construction of recombinant DNA molecules (for review see 1). Other relatively small viruses including papilloma-viruses, adenoviruses and retroviruses have also been used as vectors for a variety of reasons (for review see 2). Although genetic engineering of larger viruses such as herpes and vaccinia is more difficult, such vectors have the potential of retaining complete infectivity in a wide range of cell types and have a greater capacity for foreign DNA. For vaccinia virus there is the added incentive that recombinants expressing genes from pathogenic agents may have value as live virus vaccines.

Vaccinia virus which was widely used for smallpox vaccination is the best characterised member of the poxvirus family and as a consequence it was the poxvirus of choice to be developed as a vector. Vaccinia virus particles have a complex architecture, with a basic oval or brick shaped structure approximately 200 x 300 nm. The virus genome is large, approximately 180 kb with a coding capacity for over 150 polypeptides. One interesting feature of vaccinia virus particles is that they can synthesise functional mRNA that is capped, methylated and polyadenylated. The virus can carry out its entire life-cycle in the cytoplasm of the cell as it specifies the production of these enzymes responsible for RNA synthesis and post-transcriptional modification together with other virally coded proteins such as a DNA polymerase. Thus, poxviruses have a number of unconventional biological properties (for review see 3,4) which had to be taken into account when developing them as cloning and expression vectors.

Initially the major technical problems involved in the development of vaccinia virus as a cloning and expression vector were the insertion of DNA into the virus, efficient expression of the foreign gene and selection of the recombinant virus. The large size of the virus genome would have made the construction of recombinant genomes *in vitro* particularly difficult. Even if this were possible, isolated virus DNA is non-infectious, presumably because enzymes found in the virus particle are required for transcription of virus DNA. Thus, it would be difficult to produce infectious recombinant viruses from recombinant genomes constructed *in vitro*. Insertion of DNA into the virus genome has been achieved, in infected cells, by recombination between plasmids containing virus sequences and the homologous sequence in the virus genome.

Although vaccinia virus transcriptional control signals have not been identified, the regions upstream of several vaccinia virus genes have been found to be extremely rich in adenine and thymine differing considerably from prokaryotic or eukaryotic promoter consensus sequences (5). It would appear that the virus RNA polymerase has evolved unique regulatory sequences and presumably efficient expression of foreign genes in vaccinia virus will only be achieved by use of these sequences.

Recombinant viruses representing only a very small percentage of a virus stock can be detected by a plaque hybridisation procedure (Section 2.4.5). An alternative to the initial identification of a recombinant virus in this way is to use a selection procedure which enriches for the recombinant. This has been achieved by insertion of the foreign gene of interest into the vaccinia virus thymidine kinase (TK) gene. A TK^- recombinant is produced and can be selected with 5-bromodeoxyuridine (BUdR). Thus, the initial technical problems have been overcome and several groups have reported the construction of vaccinia virus recombinants expressing foreign genes (6–12).

2. CONSTRUCTION OF VIRUS RECOMBINANTS

2.1 Strategy

We have constructed vaccinia virus recombinants expressing foreign genes by what is essentially a two stage process. Firstly, recombinant DNA techniques are used to assemble a plasmid (insertion vector) that contains a chimeric gene flanked by vaccinia virus DNA. The chimeric gene consists of a vaccinia virus transcriptional start site and upstream regulatory sequences adjacent to the protein coding sequence of the foreign gene. Thus, when inserted into vaccinia virus, the chimeric gene should be transcribed from the normal vaccinia RNA start site and the message produced should be translated into the authentic foreign protein. The next stage is the insertion of the chimeric gene into vaccinia virus. This can be achieved by transfection of wild-type virus infected cells with the plasmid insertion vector. Homologous recombination occurs between the vaccinia sequences flanking the chimeric gene and the virus genome, thus producing a recombinant virus. A schematic representation of this is shown in *Figure 1*. The virus DNA that flanks the chimeric gene determines the site at which the foreign gene is inserted. Obviously the foreign gene must be inserted into a region of the genome that is non-essential for virus growth in tissue culture. At the time of writing,

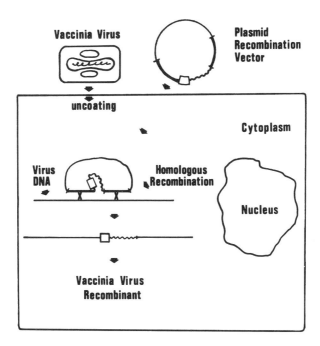

Figure 1. Generation of vaccinia virus recombinants. Cells infected with vaccinia virus are transfected with a plasmid construct. This construct contains a correctly orientated vaccinia virus promoter (—□) adjacent to a foreign gene coding sequence (〜〜) flanked on either side by virus DNA sequences. Homologous recombination occurs between the flanking sequences and virus genomic DNA with the resultant insertion of the foreign gene into virus DNA which can be packaged and produce a recombinant virus.

three regions of the virus genome have been used to insert foreign genes and these are indicated in *Figure 2* (6,9). Although any of these regions or any non-essential region could be used, we routinely insert the chimeric gene into the virus *TK* gene because recombinants will be TK^- and can be selected with BUdR.

2.2 Design of Plasmid Insertion Vectors

The sequence of several vaccinia virus genes and their promoters has been determined (5). This information has allowed the construction of several plasmid insertion vectors which with a minimal amount of manipulation can be used to insert and express foreign genes in vaccinia virus. *Figure 3* shows three such insertion vectors. These plasmids have multiple unique restriction enzyme sites engineered downstream from a vaccinia virus promoter that has been translocated within the body of the virus *TK* gene. Thus any continuous protein coding sequence can be inserted next to a vaccinia virus promoter. Potential problems arising from improper codon phasing or the generation of fusion proteins are avoided by juxtaposing the transcriptional start site of the vaccinia promoter and the translational initiation codon of the foreign gene. After transfection of wild type vaccinia virus infected cells with the insertion vector containing the foreign protein coding sequence TK^- virus is selected by plaque assay on TK^- cells in the

193

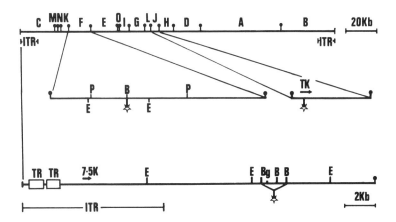

Figure 2. Location of sites of insertion (I) of foreign genes into vaccinia virus. All the *Hind*III sites (ꜜ) in the virus genome are shown and fragments labelled in order of decreasing molecular weight. The approximate location of the virus TK and the 7.5 K protein is also indicated. Details of the three sites can be found in references 6, 9 and 12. The lower line is the expanded *Hind*III C fragment. The scale shown for the *Hind*III C fragment also applies to the expanded *Hind*III F and *Hind*III J fragments. The inverted terminal repeats (ITR) and tandem repetitions (TR) in the virus DNA are indicated. Abbreviations for restriction endonuclease sites are as follows B, *Bam*HI; E, *Eco*RI; Bg, *Bgl*II; P, *Pst*I.

presence of BUdR. A high percentage of *TK⁻* viruses selected in this way are recombinants which express the foreign gene. It is possible to insert DNA into other non-essential regions of the virus genome. This can be achieved by flanking the foreign gene with the appropriate non-essential region of the virus genome. However, foreign genes inserted into these regions do not have the advantage of providing a phenotypic marker that can be selected. It is also relatively simple using synthetic linkers to create new unique restriction endonuclease sites downstream from the promoters in the insertion vectors already constructed. Thus, a new *Xho*I site was generated at the *Sma*I site of pGS20 (*Figure 3*) using synthetic linkers. This was particularly useful to insert several genes into vaccinia virus whose protein coding sequences were bounded by *Xho*I or *Sal*I sites. This strategy of translocation of a promoter, upstream of multiple restriction enzyme sites within the body of the TK gene, should also prove effective for other virus promoters as and when they are characterised. In this context, it would be particularly useful to use the promoters of the late structural genes of the virus. These are the most abundant virus coded proteins and presumably one would achieve higher levels of expression of the foreign gene if it were under control of these promoters.

2.3 Growth and Purification of Vaccinia Virus

Traditionally vaccinia virus was grown on and purified from the chorioallantoic membrane of 12 day fertile hens eggs. However, the majority of tissue culture cell lines are capable of supporting the growth of the virus. Monkey kidney cells BSc1's, CV1's and Vero's have been used extensively for plaquing the virus or growing reasonable amounts of stock virus. For large scale preparation of virus

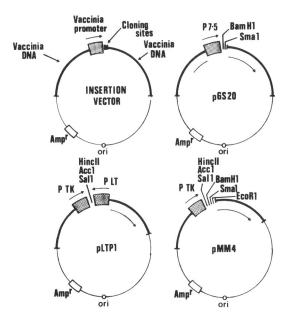

Figure 3. Plasmid insertion vectors. A generalised plasmid insertion vector is illustrated on the top left. pMM4, pGS20 and pLTP1 are specific examples of plasmid insertion vectors. The essential features of these vectors include a vaccinia virus promoter upstream from several unique restriction endonuclease sites. The promoter and cloning sites are flanked by vaccinia virus DNA taken from a non-essential region of the virus genome. The ampicillin resistance gene (*Amp*^r) and bacterial origin of replication (*ori*) enable the vector to amplify in *E. coli*. Plasmid pMM4 contains the *TK* promoter, pGS20 contains the promoter for an early virus gene coding for a 7.5 kd polypeptide and pLTP1 contains a late virus promoter found in the *Hind*III L fragment (J. Weir, personal communication). The unique endonuclease sites for cloning a foreign gene downstream of the promoter are indicated. The internal arrows indicate the virus thymidine kinase gene, the external arrows indicate the direction of transcription of the translocated promoters. For details of construction of pMM4 and pGS20 see reference 12.

HeLa S3 spinner cells seem to give the best yields although KB spinners can be used. HeLa monolayers, MRC5's, chick embryo fibroblasts, RK13, BHK and mouse L cells have all been used successfully to grow the virus.

2.3.1 *Virus Strains*

Literally hundreds of vaccinia virus strains have been documented. The more commonly used strains are IHD, Lister (Elstree) (a vaccine strain), WR, Copenhagen and Wyeth (US vaccine strain). Most of the more recent work has been done with the WR strain, however all these strains give good yields in tissue culture.

2.3.2 *Assaying Virus Stock Infectivity*

The standard method for determining the infectivity of a virus stock is a plaque assay, described in *Table 1*. Appropriate dilutions of the stock virus are incubated with confluent cell monolayers for several days. After staining the infected monolayer areas of cell degeneration and detachment (plaques) can be observed.

195

Table 1. Plaque Assay.

1.	Thaw the virus stock. Disperse any aggregated cell-debris by mild sonication (water bath sonicator; Megasonic 80 Kc).
2.	Add 0.1 volume of 2.5 mg/ml trypsin (Difco, Trypsin 1:250) and incubate at 37°C for 30 min.
3.	Make 10 fold serial dilutions[a] of virus in Hanks balanced salt solution + 0.1% BSA at 25°C.
4.	Remove media from a monolayer culture of the appropriate cell line that has just reached confluence. CV1, BS1, Vero or RK13 are all suitable host cells for the plaque assay. Wash once with PBS and drain the monolayer.
5.	Add the virus dilution to the cell monolayer in a small volume (0.5 ml for a monolayer grown in 6 cm diameter tissue culture dish or 0.15 ml for cells grown as 1.5 cm² monolayers in 24 well trays) and incubate at 37°C for 1 − 2 h.
6.	Remove the virus inoculum, add the appropriate media (depending on the cell type) containing 5% foetal calf serum (FCS). Incubate the monolayers at 37°C for 36 − 48 h.

[a]To minimise pipetting errors the dilutions can be done in duplicate and each dilution titrated on several monolayers. Monolayers with approximately 100 plaques are used to calculate the virus titre of the stock.

Table 2. Plaque Purification.

1.	Follow steps 1 − 5 in the protocol described in *Table 1* for the plaque assay.
2.	Remove the virus inoculum and overlay the monolayer with Eagle's minimal medium (MEM) containing 1% noble agar[a] and 5% FCS. If a selective media is required this can be incorporated into the agarose overlay. Incubate at 37°C for 36 − 48 h.
3.	Stain the cells by overlaying the agarose with 1% agarose containing 0.1% neutral red. Incubate at 37°C for several hours to visualise virus plaques which appear as clear areas in the monolayer since only living cells take up the netural red.
4.	Pick well isolated plaques using a Pasteur pipette with small bulb attached to it. Put the Pasteur pipette into the agarose directly over a plaque so that the pipette is in contact with the monolayer. By suction remove the agarose plug. This inevitably removes the majority of the infected monolayer in the region of the plaque.
5.	Transfer the plug and virus containing monolayer to 0.5 ml of MEM.
6.	Release the virus from the monolayer with 3 cycles of freeze thawing. This can be achieved conveniently by alternate use of dry ice and a waterbath. Between 10^4 and 10^5 p.f.u. of virus can normally be recovered from a single virus plaque.
7.	For re-plaquing follow steps 3 − 5 in *Table 1* and steps 2 − 6 above.

[a]On some cell lines with selective media we have found that 1% low gelling temperature agarose gives better results.

A knowledge of the dilution used to produce a certain number of plaques and assuming a single infectious virus particle is responsible for each plaque allows one to calculate the infectivity of a stock. This infectivity is usually expressed as plaque forming units/ml (p.f.u./ml). It should be noted that the titre of a virus stock may vary as much as 5-fold when plaqued on different cell lines.

2.3.3 *Preparation and Purification of Virus Stocks*

After the initial identification of a recombinant virus we routinely plaque purify twice before growing a small stock of the virus. A protocol for the plaque purification of virus stocks is given in *Table 2*. It is the assumption of the pro-

Table 3. Preparation of Small Virus Stocks.

1.	Use half the virus recovered from a plaque (step 6, *Table 2*) to infect a small monolayer of cells (25 cm²) which have just reached confluency (steps 4 and 5, *Table 1*).
2.	Add the appropriate medium (depending on the cell type and whether or not selection is required) containing 5% FCS. Incubate at 37°C until the majority of cells in the monolayer display viral cytopathic effect.
3.	Scrape the infected cells into the media and recover them by pelleting at low speed in a bench top centrifuge.
4.	Release the cells in 2 ml of hypotonic buffer. This may be either 10 mM Tris-HCl pH 9, 1 mM EDTA; or 5 mM citrate-phosphate buffer pH 7.4.
5.	Release the virus by three cycles of freeze-thawing. Store at −70°C. This small virus stock can then be used as required to generate larger stocks. After thawing from −70°C the cell debris is often aggregated. It can be dispersed by mild sonication before use.

tocol, that a virus plaque is the progeny of a single infectious particle. A protocol for the preparation of small stocks of virus is given in *Table 3*. For the preparation of large stocks (*Table 4*), it is advantageous to grow the virus in HeLa cells adapted for culture in Spinner flasks.

2.4 Transfection Procedures

Once a foreign gene protein coding sequence is placed under control of a vaccinia virus promoter and is flanked by non-essential vaccinia DNA, it is possible to insert the foreign gene into the virus by homogenous recombination *in vivo*. Calcium phosphate precipitates of the plasmid construct are used to transfect vaccinia virus infected CV1 cells in a manner identical to that used for marker rescue of vaccinia virus mutants (13). Conditions found to be optimal in marker rescue are used to produce recombinants. We have also generated recombinants by transfecting infected MRC5's primary chick embryo fibroblasts and human TK-143 cells. The transfection protocol is given in *Table 5*. The majority of the virus stock given by this protocol will be wild type virus but recombination will have occurred between the plasmid construct and the wild type virus to give a finite percentage of recombinant viruses. The next section describes the isolation of these recombinants.

If necessary, several parameters in this transfection procedure can be altered to increase the efficiency of recombination as described in Chapter 6 of this volume. Several workers (14−16) have found that the addition of virus DNA to the calcium phosphate precipitate increases marker rescue. Replacement of the carrier calf thymus DNA by the plasmid construct DNA may also increase the efficiency of marker rescue. Interestingly, a glycerol or dimethyl sulphoxide shock did not increase the efficiency of marker rescue. We have also used human TK^- 143 cells as an alternative to CV-1 cells. Presumably most cell lines that are susceptible to infection by vaccinia and can be efficiently transfected by the calcium phosphate method would be suitable substitutes for CV-1 cells.

2.5 Selection of Recombinants

The method used to distinguish between a recombinant virus and the wild type

Table 4. Preparation of Large Virus Stock.

A. *Crude Stocks*

1. Seed eight 150 cm² bottles each with 5 x 10⁷ HeLa spinner cells in MEM. Allow the cells to settle overnight.
2. Take half the progeny virus from a small stock (*Table 3*) and incubate with 0.1 vol of 2.5 mg/ml trypsin for 30 min at 37°C.
3. Dilute the virus stock to 16 ml with PBS + 0.1% BSA (bovine serum albumin) and add 2 ml of the virus to each bottle. Incubate at 37°C for 1 h. Add 40 ml Eagles MEM + 5% FCS per bottle and incubate for a further 48 h.
4. Recover the cells by scraping them into the media and pelleting at low speed in a bench top centrifuge. Resuspend the cells in 2 ml per monolayer (that is 16 ml in total) of MEM and subject the suspension to 3 cycles of freeze-thawing to release the virus.
5. Titrate the crude virus stock (*Table 1*). These crude stocks often have titres of 10¹⁰ p.f.u./ml.

B. *Purified Stocks*

1. Grow 2 – 10 litres of HeLa S3 spinner cells to a density of 5 x 10⁵ cells/ml.
2. Incubate the appropriate amount of crude virus stock with 0.1 vol 2.5 mg/ml trypsin at 37°C for 30 min. The amount of stock should be sufficient to infect the spinner cells at a multiplicity of 5 p.f.u./cell.
3. Concentrate the spinner cells to 5 x 10⁶ cells/ml by low speed centrifugation. Resuspend the cells in culture medium.
4. Add the trypsin-treated crude virus stock to the concentrated spinner cells and incubate in suspension for 1 h at 37°C. Dilute with culture medium to give 5 x 10⁵ cells/ml. Incubate for a further 48 h.
5. Harvest the infected cells and resuspend them in 10 mM Tris-HCl pH 9.0 at 4°C. Use 2 ml per 2 x 10⁷ infected cells from a monolayer or 2 ml per 100 ml of suspension culture. This and subsequent steps should be carried out on ice.
6. Homogenize with 15 – 20 strokes of a tight fitting Dounce homogeniser. Check for complete cell breakage by microscopy.
7. Pellet the nuclei by centrifugation at 750 g for 5 min at 4°C. Resuspend the pellet in 10 mM Tris-HCl pH 9.0. Re-centrifuge and combine the supernatants.
8. Add 0.1 volume of trypsin (2.5 mg/ml) and incubate at 37°C for 30 min with frequent vortexing.
9. Layer onto an equal volume of 36% (w/v) sucrose in 10 mM Tris-HCl pH 9.0 in a Beckman SW27 tube (1″ x 3.5″) or its equivalent. Centrifuge at 13 500 r.p.m. in the SW27 rotor (25 000 g) for 80 min at 4°C.
10. Discard the supernatant and resuspend the pellet (virus and debris) in 2 ml of 1 mM Tris-HCl pH 9.0. Add 0.1 volume 2.5 mg/ml trypsin and incubate at 37°C for 30 min.
11. Overlay 2 ml of virus suspension onto continuous sucrose gradients (15 – 40% in 1 mM Tris-HCl pH 9.0) in Beckman SW27 tubes or their equivalent. Centrifuge at 12 000 r.p.m. in the SW27 rotor (18 750 g) for 45 min at 4°C.
12. Collect the banded virus with a syringe through the side of the tube. Dilute 1:3 with 1 mM Tris-HCl pH 9.0 and pellet the virus by centrifugation at 25 000 g for 60 min at 4°C.
13. Resuspend the virus in 1 mM Tris-HCl pH 9.0. Freeze aliquots at – 70°C.
14. Thaw an aliquot and titrate following the protocol in *Table 1*. 5 litres of suspension cells should give at least 2 ml of 5 x 10¹⁰ p.f.u./ml.

virus depends on the site of insertion of the foreign gene and the nature of the foreign gene product.

2.5.1 *Selection of TK⁺ Recombinants*

The first reports of the expression of a foreign gene in vaccinia virus concerned

Table 5. Transfection Protocol for Generation of Recombinant Viruses.

1.	Infect a recently confluent monolayer of CV1 cells (25 cm²) with 0.05 p.f.u./cell of purified virus. Refer to steps 1−5 of the plaque assay in *Table 1*. At step 3 dilute the virus to give the desired multiplicity of infection.
2.	Add 1 μg of the plasmid construct to 19 μg of calf thymus DNA in 1 ml of HEPES buffered saline (see Chapter 15, Section 3.1.1).
3.	Precipitate the DNA by addition of calcium chloride to a final concentration of 125 mM. Leave at 25°C for 30 min.
4.	Two hours after infection of the monolayers, remove the virus inoculum and wash the monolayer twice with serum free medium. Add 0.5 ml of the DNA suspension and incubate at room temperature.
5.	After 30 min add 5 ml of medium + 5% FCS and incubate at 37°C for a further 3.5 h.
6.	Six hours post infection replace the media with MEM + 5% FCS.
7.	48 h post infection recover the cells by scraping them into the medium. Release the virus by three cycles of freeze-thawing.

Table 6. Generation and Selection of TK⁺ Recombinants.

1.	Infect CV1 cells at a multiplicity of 0.05 p.f.u./cell with a vaccinia virus TK^+ mutant as described in *Table 1, steps 1−5*. At step 3, dilute the virus to give the desired multiplicity.
2.	At two hours post infection add the calcium phosphate precipitated plasmid construct (see *Table 5*) to the monolayer. Leave at 25°C for 30 min. Add medium to the cells, incubate at 37°C and recover virus 48 h post infection (steps 2−7, *Table 5*).
3.	This virus stock will contain a small percentage of TK^+ recombinant viruses which can be selected by growing the virus stock in the presence of a selective media containing methotrexate (amethopterin). Plaque the virus stock on human TK^- 143 cells which have just reached confluence (steps 1−2, *Table 2*). The selective media used in the agarose overlay contains 5% FCS, 100 μM non-essential amino acids, 15 μM glycine and 1 μM methotrexate.
4.	48 h post-infection stain the cells to visualize the plaques. Pick TK^+ virus plaques and subject them to a further plaque purification in the presence of selective media (steps 4−7, *Table 2*).

the herpes simplex virus pyrimidine kinase (6,9). In this case, the expression of the foreign gene confers a TK^+ phenotype on a previously TK^- parental virus, thus the parent and the recombinant viruses are easily distinguishable. The recombinant plasmids used to generate TK^+ recombinants contain the herpes simplex type 1 pyrimidine kinase protein coding sequence adjacent to the promoter of a vaccinia gene coding for a polypeptide of 7.5 kd. This was flanked by vaccinia DNA derived from near the left hand terminus of the virus genome, known to be non-essential for virus growth. The procedure, which is given in *Table 6* could be used to generate recombinants expressing various genes linked in tandem to the herpes virus thymidine kinase.

Subsequently it was shown that TK^+ viruses have integrated the herpes virus *TK* gene into their genome and were TK^+ because they expressed the herpes gene. In theory it should be possible to insert genes into TK^- vaccinia virus in tandem with the herpes pyrimidine kinase. This approach has the advantage that all the TK^+ viruses that are generated should also contain the foreign gene. The disadvantage, however, is it necessitates constructing plasmids that are rather cumbersome and difficult to manipulate. The monolayers of human TK^- 143 cells degenerate fairly quickly in the presence of methotrexate. The integrity of the monolayer is improved when low melting point agarose is used instead of noble agar in the overlay containing the methotrexate. In an effort to see whether other cell lines were more resistant to methotrexate, we have examined mouse LMTK⁻ and BHKTK⁻ cell lines. In our hands vaccinia virus plaques poorly in LMTK⁻ cells giving very small plaques. The BHKTK⁻ line is somewhat better in this respect but the *TK*⁻ 143 cells are by far the best for plaquing the virus and consequently we have made extensive use of this cell line both for selection of *TK*⁺ and *TK*⁻ recombinant virus (see following section).

2.5.2 *Selection of Recombinant by Insertion into the Virus TK*

An alternative method for selection of recombinants is to insert the chimeric foreign gene of interest into the virus *TK* gene and to select the recombinant on the basis of its resulting *TK* phenotype. The *TK*⁻ phenotype is selectable in *TK*⁻ cells using BUdR. We have constructed the insertion vectors shown in *Figure 3* so that the virus *TK* gene is interrupted by a vaccinia promoter adjacent to multiple restriction endonuclease sites. These constructs are used as follows.

(i)   Construct recombinant plasmids in which the gene for the foreign protein is inserted at one of the endonuclease sites of the vectors shown in *Figure 3*. Introduce this DNA into vaccinia-infected cells using the standard transfection protocol outlined in *Table 5*.

(ii)   Plaque one fifth of the progeny virus from the transfection outlined in *Table 2*, on human *TK*⁻ 143 cells. At step 2 use 1% low gelling temperature agarose containing 5% foetal calf serum, Eagles MEM and 25 µg/ml BUdR.

(iii)   At 48 h post infection stain the monolayer with neutral red to visualize the *TK* virus plaques (*Table 2*, step 3).

(iv)   Pick about 24 well separated plaques and infect small monolayers (1.5 cm² 1 − 2 x 10⁵ cells) of the *TK*⁻ 143 cells in the presence of 25 µg/ml BUdR.

This can be conveniently accomplished in 24 well trays (step 6, *Table 2*; steps 4 and 5, *Table 1*). The presence of the foreign DNA in the virus genome can be tested in a simple DNA:DNA hybridisation assay or the expression of the foreign gene can be tested for by a dot blot immunoassay (see Sections 2.5.3 and 2.5.4).

Two cycles of plaquing have been found to be sufficient to give a homogeneous virus stock. In typical experiments, 20 − 90% of the initial *TK*⁻ plaques are recombinant viruses, the other viruses being spontaneous *TK*⁻ mutants. We have noticed that when very large fragments are inserted, the percentage of *TK*⁻ viruses that are recombinants decreases. Presumably the larger the foreign DNA that interrupts the flanking DNA the less efficient the homologous recombination that inserts the foreign gene into the virus.

2.5.3 *DNA Dot Blot Hybridisation*

(i)   Infect *TK*⁻ 143 cells growing in 1.5 cm² wells (∼ 1 − 2 x 10⁵ cells) with half the progeny of well separated *TK*⁻ virus plaques (Putative recombinants from a transfection, see Section 2.5.2).

(ii)   Incubate at 37°C for 36 − 48 h. Harvest the infected cells by scraping into the medium. It is convenient to use the plunger from a 1 ml syringe for this purpose.

(iii) Collect one fifth of the recovered material on nitrocellulose sheets by filtration using a microsample manifold such as the ones supplied by Schleicher and Schuell or the equivalent from other manufacturers.

(iv) Wash the nitrocellulose filter with 100 mM NaCl, 50 mM Tris-HCl pH 7.5.

(v) Place the filters for 3 min intervals on successive Whatman 3 MM papers saturated with (a) 0.5 M NaOH, (b) 1 M Tris-HCl pH 7, (c) 2 x SSC (1 x SSC is 0.15 M NaCl, 0.015 M Na citrate).

(vi) Bake the filter at 80°C in a vacuum oven for 2 h. The filter can then be hybridised with a radiolabelled probe of the foreign gene applying considerations discussed in reference 22. Hybridization using probes of a specific activity of 10^7 c.p.m./μg DNA gives a strong signal on overnight exposure of the filter to X-ray film.

(vii) Go through a further cycle of plaque purification with those virus plaques that gave a positive signal with the foreign gene. Make virus stocks as described in Sections 2.3.3 and 2.3.4.

Figure 4a shows a DNA dot blot hybridisation. In this case, VSV strain New Jersey (VSV NJ) glycoprotein cDNA was placed under control of a virus late promoter (pLTP1, *Figure 3*) translocated within the virus *TK* gene. After transfection of virus infected cells with the plasmid construct, *TK*⁻ viruses were selected in *TK*⁻ 143 cells with BUdR. Twenty four plaques were picked virus grown in

A)

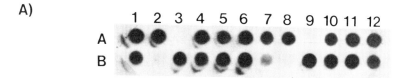

B)

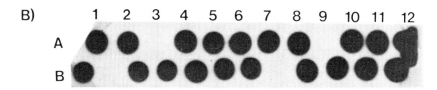

Figure 4. (a) Demonstration of VSV DNA in vaccinia recombinants. 24 *TK*⁻ virus infected small monolayers (A 1 – 12, B 1 – 12) were transferred to nitrocellulose and the DNA present probed with ³²P-labelled VSV glycoprotein cDNA (Section 2.4.3). The autoradiograph shows that 20 of the 24 monolayers were infected with a virus containing the VSV glycoprotein gene. (b) Demonstration of VSV glycoprotein expression in vaccinia recombinants. NP40 extracts of 24 small monolayers (A 1 – 12, B 1 – 12) infected in parallel with those in *Figure 4a* were spotted onto nitrocellulose and probed for VSV glycoprotein (VSVG) expression using an anti-VSVG antibody and ¹²⁵I-labelled staphylococcal A protein (Section 2.4.4). The autoradiograph shows 20 of the 24 monolayers were infected with a recombinant vaccinia virus that expresses the VSV G protein. (a) is a 6 h exposure to Kodak XAR5 X-ray film while (b) is a 48 h exposure.

small monolayers and using ^{32}P-labelled VSV cDNA tested for the presence of VSV DNA. As can be seen from *Figure 4a*, 20 of the 24 *TK$^-$* plaques picked were positive for VSV DNA.

2.5.4 *Immuno Dot Blot*

Antisera specific for the foreign gene product can also be used to detect recombinant viruses. *Figure 4b* shows the results of an experiment done in parallel with that in *Figure 4a*. The same *TK$^-$* plaques as used in *Figure 4a* were used to infect human *TK$^-$* 143 cells grown in 1.5 cm^2 wells in a 24 well tray in the presence of 25 μg/ml BUdR. The immuno dot blot procedure is carried out as follows.

(i)   Remove the media from the cells and lyse them with 100 μl of a solution containing 0.1 M Tris-HCl pH 8.0, 0.1 M NaCl, 0.5% NP40 and 0.1% Aprotinin.

(ii)   Spot 25 μl of the lysate from each monolayer onto nitrocellulose and leave it to air dry.

(iii) Pre-treat the filter with 50 ml phosphate buffered saline A (PBSA, Difco) containing 4% BSA and 0.02% sodium azide. Leave for 2 h on a rocking platform.

(iv)   Add antisera that will recognise the foreign gene product (in this example anti-VSV NJ glycoprotein) and incubate for a further 2 h on the rocking platform.

(v)   Wash the filter 5 x for 2 min with 50 ml PBSA.

(vi)   Incubate with gentle agitation for 2 h in PBSA containing 4% BSA, 0.02% sodium azide and 1 μCi of ^{125}I-labelled staphylococcal A protein.

(vii) Wash the filter 5 x for 2 min with 50 ml PBSA and allow to air dry. Expose the dot blot to X-ray film to reveal the plaques positive for foreign gene expression.

Some monoclonals that could be used as the primary antibody may not bind staphylococcal A protein. To circumvent this problem, it is possible to include a further step in the protocol. After incubation with the primary antibody and washing, a polyclonal rabbit anti-mouse IgG antibody could be used followed by the ^{125}I staphyloccal A protein step. Alternatively a peroxidase conjugated anti-mouse IgG antibody could be used with a subsequent colour development assay to detect antibody binding.

As can be seen from *Figure 4b*, the same 20 plaques that are positive for VSV DNA are also positive for expression of the foreign gene. The DNA hybridisation dot blot is somewhat more sensitive than the immuno dot blot. The immunoassay has the advantage, however, of taking less time and directly assaying for expression of the foreign gene and not solely for its insertion into the vaccinia genome.

2.5.5 *Detection of Individual Recombinant Plaques in a Monolayer*

As an alternative to the initial selection of a recombinant virus on the basis of its phenotype it is possible to detect a recombinant virus either by an *in situ* plaque DNA:DNA hybridisation protocol, or by screening virus plaques for the expression of the foreign gene using an antibody.

Virus progeny from a transfection is plaqued onto a fresh monolayer of cells at the appropriate dilution (*Table 1*) usually with an agar overlay to give approximately 100 plaques/9 cm dish. If the expected frequency of recombinants is low, then the p.f.u./monolayer can be increased. If the p.f.u./monolayer is >500 then it is preferable to shorten the time plaques are allowed to develop from 48 h to 36 h or even 24 h. Once virus plaques are produced, the monolayers are transferred to nitrocellulose as follows:

(i)   Carefully place a dry 8 cm diameter nitrocellulose filter onto a moist monolayer of cells containing virus plaques in a 9 cm tissue culture dish.

(ii)   When all of the filter is wet, place a sheet of Whatman 3 MM soaked in 2 x SSC on top of the nitrocellulose filter for several minutes.

(iii)   Press firmly and evenly on the filters and carefully remove the 3 MM paper and nitrocellulose to which the cell monolayer becomes attached. It is helpful if the cell sheet has been previously stained with 0.05% neutral red because it is then easy to see how efficiently the cells have been transferred.

(iv)   Place the nitrocellulose filter monolayer upwards sequentially for 3 min intervals onto filter papers soaked in (a) 0.5 M NaOH; (b) 1 M Tris-HCl, pH 7.5; (c) 2 x SSC. Bake the filters at 80°C for 2 h.

These filters can then be hybridised with ^{32}P-labelled DNA specific for the foreign gene using standard techniques (22). After washing and autoradiography of the filters recombinant virus plaques can be recovered from the agarose overlay corresponding to the positions of dark spots on the autoradiograph. Alternatively, recombinants can be recovered from a second nitrocellulose filter that had been pressed against the primary filter containing the cell monolayer (7). This technique has the advantage of being applicable to any cell line that can support the growth of vaccinia virus.

Antibody specific for the foreign gene product can also be used in the primary plaque screen. We have used this plaque antibody screen primarily to test the purity of our virus stocks (see *Figure 6*) but it is equally applicable to the primary identification of recombinants. The screen is performed as follows:

(i)   Fix cell monolayers having approximately 200 plaques/9 cm dish with 10 ml of cold methanol 36−48 h after infection.

(ii)   Wash with PBSA and incubate the fixed cells for one hour with 5 ml of 4% BSA, 0.02% sodium azide on a rocking platform at room temperature.

(iii)   Supplement the solution with the appropriate antibody and continue to incubate for a further hour.

(iv)   Wash the cell monolayer 5 x with PBSA.

(v)   Add 5 ml of 4% BSA, 0.02% sodium azide solution containing 0.5 μCi of ^{125}I staphylococcal A protein and incubate with rocking for a further hour.

(vi)   Wash the monolayers 10 x with PBSA. Remove the rims of the Petri dishes and expose the cell sheet to X-ray film. Dark spots on the autoradiograph represent the positions of recombinant viruses which can be recovered from the agarose overlay.

As an alternative to ^{125}I-labelled staphylococcal A protein it should be possible to use a second anti-species antibody (directed against the primary antibody) labelled with ^{125}I or coupled to an enzyme. The coupling of the second antibody to an enzyme would provide a colour reaction as described in Chapter 6 of this volume. This speeds up the assay and obviates the need for autoradiography.

3. CHARACTERISATION OF RECOMBINANTS

3.1 Analysis of Virus DNA

DNA can be extracted from the purified recombinant virus and used to analyse both the purity of the virus stock and the structure of the virus genome. However, it is possible to analyse recombinant virus DNA without first purifying the virus, by isolating total DNA from an infected monolayer and probing 'Southern blots' of endonuclease digested DNA with ^{32}P-labelled virus or foreign gene DNA. A third approach is to extract total DNA from monolayers that have been infected in the presence of [^{3}H]thymidine. Restriction endonuclease analysis followed by fluorography of the gels reveal virus specific DNA bands superimposed upon a feint background of cellular DNA. Tritium labelling of virus DNA is carried out as follows:

(i) Infect approximately 10^7 chick embryo fibroblast (CEF) cells with the virus recovered from a single plaque. Maintain the infected cells in medium supplemented with 0.5 μCi/ml [6-^{3}H]thymidine (23 Ci/mmol) until the virus cytopathic effect observed microscopically is seen to involve 80% of the cell population.

(ii) Remove the medium and scrape the infected cells into 5 ml of 0.02% EDTA in PBS. Pellet the cells by centrifugation at 500 g for 5 min.

(iii) Resuspend the infected cells in 50 mM Tris-HCl, pH 7.8, 1 mM EDTA and 30% sucrose. Lyse the cells at 4°C by adding 1% SDS and 100 mM 2-mercaptoethanol.

(iv) Leave at 4°C for 30 min and then add proteinase K to a final concentration of 500 μg/ml. Digest at 37°C for 2 h.

(v) Extract the total nucleic acids by shaking with phenol saturated with 1 M Tris-HCl pH 7.5. Remove the aqueous phase and re-extract with chloroform:isoamyl alcohol (24:1).

(vi) Remove the aqueous phase, add 1/10 vol of 5 M NaCl and 2.5 vol ethanol. Leave to precipitate overnight.

(vii) Spool out the DNA onto a glass rod. Allow to air dry and redissolve in 10 mM Tris-HCl pH 7.5, 1 mM EDTA.

The DNA may then be cleaved with restriction endonucleases and the fragments separated electrophoretically in 0.6% agarose gels. If the gels are treated for 20 min with 1 M sodium salicylate and then dried the ^{3}H-labelled DNA can be visualised by autoradiography at -70°C. *Figure 5* shows total *Hind*III cleaved [^{3}H]DNA's isolated from CEF cells infected with rabbitpox, ectromelia (an orthopoxvirus that infects mice) and a series of rabbitpox-ectromelia recombinants. Virus DNA fragments are distinctly visible above the background ^{3}H-labelled cell DNA. This may be due to the fact that CEF cells are extremely quiescent and hence little ^{3}H is incorporated into cell DNA. If this approach is tried in other cell lines such as CV-1 cells, then interference from ^{3}H-

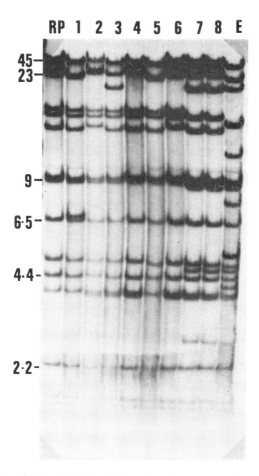

Figure 5. Fluorograph of ³H-labelled *Hind*III cleaved restriction endonuclease fragments, separated on a 0.6% Agarose gel, derived from rabbitpox virus (RP), a series of ectromelia rabbitpox recombinants (1 − 8), and ectromelia (E). The sizes of DNA fragments are indicated in kb.

cellular DNA is more of a problem. If TK^+ viruses are to be looked at, then the use of TK^- cell lines should overcome this problem. This method has the advantage of not requiring virus purification and hence is applicable for determining restriction endonuclease profiles from a large number of recombinants. It is also feasible to use this method when introducing defined deletions into the virus genome where one cannot screen for the introduction of a foreign gene.

In order to characterise the recombinant genome thoroughly, DNA should be extracted from purified virus (Section 2.3.4) as follows:

(i)   Adjust purified virus to final concentrations of 50 mM Tris-HCl pH 7.8, 1 mM EDTA, 27% sucrose, 1% SDS and 500 μg/ml Proteinase K and incubate at 37°C for 2 − 3 h.

(ii)   Deproteinise the DNA released from virus by repeated gentle extraction with 10 mM Tris-HCl pH 7.5, 1 mM EDTA saturated phenol. Care should be exercised at this stage so as not to shear the DNA.

(iii)   Extract twice with chloroform-isoamyl alcohol (24:1).
(iv)   Remove the aqueous phase, adjust to 0.5 M NaCl and precipitate with two
        volumes of ethanol.
(v)   Carefully spool out DNA, air dry and gently resuspend in 10 mM Tris-HCl
        pH 7.5, 1 mM EDTA.

Sufficient virus for DNA analysis can be obtained by using one quarter to one
half of a single plaque to infect four 150 cm^2 bottles of BSC-1's or CV-1
monolayers. It is also possible to analyse DNA from partially purified virus that
has been pelleted through a 36% sucrose cushion (steps 1 – 6, *Table 5*) without
purifying the virus further.

3.2 Characterisation of the Foreign Gene Product

As the translational initiation and termination sites of the foreign gene are maintained, authentic polypeptides are produced which can be characterised by techniques that take into account the properties of the foreign gene. Where the foreign
gene codes for an enzyme, the infected cell extracts can be assayed for the en-
zymic activity. This has been done with the herpes virus pyrimidine kinase (6,9)
and chloramphenicol acetyl transferase (12). A specific assay for the herpes virus
pyrimidine kinase based on its ability to phosphorylate ^{125}I-deoxycytidine has
been used to show that vaccinia recombinants are expressing the herpes virus
gene. CAT gene expression in cells infected with recombinant virus was
monitored by assaying the conversion of [^{14}C]chloramphenicol to its acetylated
forms (12). This enzyme is relatively stable to a single freeze thawing and this was
the method used to release the enzyme from infected cells (see also Chapter 15).
Other methods of assaying for foreign gene expression in vaccinia infected cells
include radioimmunoassay, Western blotting, binding of antibody directly to plaques (2.5.5), dot blot immunoassay (2.5.4), immunofluorescence (3.2.2) and
immunoprecipitation (3.2.1).

3.2.1 *Immunoprecipitation*

Where an antibody is available, the most generally applicable method for
characterising the foreign gene product is pulse labelling of infected cells followed
by immunoprecipitation of the foreign protein and polyacrylamide electrophoresis.

(i)   Infect monolayers of CV1 cells growing in 25 cm^2 bottles with vaccinia
        virus (wild type or recombinant) at 30 p.f.u./cell in Eagle's medium con-
        taining 0.01 mM methionine (9 parts methionine free:1 part normal
        media).
(ii)   After 2 h add 50 – 100 μCi of [^{35}S]methionine (100 Ci/mmol).
(iii) Harvest the cells at 12 h post-infection by washing the cells 3 x with PBSA
        and scraping into PBS.
(iv)   Resuspend the cells in 0.2 ml of 0.1 M NaCl, 0.1 M Tris-HCl pH 8.0,
        0.1% Aprotinin at 4°C. Add 0.05 ml of 2.5% NP40 slowly and leave at
        4°C for 10 min.

(v)   Pellet the nuclei at 800 *g* in a bench top centrifuge. Incubate the super-
        natant with 25 μl of rabbit pre-immune serum at 4°C for 15 h. Add 50 μl
        of a 20% solution of staphylococcal A protein suspension. Incubate for
        30 min at 4°C with agitation.

(vi)   Centrifuge for 2 min in microcentrifuge. Collect the supernatant and in-
        cubate for 4 h with 25 μl of the appropriate dilution of antiserum (deter-
        mined empirically) against the foreign protein. Add 50 μl of a 20% solu-
        tion of staphylococcal A protein suspension. Incubate for 30 min at 4°C
        with agitation. Spin in microcentrifuge for 2 min.

(vii)   Wash the pellet twice with 0.05 M Tris-HCl pH 7.5, 0.15 M NaCl 0.1%
        SDS, 1% Triton X-100, 1% sodium desoxycholate. Wash the pellet a fur-
        ther time with 0.4 M LiCl, 2 M urea, 10 mM Tris-HCl pH 8.0 (LUT).
        Transfer the pellet to a fresh Eppendorf tube and wash again with LUT
        buffer.

(viii) Dissolve the precipitated proteins in 100 μl of 0.06 M Tris-HCl pH 6.8, 3%
        SDS, 5% β-mercaptoethanol, 10% glycerol and 0.002% bromophenol
        blue. Leave for 15 min at 25°C and clarify by brief centrifugation. Heat
        the supernatant at 100°C for 2 min and load onto a 10% polyacrylamide
        gel (22).

The immunoprecipitate polypeptides can be detected by fluorography after
electrophoresis.

Figure 6 shows the use of three immunologically basic techniques to
characterise expression of the vesicular stomatitis virus strain New Jersey glycoprotein by a vaccinia virus recombinant. These techniques show that all the
viruses in the stock are expressing the VSV gene, that the polypeptide produced is
the same size as the authentic glycoprotein and that the expression of the gene
from the translocated promoter is regulated in a similar manner to expression
from the promoter in its native position. The glycosylation of this protein could
also have been analysed by labelling recombinant infected cells with
[^{3}H]glucosamine in the presence or absence of an inhibitor of glycosylation such
as tunicamycin (23). Immunoprecipitation and polyacrylamide electrophoresis of
the authentic protein and protein from the recombinant infected cells should
reveal any abnormalities of protein specified by the recombinant virus. The basic
protocol for labelling the carbohydrate of a foreign protein is the same as for
pulse labelling the protein with [^{35}S]methionine except that a fructose based
media supplemented with $5-10$ μCi/ml of [^{3}H]glucosamine is used (19).

3.2.2 *Immunofluorescence*

In the cases we have examined, the normal transport and subcellular localisation
of the foreign protein is maintained in recombinant infected cells. *Figure 7* shows
surface fluorescence of influenza virus haemagglutinin (HA) in MDCK cells infected with a vaccinia recombinant expressing the HA. This surface fluorescence
is detectable 2 h post infection and is the site at which one would find the HA in
cells infected with influenza virus.

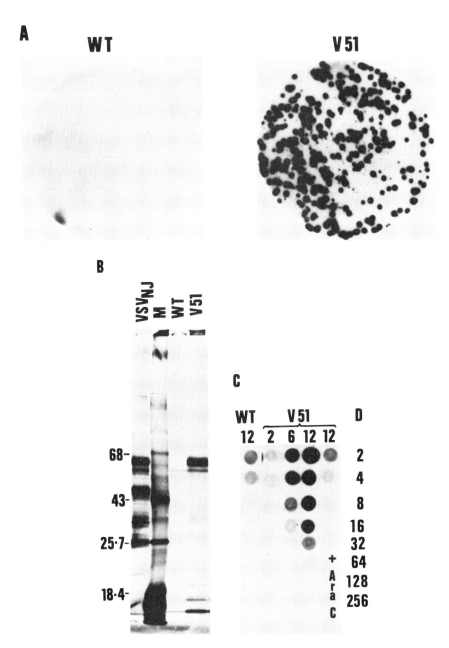

Figure 6. (a) Detection of the expression of VSV$_{NJ}$G protein by individual virus plaques. Duplicate monolayers of CV1 cells containing plaques produced by wild-type vaccinia virus (WT) or recombinant virus v51. Binding of the anti-VSV$_{NJ}$ antibodies followed by incubation with ^{125}I-staphylococcal A protein was carried out as described in Section 2.4.5. After autoradiography, the fixed monolayers were stained with crystal violet. This showed approximately the same number of virus plaques on the cells infected with wild-type virus and the cells infected with v51. It also showed that every virus in the cell monolayer infected with v51 expressed the VSV glycoprotein. (b) Characterisation of VSV polypeptides made by vaccinia virus recombinants. ^{35}S-labelled, immunoprecipitated polypeptides

from cells infected with v51 were separated on 10% polyacrylamide gels and their mobility compared with labelled proteins from VSV infected cells (VSV$_{NJ}$). Molecular weights of standards (M) are indicated in daltons x 10^{-3}. (c) Immuno dot blot extracts from cells infected with VSV vaccinia recombinants v51 or wild type vaccinia virus (WT) in the presence or absence of cytosine arabinoside (Ara C). Extracts of the infected cells and 2-fold dilutions (D) were prepared (1:2 to 1:256) and tested for the presence of VSV$_{NJ}$G by incubation with anti-VSV$_{NJ}$ antisera and ^{125}I-labelled staphylococcal A protein (Section 2.4.4). Panels a, b, c all show autoradiographs.

Figure 7 and the protocol following was kindly provided by Mike Roth of Cold Spring Harbor Labs.

(i)   Fix the cells in 2% freshly dissolved paraformaldehyde (a 1:10 dilution of a stock 37% formaldehyde solution also works) in PBS for 15 − 30 min at room temperature.

(ii)   Wash away the formaldehyde and quench the fixation reaction by adding cell culture medium. It is important to avoid the non-specific adsorption of antibodies to the monolayer and this is achieved by soaking the cells in PBS containing 0.5 − 1% BSA or gelatine for 30 min.

(iii) Incubate the fixed cells with the greatest dilution of antiserum, which still gives a good signal, for 1 h at room temperature.

(iv)   Wash away the first antiserum by five washes of 10 min in PBS.

(v)   Soak the cells in the PBS-BSA buffer prior to reaction with the fluorescent conjugated second antibody. Wash the cells 5 x in PBS. Visualise fluorescence by using the appropriate wavelength of light to fluoresce the second antibody conjugate.

Antiserum dilutions used for this procedure must be determined empirically. Thorough washing is necessary to avoid non-specific background and antisera should always be spun in a microcentrifuge for 5 − 10 min before use. For intracellular fluorescence, cells are fixed in paraformaldehyde and quenched in ammonium chloride or culture medium. They are then permeabilised with 0.1% Triton X-100 in PBS containing 0.25% BSA for 10 − 30 min at room temperature and then stained as for surface fluorescence. A particularly useful solution for washing cells which have been permeabilised in 100 mM NaCl, 50 mM Tris-HCl pH 7.5, 0.5% Nonidet P-40, 0.25% gelatin (this can be used as an immunoprecipitation buffer). The Triton X-100 method gives good preservation of cell structure and is superior to fixation with ethanol, methanol or acetone.

Vaccinia can be grown in a large number of cell types and it may be possible to study transport of selected proteins in well differentiated cell lines using vaccinia recombinants and immunofluorescence. One problem, as can be seen from *Figure 7*, is that by 4 h post infection there is a significant cytopathic effect from the virus infection.

4. CONCLUSIONS

Using the procedures described in the previous sections it is possible to insert and express a variety of foreign genes in vaccinia virus. At the time of writing, the genes that have been expressed include the prokaryotic enzyme chloramphenicol acetyl transferase (12), genes from DNA viruses such as the hepatitis virus surface

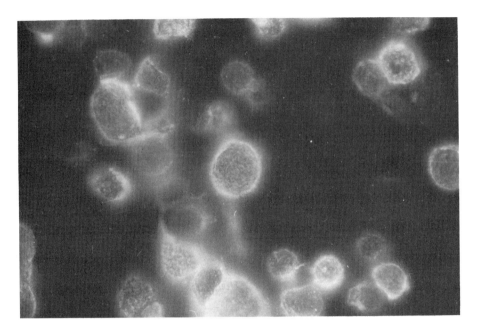

Figure. 7. Surface fluorescence of influenza HA in a vaccinia virus recombinant expressing the HA. Madin Darby canine kidney cells were infected at a multiplicity of approximately 3 p.f.u./cell with a vaccinia-HA recombinant fixed 4 h p.i. The subcellular localisation of the HA was then detected by fluorescence using a monospecific affinity purified rabbit anti-HA IgG antisera followed by an FITC conjugated goat antirabbit antisera.

antigen (HbsAg) (10), genes from RNA viruses inserted as cDNA such as the vesicular stomatitis virus glycoprotein (unpublished results) and the *Plasmodium knowlesi* gene coding for the circumsporozoite surface antigen (20). In animal model systems, recombinant viruses expressing genes from a number of pathogenic agents have been shown to protect the animal against challenge with the pathogenic agent. For example, monkeys vaccinated with a vaccinia virus recombinant expressing HbsAg were protected from disease when challenged with hepatitis virus (21). Interestingly, animals vaccinated with recombinants expressing the influenza virus HA, produce not only circulating antibody directed against HA but also cytotoxic lymphocytes that recognise the HA molecule (Bennink *et al.*, unpublished). This suggests that both the cell mediated and the humoral circulating antibody response to the foreign protein are involved in the protection of animals against challenge.

At least 25 kb of exogenous DNA can be inserted into the virus (11). It should therefore be relatively straightforward to engineer viruses that express many different foreign proteins. Viruses could then be constructed that are polyvalent vaccines, designed to vaccinate against different combinations of pathogens found in different geographical areas of the world. Presumably these vaccines would find use in both human and veterinary medicine.

5. REFERENCES

1. Subramani,S. and Southern,P.J. (1983) *Analyt. Biochem.*, **135**, 1.
2. Rigby,P.W.J. (1983) *J. Gen. Virol.*, **64**, 255.
3. Dales,S. and Pogo,B.G.T. (1981) *Virol. Monogr.*, **18**, 1.
4. Moss,B. (1978) Poxviruses, in *Molecular Biology of Animal Viruses*, Vol. **2**, Nayak,D.P. (ed.), pp. 849.
5. Weir,J.P. and Moss,B. (1983) *J. Virol.*, **46**, 530.
6. Panicalli,D. and Paoletti,E. (1982) *Proc. Natl. Acad. Sci. USA,* **79**, 4927.
7. Panicalli,D., Davis,S.W., Weinberg,R.L. and Paoletti,E. (1983) *Proc. Natl. Acad. Sci. USA,* **80**, 5364.
8. Paoletti,E., Lipinskas,B.R., Samsonoff,C., Mercer,S. and Panicalli,D. (1984) *Proc. Natl. Acad. Sci. USA,* **81**, 193.
9. Mackett,M., Smith,G.L. and Moss,B. (1982) *Proc. Natl. Acad. Sci. USA,* **79**, 7415.
10. Smith,G.L., Mackett,M. and Moss,B. (1983) *Nature,* **302**, 490.
11. Smith,G.L. and Moss,B. (1983) *Gene,* **25**, 21.
12. Mackett,M., Smith,G.L. and Moss,B. (1984) *J. Virol.,* **49**, 857.
13. Weir,J.P., Bajszar,G. and Moss,B. (1982) *Proc. Natl. Acad. Sci. USA,* **79**, 1210.
14. Sam,C.K. and Dumbell,K.R. (1981) *Ann. Viol. (Inst. Pasteur),* **132E**, 135.
15. Nakano,E., Panicalli,D. and Paoletti,E. (1982) *Proc. Natl. Acad. Sci. USA,* **79**, 1593.
16. Ensinger,M.J. (1982) *J. Virol.,* **43**, 778.
17. Villareal,L.P. and Berg,P. (1977) *Science,* **196**, 183.
18. Joklik,W.K. (1962) *Virology,* **18**, 9.
19. Payne,L.G. and Kristensson,K. (1982) *J. Virol.,* **41**, 367.
20. Smith,G.K., Godson,G.N., Nussenzweig,V., Nuzzenzweig,R., Barnwell,J. and Moss,B. (1984) *Science*, in press.
21. Moss,B., Smith,G.L., Gerin,J.L. and Purcell,R.H. (1984) *Nature,* **311**, 67.
22. Maniatis,T., Fritsch,E.F. and Sambrook,J. (1982) *Molecular Cloning. A Laboratory Manual*, published by Cold Spring Harbor Laboratory Press.

CHAPTER 8

Bovine Papillomavirus DNA: A Eukaryotic Cloning Vector

M.SAVERIA CAMPO

1. INTRODUCTION

Great advances have been made in the past few years in the development of molecular cloning vectors that allow the expression of cloned genes in bacteria, and several eukaryotic proteins can now be produced in prokaryotes.

The bacterial synthesising machine, however, imposes a number of constraints on the expression of many eukaryotic proteins. Post-translational modifications, such as glycosylation, processing and assembly, are unlikely to be carried out properly by the bacterial cell, and moreover *Escherichia coli*, the most used bacterial host, does not secrete protein into the medium. In addition, the study of several aspects of eukaryotic gene functions such as the response to hormonal or viral inducers, the effects of methylation or the interaction with specific cellular proteins is impossible in a bacterial environment.

Systems are therefore needed which allow the re-introduction of eukaryotic genes into eukaryotic cells and permit their expression in a controlled genetic environment. Most of the eukaryotic expression vectors already described are based on viral replicons. SV40 and adenoviruses have been widely used for the propagation and expression of exogeneous genes in cultured cells (for a review, see reference 1). These systems, however, present several limitations. During their productive cycle, both viruses kill their host cells, thus preventing the establishment of cell lines in which the foreign genes can be replicated and expressed. Non-permissive transformed cells contain only a few copies of the viral DNA stably integrated into the host chromosome, and therefore the amplification of gene expression that accompanies the lytic cycle does not take place in the transformed cell. Moreover, as integration is a random event, the genetic environment of the newly acquired sequences cannot be controlled. Because of the great complexity of the eukaryotic genome, the effects of neighbouring sequences or of secondary modifications, such as chromatin organisation and methylation, on the expression of the exogenous genes cannot be properly evaluated, thus making the analysis of their functions problematic.

Clearly, a vector is required that is able to replicate in a stable manner as a multicopy episome and that allows the continuous replication and expression of the cloned foreign gene, independently from chromosomal controls, and without concomitant cell lysis. Bovine papillomavirus DNA answers these requirements.

Bovine papillomavirus type 1 (BPV-1) induces epithelial and mesenchymal tumours *in vivo* and transforms bovine and rodent cells *in vitro*. In the infected or transformed cells, its genome persists exclusively as multicopy non-integrated circular DNA and this peculiarity has made it the choice vector for eukaryotic cells (for reviews, see references 2 and 3).

In this chapter, I shall describe briefly some of the salient structural and functional features of the BPV-1 genome; the recombinant BPV-1 plasmids that have been developed in several laboratories, and their transformation properties; and the already numerous eukaryotic genes that have been successfully cloned in BPV-1 vectors.

2. BPV-1 AS A VIRUS

A detailed analysis of the several aspects of the life cycle of the virus has been delayed by the lack of an *in vitro* productive system. As a consequence, our knowledge of the organisation of the viral genome, of the temporal regulation of its replication and transcription, and of the identification and functions of its products lags behind our understanding of the same phenomena in the other small DNA tumour viruses. However, the availability of the viral genome cloned in bacterial plasmids has contributed greatly to the first stages of its molecular dissection. I shall briefly consider those aspects of the virus biology which are relevant to the use of its DNA as an eukaryotic vector.

2.1 Physical and Functional Organisation of the BPV-1 Genome

The genome of BPV-1 is a double-stranded circular DNA molecule, 7954 base pairs (bp) long (4). Its physical map has been constructed in several laboratories (5 – 7) and its primary sequence has recently been determined (4).

In contrast to SV40 or polyoma, it contains open reading frames on one DNA strand only (4) and transcription is therefore unidirectional (8 – 10). Two of the four major open frames are located in the transforming region (see below); they are called E1 and E2 ('early') and are 1814 and 1229 bp long, respectively, (*Figure 1a*). The other two, L1 and L2 ('late'), are found in the structural region (see below) and are 1484 and 1406 bp long, respectively (*Figure 1a*). In addition there are several small open frames, E4 – E8, in the transforming region. Whereas L1 and L2 lie in the same reading frame, separated only by a single stop codon, the open E sequences are located on the three reading frames and are overlapping (*Figure 1a*). There is a region of ~1000 bp with no open frames, which is located between the 3' end of L1 and the 5' end of E6. This region is sensitive to DNase I (11) and contains transcriptional regulatory sequences with promoter functions (4,12,13) (*Figure 1b*), and thus appears to play a major role in the life cycle of the virus. An enhancer element has been located at the distal end of the early region (12) (*Figure 1b*) which is necessary for BPV-mediated transformation and can be substituted for by the enhancer sequences of SV40 or polyoma (14). Polyadenylation sites are found at position 4179 at the 3' end of the early region and at position 7091 and 7155 at the 3' end of the late region (4) (*Figure 1b*).

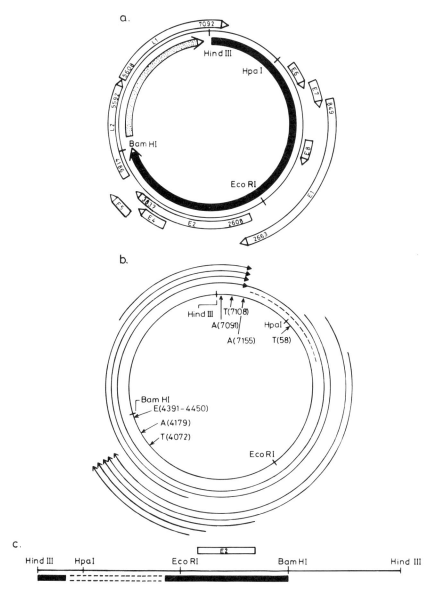

Figure 1. Organisation of the BPV-1 genome. **a**. The solid block is the *Hind*III-*Bam*HI 69% transforming fragment. The stippled block is the *Bam*HI-*Hind*III 31% structural fragment. Open blocks are 'early' (E1 – E8) or 'late' (L1 – L2) open reading frames. The previously described E2 and E3 frames are one continuous open frame E2, as a result of the recent addition of a G residue at position 3444 (A.Stenlund, personal communication). The arrows indicate the direction of transcription, and the numbers the positions of the start and stop codon. The first nucleotide of the single *Hpa*I site is nucleotide 1, and the *Hpa*I site is 0/1.0 map unit. **b**. The viral transcripts detected in transformed cells are indicated by open arrows. Closed arrows indicate the viral RNA species found in productive warts. The dashed line indicates the putative 5′ leader sequences. T = TATA box; E = enhancer element; A = polyadenylation site. **c**. The thin line represents the linearised BPV-1 genome. The shaded boxes indicate the segments required in *cis* for *in vitro* cell transformation. The dashed box indicates the deletion. The E2 open reading frame is shown above. Compiled from references 4,8,9,10,12,13,14 and 15.

2.2 **The Transformation Function of the BPV-1 Genome**

The cell transformation ability of BPV-1 DNA and the function for maintenance as a free episome have been localised in the 5.4 kilobase (kb) *Hind*III-*Bam*HI fragment, which represents 69% of the full genome length (15) (*Figure 1a* and *1c*). This fragment will be refered to as the 69T fragment. By the use of deletion mutants, Nakabayashi and colleagues (13) have found that two discontinuous segments of the 69T fragment are required in *cis* for cell transformation: one is at the 5' end of the 69T fragment and corresponds to the region where the transcriptional control elements have been mapped (12); the other is a 2.3 kb segment at the 3' end of the 69T fragment and includes the E2 open frame (*Figure 1c*). In cells transformed by the deletion mutant containing the above segments, only a few copies of integrated viral DNA are present, showing that episomality and transformation are encoded in different portions of the viral genome, and suggesting that the E1 open frame is responsible for the maintenance of the episomal state.

2.3 **Transcription of the BPV-1 Genome**

BPV-1 specific transcripts have been identified in productive warts and in non-productive transformed cells (8 − 10). Five mRNA species are present in transformed cells, which predictably all map within the 69T fragment. They are 4050, 3800, 1700, 1150 and 1050 bases long, and have a 3' co-terminus at the polyadenylation site at position 4179 (8,9). The 5' end of their bodies has been mapped at 0.03, 0.09, 0.34, 0.39 and 0.41 map unit, respectively, (9). Remote 5' leader sequences have been mapped around unit 0.9 to 0.1 indicating that the early transcripts are generated by the differential splicing of a common leader onto several main bodies.

In addition to the RNA species found in transformed cells, five more viral transcripts of 8000, 6700, 3800, 3700 and 1700 bases are found in the productive tumours. The 3700 bases long RNA species has not been precisely mapped and will not be discussed further. The other four wart-specific transcripts are 3' co-terminal at map unit 0.9, at the polyadenylation sites at position 7091 and 7155, and the 5'ends map at 0.9, 0.01, 0.42 and 0.71 unit, respectively, (10). It is possible that the two largest RNA species are non-spliced precursors both for the early and the late transcripts. In this case, mechanisms must exist, operating at different stages during the viral cycle, that control the choice of both splicing and termination signals. Whether these mechanism are under viral control or are provided by the differentiating keratinocytes is far from being understood.

3. BPV-1 AS A VECTOR

Methods that have been followed for the construction of recombinant BPV-1 plasmids are standard and are described in other cloning manuals (16). This chapter will concentrate upon a description of these vectors and their transformation abilities. This will be followed by the description of BPV-1 plasmids carrying other dominant selectable markers, and by an account of their successful use for the cloning, propagation and analysis of eukaryotic genes. Examples of cloning strategies to produce BPV recombinants that express exogenous genes may be found in the legends to *Figures 6–12*.

3.1 Recombinant BPV-1 Plasmids

The bacterial plasmids pBR322 (17) and its deletion derivatives pAT152 (18), pBR327 (19) and pML2 (20) have all been used as cloning vehicles for either full length BPV-1 DNA or the 69T fragment. With the exception of pBR322, they all lack most of the 'poison sequence', mapping between nucleotides 2067 and 2533, that has been shown to inhibit SV40 DNA replication in monkey cells (20) (*Table 1*), and can thus be used for BPV-1 DNA-mediated cell transformations without prior separation of viral and plasmid DNA (see below).

Table 2 lists the current BPV-1 plasmids and *Figure 2* shows their general structure. These plasmids have been used to transform both C127 and NIH3T3 mouse fibroblasts (15,21,22,23) and FR3T3 rat fibroblasts (24) by DNA transfection. Their transformation efficiencies are listed in *Table 3*. A protocol for transfection with selection for morphological transformation is given in *Table 4*. It is clear that, when linked to pBR322, both BPV-1 DNA and the 69T fragment, lose their transformation ability, whereas cleavage of the viral sequences from

Table 1. Deletion Mutants of pBR322 used to Clone BPV-1 DNA.

Plasmid	Deletion[a]	Length[b]	Reference
pBR322	–	4.4	17
pAT153	1646–2451	3.7	18
pBR327	1440–2500	3.3	19
pML2	1095–2485	3.0	20

[a]Nucleotides bracketing the deleted fragment (29).
[b]Length in kilobases approximated to the nearest first decimal figure.

Table 2. Recombinant BPV-1 Plasmids.

Recombinant	Vector	BPV-1 insert	Site	Reference
pBPV-1 8-2	pBR322	Full length	*Bam*HI	25
pBPV-1 9-1	pBR322	Full length	*Hind*III	25
pBPV69T	pBR322	69T fragment	*Hind*III-*Bam*HI	15
pBV1	pAT153	Full length	*Hind*III	26
pBV1-DI	pAT153	69T fragment	*Hind*III-*Bam*HI	22
pd1BPV69T	pML2	69T fragment	*Hind*III-*Bam*HI	25
pdBPV69T	pML2	69T fragment	*Bam*HI[a]	25
pdBPV-1	pML2d[b]	Full length	*Bam*HI[c]	25
pB2	pBR322	Full length	*Bam*HI	24
pMB2	pML2	Full length	*Bam*HI	24
pMH4	pML2	Full length	*Hind*III	24
pM69	pML2	69T fragment	*Hind*III-*Bam*HI	24

[a]pdBPV69T is derived from pd1BPV69T by converting the single *Hind*III site to a *Bam*HI site.
[b]pML2d was recovered by cleavage of pdBPV69T with *Bam*HI, followed by circularization.
[c]pdBPV-1 was generated by inserting *Bam*HI-linear BPV-1 DNA in the single *Bam*HI site of pML2d.

217

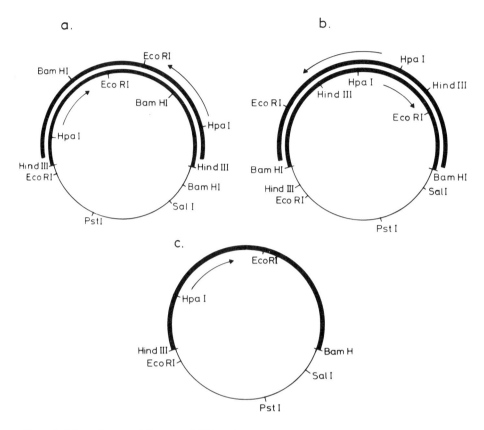

Figure 2. Schematic maps of the parental BPV-1 plasmids. In all cases the thin line represents prokaryotic vector DNA and the thick line viral DNA. The vector can be pBR322, pAT153 or pML2. **a.** Full length BPV-1 DNA cloned at the *Hind*III site. Both orientations of the BPV-1 DNA with respect to the vector are indicated. **b.** Full length BPV-1 DNA cloned at the *Bam*HI site. The viral DNA is shown in either orientation. **c.** *Hind*III-*Bam*HI 69T fragment cloned in the large *Hind*III-*Bam*HI fragment of the vector. Because of the two different cloning sites, only one orientation of the viral DNA is possible. In this and in all the following figures the maps are not shown to scale, only some of the restriction sites are indicated and the arrows indicate the direction of transcription.

the plasmid DNA restores it. This *cis* inhibition is ascribed to the presence in pBR322 of a poison sequence which suppresses SV40 DNA replication in monkey cells (20). In contrast, neither pAT153 nor pML2, which have undergone deletion of this sequence (*Table 1*), show this inhibitory effect on full length BPV-1 DNA-mediated transformation of either NIH3T3 or C127 mouse fibroblasts. Surprisingly, however, uncleaved pMB2 (BPV-1 DNA in pML2) does not transform FR3T3 rat fibroblasts (24) (*Table 3*). It is assumed that the different genetic background of the host cells may be responsible for this discrepancy. Another difference in transformation behaviour is found between pBV1-D1 and pdBPV69T, which carry the 69T fragment in pAT153 and pML2, respectively, (*Table 2*). Whereas pBV1-D1 transforms NIH3T3 cells equally efficiently before and after cleavage (22), pdBPV69T is capable of transforming C127 cells only when cleaved

Table 3. Transformation Efficiency of Recombinant BPV-1 Plasmids.

Plasmid	Enzyme	Cells	No. foci[a]	Reference
pBPV-1 8-2	None	C127	2	25
	BamHI	C127	246	
pBPV-1 9-1	None	C127	1	25
	HindIII	C127	197	
pBPV69T	None	C127	1	25
	HindIII + BamHI	C127	107	
pBV2[b]	None	NIH3T3	384	22
	HindIII	NIH3T3	667	
pBPV-1-DI	None	NIH3T3	300	22
	HindIII + BamHI	NIH3T3	280	
pdBPV-1	None	C127	198	25
	BamHI	C127	200	
pdBPV69T	None	C127	0	25
	BamHI	C127	126	
pB2	None	FR3T3	1	24
	BamHI	FR3T3	250	
pMB2	None	FR3T3	1	24

[a]DNA-mediated cell transformation was achieved by using the calcium precipitation method (27), followed by glycerol shock (28) in reference (25). Transformation efficiency is calculated as number of foci per 10^6 cells per μg of recombinant DNA (22,24) or per μg of viral DNA (25).
[b]pBV1 and pBV2 have the same transformation efficiency both before and after cleavage with HindIII (Campo and Spandidos, unpublished results). Shown here are the published data for pBV2.

(25) (*Table 3*). The reason for this disparity is not understood; it may be due to the different origin of NIH3T3 and C127 cells. This, however, seems unlikely, as uncleaved pBV1-D1 is capable of inducing transformation of C127 cells also, with an efficiency of ~100 foci per μg of DNA (D.DiMaio, personal communication). Thus, it would appear that, although both vectors have undergone similar deletions (*Table 1*), pAT153 is a better vehicle for the 69T fragment.

The transformation efficiency of the 69T fragment is ~30% that of the full length genome (15,13). It is concluded that, although unnecessary for cellular transformation, the 31% non-transforming fragment has a facilitative role, possibly due to the presence of additional transcriptional control elements.

Cells transformed by recombinant BPV-1 DNA exhibit the same morphology and the same behaviour (growth in agar, tumourigenicity, etc.) as their counterparts transformed by the virus (15,21,22), and, as in virus-transformed cells, the viral DNA is found in a non-integrated form. In cells transformed by linear BPV-1 DNA recircularisation takes place and the viral genome is found as monomeric episomes (23). In cells transformed by the 69T fragment, circularisation of the DNA is accompanied by the acquisition of new (cellular?) sequences or by duplication of viral ones (23). This last observation suggests that an optimum genome length exists for DNA replication, irrespective of the source or nature of the acquired sequences, and may explain the greater transformation ability of full-length BPV-1 DNA as compared with that of the 69T fragment.

In mouse cells transformed by intact BPV-1 plasmids, the hybrid molecules

Table 4. DNA Transfection and Selection for Morphological Transformants.

A. *Transfection with recombinant BPV DNA (see Chapter 6, Section 5)*

1. Cleave pBR322 from the BPV recombinant with the appropriate restriction enzyme to eliminate the poison sequences, or use intact supercoiled recombinant plasmid if the bacterial vector is either pAT153, pBR327 or pML2.
2. One day before transfection, plate the cells to be used at a density of 10^4 cells per cm^2 in medium containing 15% fetal calf serum.
3. Dissolve the donor DNA in 0.1 mM EDTA, 1 mM Tris-HCl pH 8.0, and add 2.5 M CaCl$_2$ to a final concentration of 125 mM. The concentration of DNA should be 20 μg/ml. If less than 1 μg of transforming DNA is used, add carrier DNA.
4. Add the DNA solution to an equal volume of 2 x HBS buffer[a] and pipette the DNA-CaPO$_4$ precipitate to the cells.
5. Incubate the cells at 37°C for 3−4 h and then proceed with the 'glycerol-shock' treatment (see Chapter 15, section 3.2) which sometimes improves the transfection efficiency. Alternatively, allow the cells to remain in contact with the DNA-CaPO$_4$ precipitate for 18 h and omit the glycerol-shock.
6. Remove the medium, wash the cells with serum-free medium, and replace with fresh medium containing 15% fetal calf serum. Incubate for 24 h at 37°C.

B. *Selection for morphological transformants*

1. Remove the medium and replace with medium containing only 5% fetal calf serum. Change the medium every three days for up to three weeks. Use 5% fetal calf serum throughout. At this time, morphologically transformed foci should be visible. Count foci, if required.
2. Place stainless steel cloning rings over individual foci and trypsinize the cells by incubating for 2−3 min at room temperature with 0.1 ml of 0.25% trypsin in sterile 1 mM EDTA, phosphate-buffered saline (PBS)[b]. The lower rim of the rings must be coated with a layer of silicon grease to achieve a watertight seal between the ring and the culture dish. Pipette off the trypsinized cells and pellet them by centrifugation. Resuspend the cells in 0.2 ml of medium.
3. Mix the cell suspension with 20 ml of medium containing 0.9% methocel MC 4000 CP (from Fluke) and 20−30% fetal calf serum, and plate on bacteriological plates. High serum concentrations increase plating efficiency.
4. Incubate the plates at 37°C for 10 days. At this time, colonies should be seen on the surface of the semi-solid medium. Count colonies if required.
5. Pick individual colonies using a Pasteur pipette and grow them in fresh medium. This selection procedure allows the isolation of cell lines which have lost contact-inhibition and are anchorage-independent. The tumorigenicity of the lines can be tested by subcutaneous injection of 3−5 x 10^6 cells into nude mice.

[a]For the composition of HBS, see Chapter 6, Section 5.1.
[b]PBS contains 8.0 g NaCl, 0.2 g KCl, 1.15 g Na$_2$HPO$_4$, 0.2 g KH$_2$PO$_4$ per litre.

are present as multiple copies of unrearranged non-integrated monomers (22,25). Rearrangements of the recombinant plasmids are more common in rat fibroblasts and appear to be mostly deletions (24).

Unrearranged BPV-1 plasmids can be rescued in *E. coli* by transfection of bacteria with low molecular weight DNA from transformed cells. A protocol for the extraction of low molecular weight DNA from transformed cells is given in *Table 5*. The recovered plasmids have the same structure and the same transformation efficiency as the original DNA (24,25). This finding has been a fundamental step in the development of BPV-1-based vectors. The ability of BPV-1 plasmids to behave as independent replicons both in prokaryotic and eukaryotic cells makes them extremely suitable as vectors for shuttling genes between mammalian and

Table 5. Extraction of low molecular Weight DNA from Transformed Cells (adapted from reference 30).

1.	Remove the growth medium and rinse the cells with 0.15 M NaCl, 10 mM Tris-HCl pH 7.5, taking care not to disturb the monolayer.
2.	Add 0.6% SDS 10 mM EDTA, 10 mM Tris-HCl pH 7.5 (20 ml to a 75 cm² flask). Spread the solution evenly and incubate at room temperature for 10−20 min.
3.	Pour the cell lysate into a flask and add 5 ml NaCl (final salt concentration is 1 M). A precipitate will form instantly. Leave overnight or longer at 4°C. High molecular weight DNA will precipitate, whereas low molecular weight DNA will remain in solution.
4.	Centrifuge at 25 000 r.p.m. for 1 h at 4°C. Keep the supernatant.
5.	Extract the supernatant with an equal volume of phenol saturated with 10 mM NaCl, 10 mM EDTA, 50 mM Tris-HCl pH 8.0.
6.	Collect the aqueous layer by centrifugation at 10 000 r.p.m. for 10 min, and extract with an equal volume of phenol-chloroform (1:1, v/v).
7.	Collect the aqueous layer as above. Precipitate the DNA by adding three volumes of ice-cold ethanol and leave overnight or longer at −20°C.
8.	Pellet the DNA by centrifugation at 10 000 r.p.m. for 20 min.
9.	Wash the pellet twice with 70% ethanol, 0.2 mM NaCl, and once with 70% ethanol.
10.	Dry the pellet and resuspend the DNA in 0.2 M NaCl, 1 mM EDTA, 10 mM Tris-HCl pH 7.5.
11.	Add RNase to a final concentration of 20 μg/ml and incubate at 37°C for 1 h.
12.	Extract with an equal volume of phenol-chloroform.
13.	Repeat steps 7−9.
14.	Dry the pellet and resuspend the DNA in 1 mM EDTA, 10 mM Tris-HCl pH 7.5. The DNA is ready for transfection into bacteria.

bacterial cells. They have already been employed on the analysis of gene regulation (see below) and will prove invaluable in the isolation of specific genes. The genes can be selected on the basis of phenotype conversion of mammalian cells, and then recovered in large amounts by rescue in *E. coli* thus lending themselves to detailed analysis.

3.2 Hybrid Plasmids Expressing Selectable Markers

One major limitation of the BPV-1 vector systems described in the preceding section is that identification of transformants depends on the morphological transformation of contact-inhibited cells as a selective marker. This restricts the host range for BPV-1 vectors to those cells which are capable of focus formation. The introduction of dominant selectable markers into BPV-1 plasmids has allowed the expansion of the host range of BPV-1 to cells which do not express the transformed phenotype but may be capable of supporting the episomal replication of BPV-1 DNA.

The biochemical markers which have already been incorporated into BPV-1 vectors are the thymidine kinase (*tk*) gene from herpes simplex virus type 1 (HSV-1), the xanthine-guanine phosphoribosyl-transferase (*gpt*) gene from *E. coli*, and the aminoglycoside-3′-phosphotransferase (*aph*) gene (neomycin resistance) from transposon Tn5.

3.2.1 *BPV-tk Hybrid Plasmids*

These are plasmids containing the HSV-1 *tk* gene linked to the 69T fragment of BPV-1 (*Figure 3*). They are capable of transforming several TK⁻ cell lines to the TK⁺ phenotype, including mouse L, BHK21, Rat 2 and human 143

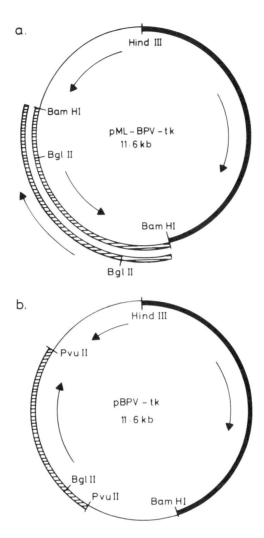

Figure 3. BPV-1-tk hybrid plasmids. **a.** pML-BPV-tk was constructed by inserting the 3.4 kb *Bam*HI fragment containing the HSV-1 *tk* gene (28) into the *Bam*HI site of pML-BPV69T. The two possible orientations of the *tk* gene are shown. In pML-BPV-tk4 the direction of transcription of the *tk* gene is the same as that of the 69T fragment, whereas in pML-BPV-tk2, it is opposite. **b.** pBPV-tk was constructed by ligating the 69T fragment to the large *Hind*III-*Bam*HI fragment of pAGO, a plasmid which consists of a 2 kb *Pvu*II fragment containing the HSV-1 *tk* gene and pBR322 (29). As unique *Hind*III and *Bam*HI sites were used, the orientation of the ligated fragment was predetermined, and the direction of transcription is the same for both the *tk* gene and the 69T fragment. ■, BPV69T fragment; ▥, HSV-1 *tk*; −, plasmid.

(14,33). pMLBPVTK2 and pMLBPVTK4 transform both Rat 2 and human 143 cells with higher efficiency than the parental plasmid pMLTK (*Table 4*), although the increase is less marked in the human cell line. This increase is due to the presence of the BPV-1 enhancer sequence (12,14), and is independent of both its position and its orientation relative to the direction of transcription of the *tk* gene. This effect is observed also in the case of pBPV-TK in hamster BHK21

Table 6. Transformation Efficiency of BPV-tk Plasmids.

Plasmid	Rat 2[a]	Human 143[b]	Mouse L[c]	BHK21[c]	Reference
pMLTK	160	51			14
pMLBPVTK2	1300	131			14
pMLBPVTK4	400	—			14
pAGO			25	2.8×10^{-2}	33
pBPV-TK			26	7.8×10^{-2}	33

[a]Number of colonies per μg of recombinant plasmid.
[b]Number of colonies induced by 5 μg of recombinant plasmid.
[c]Number of colonies per ng of recombinant plasmid.

cells but not in mouse L cells (*Table 6*). This last result is surprising, as the BPV enhancer element alone increases the efficiency of transformation of the *tk* gene in mouse L cells by ~7-fold, even when inserted ~800 bp upstream from the *tk* promoter (12). The reasons for this difference are not known but may be due to the different types of plasmids used.

In the transformed cells the great majority of the recombinant plasmid DNA is integrated into the cell genome and only a small proportion is present as episomes. The extrachromosomal plasmids can be rescued in bacteria and used in a second round of cell transformation with the same efficiency as that of the initial plasmids (33). However, as only a small proportion of these plasmids is maintained in an extrachromosomal form, they seem to be of limited value as shuttle vectors.

3.2.2 *BPV-gpt Hybrid Plasmids*

In these plasmids, the dominant selectable marker is the *E. coli gpt* gene, which operates under the control of the SV40 early-region transcriptional unit (34). Cells incorporating and expressing the bacterial gene are able to grow in HAT (hypoxanthine-aminopterine-thymidine) media containing xanthine and mycophenolic acid. Protocols for this scheme of dominant selection are given in Chapter 15.

In pBPV69T-SV2gpt, the SV2gpt gene was linked to the 69T fragment of BPV-1 in either orientation (35) (*Figure 4*). These plasmids induce morphological transformation in C127 mouse cells and permit them to grow in selective media. The two phenotypic markers are co-expressed in the great majority of the transformed cell lines. Cells selected for BPV-induced focal transformation can grow in the presence of mycophenolic acid, and *vice versa* cells selected for their resistance to the acid are not contact-inhibited and can grow in soft agar. The high incidence of co-expression shows that the linkage between the two genes has been maintained, irrespective of their relative orientation.

However, as in the case of the BPV-*tk* plasmids, selection for the expression of the *gpt* gene results in rearrangements and integration of the BPV-*gpt* plasmids, thus making them unsuitable as shuttle vectors.

3.2.3 *BPV-Neomycin Resistance Plasmids*

The neomycin resistance (*neo*ʳ) gene carried by Tn5 confers kanamycin resistance to bacterial cells and, if under the control of an eukaryotic transcriptional pro-

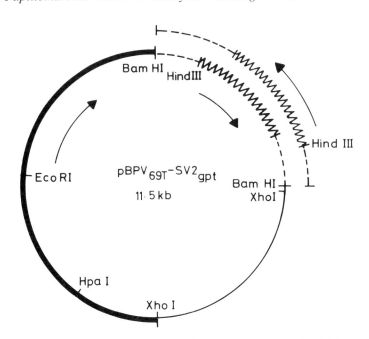

Figure 4. BPV-1-gpt hybrid plasmids. pBPV69T-SV2gpt was constructed by first joining the 69T fragment of BPV-1 to the 3.7 kb *Eco*RI-*Sal*I fragment of pBR322 by means of *Xho*I linkers. The *Pvu*II site
of pSV2gpt (31) was converted to a *Bam*HI site, and the *Bam*HI fragment containing the *gpt* gene was
inserted in the single *Bam*HI site of the BPV69T plasmid. In pBPV69T-SV2gpt (57-5) the 69T fragment
and the SV2gpt fragment are in the same transcriptional orientation, and in pBPV69T-SV2gpt (57-1) they
are in the opposite orientation. ■, BPV69T fragment; - - - -, SV40 DNA; ⋀⋀, *gpt*; −, pBR322.

moter, resistance to the amino-glycoside G418 to mammalian cells (36). The *neo*^{r}
gene, flanked by eukaryotic promoter and termination signals, has been linked
to either the complete BPV-1 genome (37,38) or to the 69T fragment (39) (*Figure
5*).

In pdBPV-MMTneo (342-12), the *neo*^{r} gene, flanked by the mouse metallothionein I gene promoter element (40) at the 5' end and by the SV40 small t
antigen splicing and termination signals (41) at the 3' end, is linked to a derivative
of pdBPV-1 (25) (*Figure 5a*).

pCGBPV is a cosmid carrying the ColE1 replication origin and the λ *cos* region,
linked to the entire genome of BPV-1. The *neo*^{r} gene is under the transcriptional
control of both a eukaryotic promoter, that of the HSV-1 *tk* gene, and a prokaryotic promoter, the P1 promoter of pBR322 (42), and hence can be selected
for both in bacterial and in mammalian cells. The polyadenylation signals are
provided by the 3' end of the HSV-1 *tk* gene (*Figure 5b*).

The plasmid pV69 carries a similar *neo*^{r} transcriptional cassette, but the P1
promoter has been substituted by the homologous to *neo*^{r} promoter of Tn5 (36).
The *neo*^{r} gene is linked to the 69T fragment of BPV-1 (*Figure 5c*).

When these plasmids are transfected into C127 mouse cells, they induce the
formation both of colonies resistant to G418, and of morphologically transformed
foci (*Table 7*). The efficiency of transformation to G418-resistance is higher than

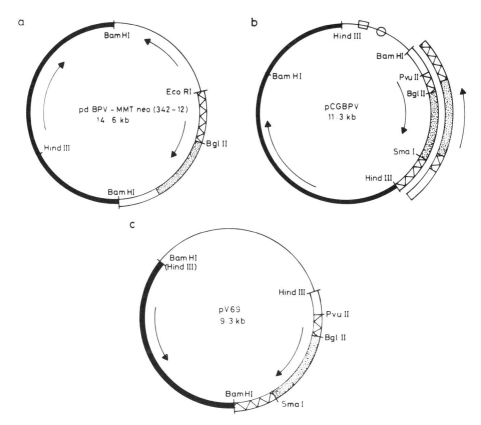

Figure 5. BPV-1 neomycin reistance plasmids. **a**. pdBPV-MMTneo contains: the *Bam*HI-*Hind*III fragment of pML2 (with an inverted duplication of the 29 bp *Eco*RI-*Hind*III segment) carrying the *amp*r gene and the origin of DNA replication; the *Eco*RI-*Bgl*II transcriptional promoter of the mouse metallothionein I gene (40); the *Bgl*II-*Xho*I body of the *neo*r gene in which the *Sal*I site has been abolished; the 850 bp fragment of SV40 carrying the small t antigen splicing signals and 3′ transcriptional terminator signals (41); and the complete *Bam*HI-*Bam*HI BPV-1 genome. ■, BPV-1; ◩, MMT promoter; ▦, *neo*r gene; □, SV40; −, pML2. **b**. pCGBPV contains: a DNA segment carrying the ColE1 replication origin and the λ *cos* region; the P1 promoter of pBR322 (42); the HSV-1 *tk* promoter and polyadenylation signals (31) flanking the Tn5 *neo*r gene; and the complete *Hind*III-*Hind*III BPV-1 genome. In pCGBPV7 the direction of transcription of the *neo*r gene is the same as that of BPV-1; in pCGBPV9 it is opposite. ■, BPV-1; ○, ColE1 origin of replication; □, λ *cos* region; ▭, P1 promoter of pBR322; ▧▧, HSV-1 *tk* promoter and polyadenylation signals; ▦, *neo*r gene. **c**. pV69 contains: the *amp*r gene and the replication origin of pBR322; the *neo*r promoter element of Tn5 (36); the HSV-1 *tk* promoter and polyadenylation signals, flanking the *neo*r gene; and the 69T fragment of BPV-1. The *Hind*III site of BPV-1 DNA has been converted to a *Bam*HI site. Arrows indicate the direction of transcription of BPV-1 and the *neo*r gene. ■, BPV-169T fragment; −, pBR322; ▭, *neo*r promoter; ▧▧, HSV-1 *tk* promoter and polyadenylation sequences; ▦, *neo*r gene.

that of their respective parental plasmids, and this effect is probably due to the presence of the enhancer element of BPV-1 (12,14). The number of G418-resistant colonies is always higher than the number of transformed foci (*Table 7*) and most of them maintain the phenotype of normal fibroblasts. However, with further expansion in culture, these cells eventually acquire a fully transformed phenotype. A protocol for G418-resistance selection is given in Chapter 6.

Table 7. Transformation of Mouse C127 Cells by BPV-neo[r] Plasmids.

Plasmid	G418[r] colonies	Foci	Reference
pdMMTneo (302-3)	9[a]	0[a]	37
pdBPV-MMTneo (342-12)	30[a]	6[a]	37
pAG60	510[b]	0[b]	38
pCGBPV7	3441[b]	176[b]	38
pCGBPV9	4529[b]	573[b]	38
pAG60	392[a]	–	39
pV69	550[a]	20[c]	39

[a]Transformation efficiency expressed as number of G418-resistant colonies and of transformed foci per μg plasmid DNA.
[b]Transformation efficiency expressed as number of G418-resistant colonies and of transformed foci per pmol plasmid DNA.
[c]Number of foci obtained after seeding in normal medium 500 G418-resistant cells together with 5×10^5 untransformed C127 cells.

Cells transformed with pdBPV-MMTneo and selected for the transformed phenotype, are resistant to G418 (37), whereas many foci induced by both pCGBPV and pV69 do not exhibit resistance to the drug and contain deleted forms of plasmid DNA (see below) (38,39). The reason for this discrepancy is unknown.

In cell lines selected for their resistance to G418 and, in the case of pdBPV-MMTneo also, for morphological transformation, the plasmids are maintained as multiple monomeric episomes, ranging from 20 to 100 copies per diploid gene. These plasmids can be rescued in bacteria and their structure is the same as that of the input plasmid. Interestingly, this is not the case for pCGBPV7 which is found mainly in the high molecular weight. DNA fraction of the transformed cells (38). There is as yet no explanation for the different behaviour of pCGBPV7 and pCGBPV9, but it may be relevant that the direction of transcription of BPV-1 DNA and the *neo*[r] gene is the same in pCGBPV7 and opposite in pCGBPV9 (*Figure 5b*).

In cell lines selected for morphological transformation, the transforming plasmids, although present in multiple extrachromosomal copies, have undergone rearrangements and deletions, affecting preferentially the bacterial sequences (38,39). As mentioned above, pdBPV-MMTneo is exceptional as it is maintained unrearranged in morphologically transformed cells and can be rescued in bacteria (37).

Because of their extrachromosomal maintenance, the BPV-neo plasmids provide by far the most promising eukaryotic shuttle vectors with a dual selection system. The delayed acquisition of the characteristics of full malignant transformation by the G418-resistant cells is interesting: it may reflect the need to reach a threshold level either in BPV-1 genome copy number and/or in BPV-1 transforming products. These cells would be at an early stage of transformation and would progress further towards full malignancy when enough viral products have accumulated or when an interaction has been established with as yet unknown cellular factors.

Table 8. Eukaryotic Genes Cloned in BPV Vectors.

Recombinant plasmids	BPV	Eukaryotic gene	Size of cloned gene (kb)	pBR sequences[a]	Figure	References
pBPV69T-rI1	69T	Rat preproinsulin	1.62	−	6	43
pBPV-β	69T	Human β-globin	7.6	+	7	46
pGP	69T	Rat growth hormone	5.8	+	8a	48
pBPVMG	69T	Human growth hormone plus mouse metallothionein promoter element	3.5	−	8b	49
pBPV69T-IFNβ1	69T	Human β-interferon	1.6	−	9a	52
pBPV-IF	69T	Human β-interferon	1.6	−	9b	53
pM18/19	69T	MMTV LTR + H-MSV v-ras	3.7	−	10	55
pBPV-βHLA	69T	Human HLA heavy chain plus human β-globin	8.5 + 7.6	+	11	56
pM2 6V4	Whole genome	HBV surface antigen	2.7	−	12	57

[a]Indicates whether the plasmid sequences were maintained (+) or removed (−) for cell transformation.

3.3 Eukaryotic Genes Cloned in BPV Vectors

To be useful cloning vectors, BPV plasmids must answer a number of requirements. They should be capable of propagation as episomes, recovery when transferred into bacteria, and they should allow the regulated expression and faithful transcription of the cloned genes. To date, several eukaryotic genes have been successfully cloned and propagated in BPV vectors (*Table 8*). These genes are maintained in the transformed cells mostly as episomes, and are correctly transcribed and expressed, although not all of the required characteristics have always been met.

3.3.1 *Physical Organisation*

The BPV69T fragment has been used in all cases, except for the cloning of the hepatitis B virus surface antigen (HBVsAg) gene (*Table 8*). The size of the cloned eukaryotic sequences ranges from 1.6 kb for the rat preproinsulin (rI1) gene and the human β-interferon (IF) gene, carried in pBPV69T-rI1 and pBPV-IF, respectively, (43,52,53) (*Figures 6* and *9*), to 16.1 kb, this being the combined size of the 7.6 kb human β-globin gene (*Figure 7*) and a 8.5 kb human HLA gene, carried in pBPV-βHLA (56) (*Figure 11*).

Most genes have been cloned in both transcriptional orientations relative to the direction of transcription of the BPV sequences; some genes were cloned in only one orientation, predetermined by the choice of restriction enzyme sites (see *Figures 6, 8a* and *9b*).

3.3.2 *Efficiency of Transformation*

As the bacterial plasmid sequences seem to inhibit cell transformation by the BPV69T fragment (see paragraph 3.1 and *Table 3*), they were removed by restric-

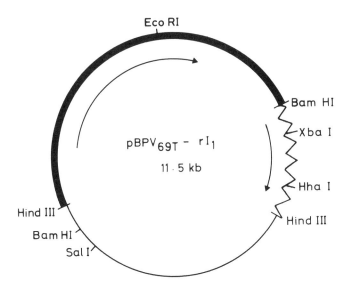

Figure 6. The rat preproinsulin gene in a BPV plasmid. The *Hinc*II site of the 1.62 kb *Bam*HI-*Hinc*II fragment containing the coding sequences of the rat preproinsulin gene (rI1), its intervening sequences and the 5′ and 3′ regulatory signal (44), was converted to a *Hind*III site by the use of *Hind*III linkers. The modified fragment was ligated to the BPV69T fragment, and, after digestion with *Hind*III, the BPV69T-rI1 sequences were inserted into the *Hind*III site of pBR322. Before transfection into C127 cells, pBR322 was removed from pBPV69T-rI1 by *Hind*III digestion. The transformation efficiency of the hybrid DNA was ~200 foci per μg of DNA. ■, BPV69T fragment; −, pBR322; ∿, rI1 gene.

tion enzyme cleavage before transfection of the hybrid DNA into C127 cells. The hybrid sequences were transfected either as linear molecules, or were re-circularised by ligation and then transfected, as in the case of the BPVMG plasmids (49) (*Figure 8b*). When transfected as linear molecules, the plasmids recircularise within the cell (see below, section 3.3.3), and often lose the cleavage restriction site, as in the case of BPV69T-rI1 (43) (see *Figure 6*).

Removal of the bacterial plasmid sequences is not necessary for efficient cell transformation in the case of pBPV-β (46) and pGP (48), which carry the human β-globin gene and the rat growth hormone (GH) gene, respectively. This is ascribed to an ill-defined stabilizing effect of the inserted DNA.

The efficiency of transformation varies from 10 foci per μg of DNA, in the case of pM2 6V4 (57), to 400 foci per μg of DNA in the case of pBPV-β (46). It is not clear whether such variability results from intrinsic structural features of the plasmids, or to differences in transfection protocols and competence of the recipient cells.

3.3.3 *Chromosomal State*

The majority of the BPV recombinants is present in the transformed cells as multi-copy circular episomes, ranging from 100 to 200 copies in different cell lines. pBPV-β and pGP, which have retained their plasmid sequences, can be rescued in *E. coli* as unrearranged monomeric episomes (46,48). Some chimeric plasmids

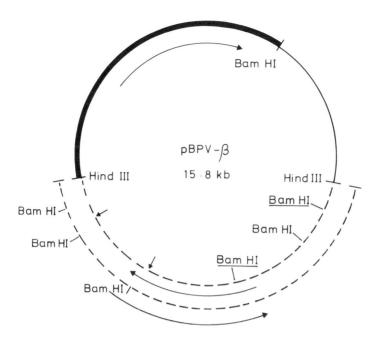

Figure 7. The human β-globin gene in a BPV plasmid. The 7.6 kb *Hind*III fragment of the human β-globin gene cluster (45) was inserted in either orientation into the *Hind*III site of pBPV-H11, a plasmid which carries the BPV69T fragment in pML2, to generate pBPV-β. In pBPV-β1 the transcriptional direction of BPV DNA and the globin gene is the same, and is opposite in pBPV-β3. The transformation efficiency of pBPV-β is ~400 foci per μg of viral DNA. ■, BPV69T fragment; −, pML2; - - -, β-globin gene sequences. The two underlined *Bam*HI sites and the two little arrows in the β-globin sequences define the deletion end-points of a mutant used to clone the human β-interferon gene (see text).

have, however, undergone either rearrangements or integration into the host chromosome or both.

For instance, although propagated as episomes, BPV69T-IFNβ1 and BPV-IF, both of which carry the human β-interferon gene (*Figure 9*), have acquired foreign DNA sequences either from the host genome or from the carrier DNA, and, in some cell lines, are present exclusively as multimeric forms (52,53). Also, pM2 6V4, which carries the sAg gene of HBV (*Figure 12*), has suffered rearrangements involving deletions and insertions (57).

In contrast to the other members of the same series (see below, section 3.3.4(ii)), the plasmids pM20 and pM21 (*Figure 10c*) are found almost always rearranged and integrated, and their copy number is lower than that of their sister plasmids (55). The reasons for their different behaviour are not understood, but the presence of the 72 bp enhancer element of Harvey mouse sarcoma virus (H-MSV) may be responsible.

Also, the plasmid pBPV-βHLA (*Figure 11*) is maintained in the transformed cells as high molecular weight structures made up of tandemly integrated sequences. Integration may be due to the large size of the plasmid, to homologous recombination with the mouse H2 genes, or to other unknown factors.

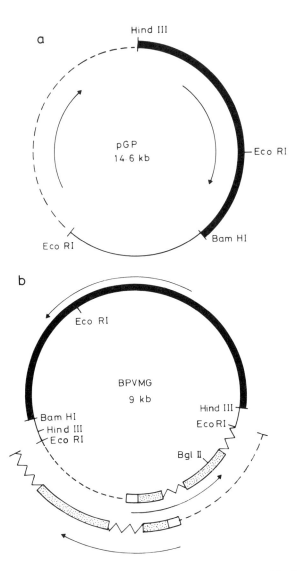

Figure 8. Growth hormone genes in BPV-based vectors. **a**. pGP was constructed by linking the 5.8 kb *EcoRI-Hind*III fragment of rat DNA containing the growth hormone gene (47) to the large *EcoRI-Hind*III fragment of pBR327, and by subsequent insertion of the BPV69T fragment between the *Hind*III and *Bam*HI sites. Its transformation efficiency is ~25 foci per μg of DNA. ■, BPV69T fragment; −, pBR327; - - -, rat growth hormone gene sequences. **b**. BPVMG was constructed using a *Bam*HI fragment containing partly genomic DNA and partly cDNA of the human growth hormone gene (50). This was inserted into the *Bgl*II site of the 5′ untranslated region of the mouse metallothionein gene (MT) (51). The hybrid MT-GH gene was recovered as a *Hind*III fragment and inserted into *Hind*III-digested pdBPV69T. The plasmid sequences were removed by *Bam*HI digestion, and the BPV-MT-GH sequences were recircularised by ligation before transfection into C127 cells. BPVMG thus contains, in addition to the 69T fragment, 1.9 kb of MT 5′ flanking sequences, 68 bp of MT 5′ untranslated sequences, the 70 bp first exon of GH, a 250 bp GH intron, 750 bp of GH structural sequences and 450 bp of GH 3′ untranslated sequences. The direction of transcription of BPV and GH is the same in BPVMG7 and opposite in BPVMG6. ■, BPV69T fragment; - - -, MT 5′ flanking sequences; □, MT 5′ untranslated leader sequence; ▨, GH exons; ∿, GH introns and 3′ flanking sequences.

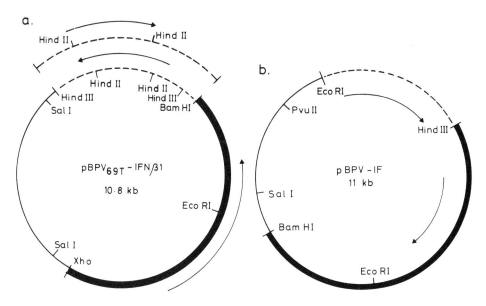

Figure 9. The human β-interferon gene in BPV plasmids. **a.** pBPV69T-IFNβ1; a 1.6 kb *Hind*III DNA fragment containing the β-interferon gene, and the 28 bp *Eco*RI-*Hind*III fragment of pBR322 was inserted in either orientation into the *Hind*III site of a pBPV69T vector containing a modified 3.7 kb *Sal*I-*Sal*I fragment of pBR322. In pBPV69T-IFNβ1 117-13 the transcriptional orientation of the BPV and the interferon sequences is the same, and opposite in pBPV69T-IFNβ1 117-211 (52,53). **b**; pBPV-IF; the 1.6 kb human β-interferon gene was cloned as an *Eco*RI-*Hind*III fragment into a deletion mutant of pBR322 lacking the poison sequence, thus generating pIFR. The *Hind*III-*Bam*HI 69T fragment was ligated to *Hind*III-*Bam*HI digested pIFR, thus generating pBPV-IF. The direction of transcription of the interferon gene is the same as that of BPV DNA. Before transfection, the plasmid sequences were removed by digestion with *Sal*I in the case of pBPV69T-IFN, and with *Sal*I-*Pvu*II in the case of pBPV-IF. ■, BPV69T fragment; - - -, β-interferon gene; —, pBR322.

3.3.4 *Expression of Cloned Genes*

All the eukaryotic genes cloned in BPV vectors are under the control of either their own transcriptional promoter or of an heterologous one. None is controlled by the BPV promoter elements.

(i) *Non-inducible Genes.* The rat preproinsulin gene, the human β-globin and HLA genes and the HBVsAg gene are all correctly transcribed and expressed in cells transformed by the respective BPV hybrid plasmids. The insulin mRNA is correctly translated into preproinsulin, which is processed to proinsulin and is secreted into the medium. However, the second maturation step, from proinsulin to insulin, does not take place in C127 cells, probably because of the absence of the processing enzymes (43). The β-globin mRNA is correctly spliced and polyadenylated, but β-globin has not been detected in the transformants, possibly because of the instability of globin proteins in non-erythroid cells (46). The HLA gene is also accurately and efficiently transcribed. In this case high levels of human heavy chain are produced, and some of the human protein is found at the cell surface associated with mouse β2 microglobulin (56). Despite the extensive rearrangements suffered by the BPV-HBV hybrid DNA, the sAg gene is

231

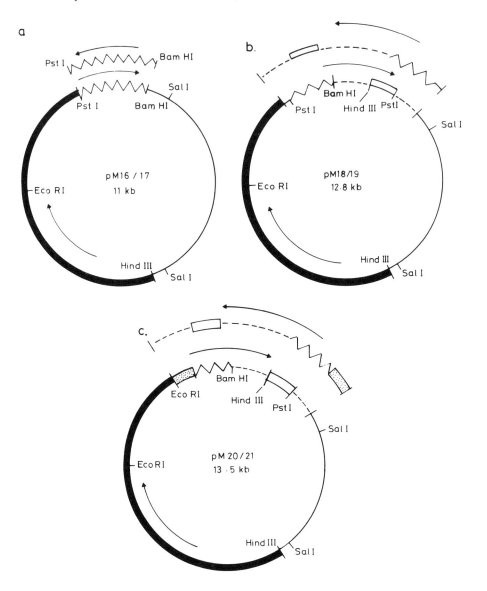

Figure 10. BPV-MMTV LTR hybrid plasmids. All the BPV-MMTV LTR plasmids contain the BPV69T fragment and a modified pBR322 in which the *Eco*RI site has been converted into a *Sal*I site. **a**. pM16 and pM17 contain the 1.3 kb MMTV LTR in the same and in the opposite transcriptional orientation of BPV DNA, respectively. **b**. pM18 and pM19 contain the v-ras gene sequences of Harvey MSV fused 3' to MMTV LTR. In pM18 the direction of transcription of the v-ras gene is the same as that of BPV DNA and opposite in pM19. **c**. pM20 and pM21 carry a 700 bp fragment containing the 72 bp repeat enhancer sequences of Harvey MSV linked 5' to MMTV LTR. The direction of of transcription of v-ras and BPV sequences is the same in pM20, and opposite in pM21. After removal of pBR322 by *Sal*I digestion, the hybrid DNA was either transfected as linear molecules, or recircularised by ligation and then transfected. The efficiency of tranformation of recircularized DNA is between 3000 and 7000 foci per pmol of DNA. ■, BPV69T; −, pBR322; ⋀⋀, MMTV LTR; □, Harvey MSV v-ras gene; - - -, 5' and 3' flanking sequences; ▒, Harvey MSV enhancer sequences.

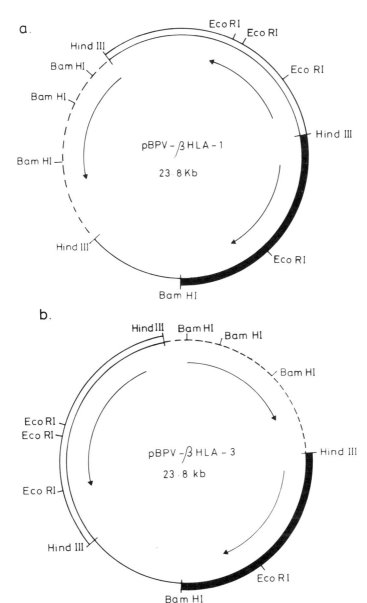

Figure 11. The human HLA heavy chain gene in a BPV-β-globin vector. A 8.5 kb *Hind*III fragment containing the human HLA heavy chain gene was ligated to *Hind*III digested pBPV-H11. The resulting plasmid was partially digested with *Hind*III, and full-length linear molecules were ligated to the 7.6 kb *Hind*III fragment of the human β-globin gene cluster to generate pBPV-β HLA-1, in which the direction of transcription of the HLA gene is the same as that of the β-globin gene, but opposite to that of BPV DNA; and pBPV-βHLA-3, in which the HLA gene is in the opposite orientation to both the β-globin gene and BPV DNA. The human β-globin gene fragment was introduced into the recombinant BPV-HLA plasmids because of its stabilising effect upon pBPV plasmids (46; see Section 3.2.2). The transformation efficiency of pBPV-βHLA is ~150 foci per μg of DNA. ■, BPV69T fragment; □, HLA heavy chain sequences; - - -, β-globin sequences; −, pBR322.

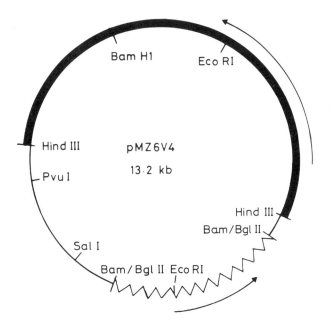

Figure 12. The HBV surface antigen gene in a BPV plasmid. A 2.75 kb *Bgl*II fragment of HBV DNA containing the surface antigen gene (sAg) and its putative promoter sequences (58) was inserted in the *Bam*HI site of pML2, and the *Hind*III cleaved full-length linear BPV-1 DNA was inserted into the *Hind*III site. In the resulting recombinant plasmid pM2 6V4 the direction of transcription of the sAg gene is the same as that of BPV DNA. Before transfection, the plasmid sequences were removed by digestion with *Sal*I-*Pvu*II. The transformation efficiency is ~ 10 foci per μg of DNA. ■, BPV DNA; ∿, HBsAg sequences; −, pML2.

correctly expressed (57). The transformed cells produce and secrete HBsAg particles which are indistinguishable from the 22 nm particles present in the serum of HBV patients, thus showing that the protein has been accurately synthesised, glycosylated, assembled and secreted. The structure of sAg mRNA, however, is not known.

(ii) *Inducible Genes.* BPV vectors have been used to analyse the expression of genes regulated by specific inducers. The rat growth hormone gene contained in pGP (*Figure 8*) is, however, constitutively expressed and does not respond to glucocorticoid treatment in mouse cell transformants (48). The loss of inducibility may be due to incorrect transcription initiation since the mRNA made in the transformants is approximately 200 nucleotides longer than authentic GH mRNA. The promoter sequences of the MT gene have been fused to a human GH mini-gene in plasmid pBPVMG (49) (*Figure 8b*). Treatment of the BPVMG-transformed cell lines with cadmium increases both the level of hybrid MT-GH mRNA (2- to 7-fold) and the amount of secreted GH (2-fold). In contrast, treatment with dexamethasone failed to do so. It appears therefore that the regulation of expression of the MT gene by metals and hormones proceeds *via* different mechanisms, one of which at least is not operative on the cloned regulatory sequences. When cells transformed by the BPV-MMTV LTR chimeras (*Figure*

10) are grown in dexamethasone, the amount of v-ras mRNA, whose transcription initiates in the glucocorticoid responsive LTR, increases from 30 to 100 times compared with the RNA of uninduced cells. This observation confirms the previous finding that the LTR of MMTV is the target site for glucocorticoids induction, and extends its validity to a situation in which the LTR element, being part of an episomal structure, is removed from the possible control of cellular flanking sequences.

Cells transformed by BPV-IF hybrid plasmids (52,53) (*Figure 9*) produce human β-interferon constitutively and respond to induction by both inactivated Newcastle disease virus and poly(I).poly(C). Induction increases the levels of human β-IF mRNA up to 32-fold, and the level of human β-interferon up to 120-fold. The constitutive transcripts are probably incorrectly initiated. Recently two elements which regulate the expression of human β-interferon gene have been localized and identified (54). One is located between positions -77 and -19 from the *cap* site and is required for both constitutive and induced β-IF expression. The other is located between -210 and -107 and, when deleted, the constitutive level of β-IF mRNA increases approximately 10-fold, whereas the amount of induced mRNA is unaffected.

4. CONCLUSIONS

Despite its successful and increasing use as an eukaryotic cloning vector, several features of the biology of BPV-1 DNA remain obscure.

BPV-1 DNA, or its vector derivatives, establishes itself in transformed cells as multiple episomes, the number of which varies in different cell lines, approximately from 20 to 300 copies. However, in a given cell line, their number remains constant over very long periods of time. As the BPV minichromosome does not contain a centromere, its segregation to the daughter cells is likely to be a random event, and the copy number is therefore an average. The factors, if any, that determine the number of episomes in a given cell, and that maintain a constant average number in a given cell line, are unknown. Nor is it known whether BPV DNA replicates at a particular stage of the cell cycle, or whether its synthesis can occur throughout.

Also unclear are the reasons why, in order to achieve efficient cell transformation, the 69T fragment has to be separated from its vector (but not from pAT153), and why the insertion of either the human β-globin gene fragment or the rat growth hormone gene fragment confers cell transformation ability to the undivided plasmids.

When dominant markers are used in BPV vectors and cells are selected for *tk* or *gpt* expression, the BPV plasmids are found in an integrated form, whereas they are maintained as episomes in cells selected for neomycin resistance. Is this due to the need to regulate the expression of the *tk* and *gpt* genes through the action of controlling flanking cellular sequences? An analysis of the cellular sequences at the insertion sites may provide an answer to the question.

Another puzzling feature of the BPV plasmids is their ability to acquire new DNA sequences, especially when only the 69T fragment is used. This may reflect the requirement for a minimum size for effective DNA replication, but on the other hand it may be a random consequence of the recombination mechanisms of the cell.

Even more puzzling is the observation that very similar hybrid plasmids display different behaviour. The best example is given by pCGBPV7 and pCGBPV9: the former integrates in the DNA of transformed cells, whereas the latter persists as an autonomous replicon. The difference between the two plasmids resides in the relative transcriptional orientation of their component parts. In the former plasmid, transcription of the Tn5 and the BPV sequences proceeds unidirectionally, whereas in the latter it is bidirectional. As the structure of the mRNAs transcribed from these plasmids is not known, it is not yet possible to say if and how the transcriptional pattern can affect their chromosomal behaviour.

Despite these difficulties, the BPV vectors have made possible the cloning of several genes, and the unequivocal identification of the regulatory elements of inducible genes, as in the elegant work conducted on the human β-interferon gene. It should now be possible to study the interaction of controlling factors with these elements, and thus gain a better knowledge of the mechanisms of gene induction.

The correct expression of the cloned genes makes the BPV vector a suitable system not only for analytical purposes but also for practical applications. The possibility of introducing several copies of the HBV surface antigen gene in mammalian cells, where the protein is accurately synthesized and modified, can be a very important factor in the development of a vaccine against the virus.

As mentioned before, not all the chimeric plasmids are maintained episomally, and some integrate possibly as a result of their large size, as in the case of the HLA plasmids. The ability of most of the BPV-neo plamids to persist as episomes and to shuttle between eukaryotic and prokaryotic cells will make them very useful. Neomycin-resistance plasmids can be constructed where only the BPV functions needed for autonomous replications are maintained. Such smaller plasmids should be able to incorporate larger DNA fragments, and to propagate autonomously in G418-resistant cells.

The growing understanding of the functional organisation of the BPV genome will undoubtedly benefit not only the investigations of those of us who work on the biology of the virus, but also the development of newer and better eukaryotic vectors.

5. ACKNOWLEDGEMENTS

Thanks are due to Dr D.DiMaio for providing unpublished results, to Dr K.Smith for his critical reading of the manuscript, to Miss R.Forbes for excellent secretarial assistance, and to Mr D.Tallach for skilful photographic work. The author's research is supported by the Cancer Research Campaign (CRC), and the author is a Fellow of the CRC.

6. REFERENCES

1. Rigby,P.W.J. (1982) in *Genetic Engineering,* Vol. 3, Williamson,R. (ed.), Academic Press, p. 83.
2. Lancaster,W.D. and Olson,C. (1982) *Microbiological Reviews,* **46**, 191.
3. Pfister,H. (1984) *Rev. Physiol. Biochem. Pharmacol.,* **99**, 112.
4. Chen,E.Y., Howley,P.M., Levinson,A.D. and Seeburg,P.H. (1982) *Nature, 299,* 529.

5. Lancaster,W.D. (1979) *J. Virol.*, **32**, 684.
6. Campo,M.S., Moar,M.H., Laird,H.M. and Jarrett,W.F.H. (1981) *Virology*, **113**, 323.
7. Morgan,D.M., Murphy,M.F., Abraham,J.M. and Meinke,W.J. (1981) *J. Gen. Virol.*, **56**, 213.
8. Amtmann,E. and Sauer,G. (1982) *J. Virol.*, **43**, 59.
9. Heilman,C.A., Engel,L.W., Lowy,D.R. and Howley,P.M. (1982) *Virology*, **119**, 22.
10. Engel,L.W., Heilman,C.A. and Howley,P.M. (1983) *J. Virol.*, **47**, 516.
11. Rosl,F., Waldek,W. and Sauer,G. (1983) *J. Virol.*, **46**, 567.
12. Campo,M.S., Spandidos,D.A., Lang,J. and Wilkie,N.M. (1983) *Nature*, **303**, 77.
13. Nakabayashi,Y., Chattopadhyay,S.K. and Lowy,D.R. (1983) *Proc. Natl. Acad. Sci. USA*, **80**, 5832.
14. Lusky,M., Berg,L., Weiher,H. and Botchan,M. (1983) *Mol. Cell. Biol.*, **3**, 1108.
15. Lowy,D.R., Dvoretzky,I., Shober,R., Law,M.F., Engel,L. and Howley,P.M. (1980) *Nature*, **287**, 72.
16. Maniatis,T., Fritsch,E.F. and Sambrook,J. (1982) *Molecular Cloning: A Laboratory Manual*, published by Cold Spring Harbor Laboratory Press.
17. Bolivar,F., Rodriguez,R., Green,P.J., Betlach,M., Heyneker,H.L., Boyer,H.W., Crosa,J. and Falkow, S. (1977) *Gene*, **2**, 95.
18. Twigg,A.J. and Sherratt,D. (1980) *Nature*, **283**, 216.
19. Soberon,X., Covarrubias,L. and Bolivar,F. (1980) *Gene*, **9**, 287.
20. Lusky,M. and Botchan,M. (1981) *Nature*, **293**, 79.
21. Howley,P.M., Law,M.F., Heilman,C., Engel,L., Alonso,M.C., Israel,M.A., Lowy,D.R. and Lancaster,W.D. (1980) in *Viruses in Naturally Occurring Cancers*, Essex,M., Todaro,G., zur Hans,H. (eds.), Cold Spring Harbor Laboratory Press, p.233.
22. Campo,M.S. and Spandidos,D.A. (1983) *J. Gen. Virol.*, **64**, 549.
23. Law,M.F., Lowy,D.R., Dvoretzky,I. and Howley,P.M. (1981) *Proc. Natl. Acad. Sci. USA*, **78**, 2727.
24. Binetruy,B., Meneguzzi,G., Breathnach,R. and Cuzin,F. (1982) *EMBO J.*, **1**, 621.
25. Sarver,N., Byrne,J.C. and Howley,P.M. (1982) *Proc. Natl. Acad. Sci. USA*, **79**, 7147.
26. Campo,M.S. and Coggins,L.W. (1982) *J. Gen. Virol.*, **63**, 255.
27. Graham,F.L. and Van Der Eb,A.J. (1973) *Virology*, **54**, 536.
28. Frost,E. and Williams,J. (1978) *Virology*, **91**, 39.
29. Sutcliffe,J.G. (1978) *Nucleic Acids Res.*, **5**, 2721.
30. Hirt,B. (1967) *J. Mol. Biol.*, **26**, 365.
31. McKnight,S.L., Gavis,E.R., Kingsbury,R. and Axel,R. (1981) *Cell*, **25**, 385.
32. Colbere-Garapin,F., Chousterman,S., Horodiceanu,F., Kourilsky,P. and Gerapin,A.C. (1979) *Proc. Natl. Acad. Sci. USA*, **76**, 3755.
33. Sekiguchi,T., Nishimoto,T., Kai,R. and Sekiguchi,M. (1983) *Gene*, **21**, 267.
34. Mulligan,R.C. and Berg,P. (1981) *Proc. Natl. Acad. Sci. USA*, **78**, 2072.
35. Law,M.F., Howard,B., Sarver,N. and Howley,P.M. (1982) in *Eukaryotic Viral Vectors*, Gluzman,Y. (ed.), Cold Spring Harbor Laboratory Press, p.79.
36. Colbere-Garapin,F., Horodiceanu,F., Kourilsky,P. and Gerapin,C.A. (1981) *J. Mol. Biol.*, **105**, 1.
37. Law,M.F., Byrne,J.C. and Howley,P.M. (1983) *Mol. Cell. Biol.*, **3**, 2110.
38. Mathias,P.D., Bernard,H.V., Scott,A., Brady,G., Hashimoto-Gatoh,T. and Schutz,G. (1983) *EMBO J.*, **2**, 1487.
39. Meneguzzi,G., Binetruy,B., Grisoni,M. and Cuzin,F. (1984) *EMBO J.*, **3**, 365.
40. Pavlakis,G.N. and Hamer,D.H. (1983) *Proc. Natl. Acad. Sci. USA*, **80**, 397.
41. Mulligan,R.C. and Berg,P. (1980) *Science (Wash.)*, **209**, 1422.
42. Stuber,D. and Mujard,H. (1981) *Proc. Natl. Acad. Sci. USA*, **78**, 167.
43. Sarver,N., Gruss,P., Law,M.F., Khoury,G. and Howley,P. (1981) *Mol. Cell. Biol.*, **1**, 486.
44. Lomedico,P., Rosenthal,N., Efstratiadis,A., Gilbert,W., Kolodner,R. and Tizard,R. (1979) *Cell*, **18**, 545.
45. Fritsch,E.F., Lawn,R.M. and Maniatis,T. (1980) *Cell*, **19**, 959.
46. DiMaio,D., Treisman,R. and Maniatis,T. (1982) *Proc. Natl. Acad. Sci. USA*, **79**, 4030.
47. Page,G.S., Smith,S. and Goodman,H.M. (1981) *Nucleic Acids Res.*, **9**, 2087.
48. Kushner,P.J., Levinson,B.B. and Goodman,H.M. (1982) *J. Mol. Appl. Genet.*, **1**, 527.
49. Pavlakis,G.N. and Hamer,D.H. (1983) *Proc. Natl. Acad. Sci. USA*, **80**, 397.
50. Martial,J.A., Hallewell,R.A., Baxter,J.D. and Goodman,H.M. (1979) *Science (Wash.)*, **205**, 602.
51. Hamer,D.H. and Walling,M.J. (1982) *J. Mol. Appl. Genet.*, **1**, 273.
52. Mitrani-Rosenbaum,S., Maroteaux,L., Mory,Y., Revel,M. and Howley,P.M. (1983) *Mol. Cell. Biol.*, **3**, 233.
53. Zinn,K., Mellon,P., Ptashne,M. and Maniatis,T. (1982) *Proc. Natl. Acad. Sci. USA*, **79**, 4897.
54. Zinn,K., DiMaio,D. and Maniatis,T. (1983) *Cell*, **34**, 865.
55. Ostrowski,M.C., Richard-Foy,H., Wolford,R.G., Berard,D.S. and Hager,G.L. (1983) *Mol. Cell. Biol.*, **3**, 2045.

56. DiMaio,D., Corbin,V., Sibley,E. and Maniatis,T. (1984) *Cell. Mol. Biol.,* **4**, 340.
57. Stenlund,A., Lamy,D., Moreno-Lopez,J., Ahola,H., Pettersson,V. and Tiollais,P. (1983) *EMBO J.,* **2**, 669.
58. Pourcel,C., Louise,A., Gervais,M., Chenciner,N., Dubois,M.F. and Tiollais,P. (1982) *J. Virol.,* **42**, 100.

INDEX

244

Published in the Practical Approach series

Nucleic acid hybridisation
a practical approach

Edited by B D Hames and S J Higgins,
University of Leeds

AN INVALUABLE GUIDE TO ALL ASPECTS OF HYBRIDISATION

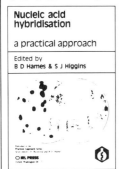

Nucleic acid hybridisation

a practical approach

Edited by
B D Hames & S J Higgins

November 1985;
256pp; softbound:
0 947946 23 3
hardbound:
0 947946 61 6

Hybridisation techniques have established themselves as invaluable in recombinant DNA technology – genetic engineering – and throughout molecular biology in general. *Nucleic acid hybridisation* is unique in this field as a source for the techniques' major applications at both the theoretical and practical levels. It encompasses the theory behind experiments, their rationale and design, and the detailed protocols for researchers and clinicians to identify and analyse specific sequences.

Like previous books in the Practical Approach series, *Nucleic acid hybridisation* is a laboratory-bench manual which acts as a source-book of ideas and experimental protocols for researchers from senior-undergraduate and post-graduate levels upwards, in industry, universities and research institutes. In addition, the usefulness of *Nucleic acid hybridisation* is enhanced by appendices containing essential data on restriction enzymes, nucleic acid size markers, sources of equipment and chemicals and a computer program for the analysis of hybridisation data.

Contents

Introduction *E M Southern* ● Hybridisation strategy *R J Britten and E H Davidson* ● Preparation of nucleic acid probes *J E Arrand* ● Quantitative analysis of solution hybridisation *B D Young and M L M Anderson* ● Quantitative filter hybridisation *M L M Anderson and B D Young* ● Hybridisation in the analysis of recombinant DNA *P J Mason and J G Williams* ● Hybridisation in the analysis of RNA *J G Williams and P J Mason* ● Electron microscopic visualisation of nucleic acid hybrids *P Oudet and C Schatz* ● *In situ* hybridisation *M L Pardue* ● Appendices ● Index

For details of price and ordering consult our current catalogue or contact:

IRL Press Ltd,
PO Box 1, Eynsham,
Oxford OX8 1JJ, UK

IRL Press Inc,
PO Box Q, McLean
VA 22101 USA

IRL PRESS
Oxford · Washington DC